Undergraduate Texts in Mathematics

Kai Lai Chung

Elementary Probability Theory with Stochastic Processes

Springer-Verlag
New York Heidelberg Berlin

Kai Lai Chung
Department of Mathematics
Stanford University
Stanford, California 94305
USA

AMS Subject Classifications: 6001, 60C05, 60J10, 60J15, 60J20, 60J75

Library of Congress Cataloging in Publication Data
Chung, Kai Lai,
Elementary probability theory with stochastic processes.

(Undergraduate texts in mathematics)
Bibliography: p.
Includes index.
1. Probabilities. 2. Stochastic processes.
I. Title.
QA273.C5774 1979 519.2 79-1193

Third Edition.

Printed in the United States of America.

9 8 7 6 5 4 3 2 1

ISBN 0-387-90362-3 Springer-Verlag New York
ISBN 3-540-90362-3 Springer-Verlag Berlin Heidelberg

'Tis the good reader that makes the good book.
Ralph Waldo Emerson

Preface to the Third Edition

A new feature of this edition consists of photographs of eight masters in the contemporary development of probability theory. All of them appear in the body of the book, though the few references there merely serve to give a glimpse of their manifold contributions. It is hoped that these vivid pictures will inspire in the reader a feeling that our science is a live endeavor created and pursued by real personalities. I have had the privilege of meeting and knowing most of them after studying their works and now take pleasure in introducing them to a younger generation. In collecting the photographs I had the kind assistance of Drs Marie-Hélène Schwartz, Joanne Elliot, Milo Keynes and Yu. A. Rozanov, to whom warm thanks are due.

A German edition of the book has just been published. I am most grateful to Dr. Herbert Vogt for his careful translation which resulted also in a considerable number of improvements on the text of this edition. Other readers who were kind enough to send their comments include Marvin Greenberg, Louise Hay, Nora Holmquist, H.-E. Lahmann, and Fred Wolock. Springer-Verlag is to be complimented once again for its willingness to make its books "immer besser."

K.L.C.
September 19, 1978

Preface to the Second Edition

A determined effort was made to correct the errors in the first edition. This task was assisted by: Chao Hung-po, J. L. Doob, R. M. Exner, W. H. Fleming, A. M. Gleason, Karen Kafador, S. H. Polit, and P. van Moerbeke. Miss Kafador and Dr. Polit compiled particularly careful lists of suggestions. The most distressing errors were in the Solutions to Problems. All of them have now been checked by myself from Chapter 1 to 5, and by Mr. Chao from Chapter 6 to 8. It is my fervent hope that few remnant mistakes remain in that sector. A few small improvements and additions were also made, but not all advice can be heeded at this juncture. Users of the book are implored to send in any criticism and commentary, to be taken into consideration in a future edition. Thanks are due to the staff of Springer-Verlag for making this revision possible so soon after the publication of the book.

K. L. C.

Preface to the First Edition

In the past half-century the theory of probability has grown from a minor isolated theme into a broad and intensive discipline interacting with many other branches of mathematics. At the same time it is playing a central role in the mathematization of various applied sciences such as statistics, operations research, biology, economics and psychology—to name a few to which the prefix "mathematical" has so far been firmly attached. The coming-of-age of probability has been reflected in the change of contents of textbooks on the subject. In the old days most of these books showed a visible split-personality torn between the combinatorial games of chance and the so-called "theory of errors" centering in the normal distribution. This period ended with the appearance of Feller's classic treatise (see [Feller 1]†) in 1950, from the manuscript of which I gave my first substantial course in probability. With the passage of time probability theory and its applications have won a place in the college curriculum as a mathematical discipline essential to many fields of study. The elements of the theory are now given at different levels, sometimes even before calculus. The present textbook is intended for a course at about the sophomore level. It presupposes no prior acquaintance with the subject and the first three chapters can be read largely without the benefit of calculus. The next three chapters require a working knowledge of infinite series and related topics, and for the discussion involving random variables with densities some calculus is of course assumed. These parts dealing with the "continuous case" as distinguished from the "discrete case" are easily separated and may be postponed. The contents of the first six chapters should form the backbone of any meaningful first introduction to probability theory. Thereafter a reasonable selection includes: §7.1 (Poisson distribution, which may be inserted earlier in the course), some kind of going over of §7.3, 7.4, 7.6 (normal distribution and the law of large numbers), and §8.1 (simple random walks which are both stimulating and useful). All this can be covered in a semester but for a quarter system some abridgment will be necessary. Specifically, for such a short course Chapters 1 and 3 may be skimmed through and the asterisked material omitted. In any case a solid treatment of the normal approximation theorem in Chapter 7 should be attempted only if time is available as in a semester or two-quarter course. The final Chapter 8 gives a self-contained elementary account of Markov chains and is an extension of the main course at a somewhat more mature level. Together with the asterisked sections 5.3, 5.4 (sequential sampling and Pólya urn scheme) and 7.2 (Poisson process), and perhaps some filling in from the Appendices, the material provides a gradual and concrete passage into the domain of sto-

† Names in square brackets refer to the list of General References on p. 307.

chastic processes. With these topics included the book will be suitable for a two-quarter course, of the kind that I have repeatedly given to students of mathematical sciences and engineering. However, after the preparation of the first six chapters the reader may proceed to more specialized topics treated e.g. in the above mentioned treatise by Feller. If the reader has the adequate mathematical background, he will also be prepared to take a formal rigorous course such as presented in my own more advanced book [Chung 1].

Much thought has gone into the selection, organization and presentation of the material to adapt it to classroom uses, but I have not tried to offer a slick package to fit in with an exact schedule or program such as popularly demanded at the quick-service counters. A certain amount of flexibility and choice is left to the instructor who can best judge what is right for his class. Each chapter contains some easy reading at the beginning, for motivation and illustration, so that the instructor may concentrate on the more formal aspects of the text. Each chapter also contains some slightly more challenging topics (e.g., §1.4, 2.5) for optional sampling. They are not meant to deter the beginner but to serve as an invitation to further study. The prevailing emphasis is on the thorough and deliberate discussion of the basic concepts and techniques of elementary probability theory with few frills and minimal technical complications. Many examples are chosen to anticipate the beginners' difficulties and to provoke better thinking. Often this is done by posing and answering some leading questions. Historical, philosophical and personal comments are inserted to add flavor to this lively subject. It is my hope that the reader will not only learn something from the book but may also derive a measure of enjoyment in so doing.

There are over two hundred exercises for the first six chapters and some eighty more for the last two. Many are easy, the harder ones indicated by asterisks, and all answers gathered at the end of the book. Asterisked sections and paragraphs deal with more special or elaborate material and may be skipped, but a little browsing in them is recommended.

The author of any elementary textbook owes of course a large debt to innumerable predecessors. More personal indebtedness is acknowledged below. Michel Nadzela wrote up a set of notes for a course I gave at Stanford in 1970. Gian-Carlo Rota, upon seeing these notes, gave me an early impetus toward transforming them into a book. D. G. Kendall commented on the first draft of several chapters and lent further moral support. J. L. Doob volunteered to read through most of the manuscript and offered many helpful suggestions. K. B. Erickson used some of the material in a course he taught. A. A. Balkema checked the almost final version and made numerous improvements. Dan Rudolph read the proofs together with me. Perfecto Mary drew those delightful pictures. Gail Lemmond did the typing with her usual efficiency and dependability. Finally, it is a pleasure to thank my old publisher Springer-Verlag for taking my new book to begin a new series of undergraduate texts.

K. L. C.
March 1974.

Borel

Lévy

Keynes

Feller

Doob

Pólya

Kolmogorov

Cramér

CONTENTS

Chapter 1

Set

1.1. Sample sets

These days school children are taught about sets. A second grader* was asked to name "the set of girls in his class." This can be done by a complete list such as:

"Nancy, Florence, Sally, Judy, Ann, Barbara, . . ."

A problem arises when there are duplicates. To distinguish between two Barbaras one must indicate their family names or call them B_1 and B_2. The same member cannot be counted twice in a set.

The notion of a set is common in all mathematics. For instance in geometry one talks about "the set of points which are equi-distant from a given point." This is called a circle. In algebra one talks about "the set of integers which have no other divisors except 1 and itself." This is called the set of prime numbers. In calculus the domain of definition of a function is a set of numbers, e.g., the interval (a, b); so is the range of a function if you remember what it means.

In probability theory the notion of a set plays a more fundamental role. Furthermore we are interested in very general kinds of sets as well as specific concrete ones. To begin with the latter kind, consider the following examples:

(a) a bushel of apples;
(b) fifty five cancer patients under a certain medical treatment;
(c) all the students in a college;
(d) all the oxygen molecules in a given container;
(e) all possible outcomes when six dice are rolled;
(f) all points on a target board.

Let us consider at the same time the following "smaller" sets:

(a′) the rotten apples in that bushel;
(b′) those patients who respond positively to the treatment;
(c′) the mathematics majors of that college;
(d′) those molecules which are traveling upwards;
(e′) those cases when the six dice show different faces;
(f′) the points in a little area called the "bull's eye" on the board.

* My son Daniel.

We shall set up a mathematical model for these and many more such examples that may come to mind, namely we shall abstract and generalize our intuitive notion of "a bunch of things." First we call the things points, then we call the bunch a space; we prefix them by the word "sample" to distinguish these terms from other usages, and also to allude to their statistical origin. Thus a *sample point* is the abstraction of an apple, a cancer patient, a student, a molecule, a possible chance outcome, or an ordinary geometrical point. The *sample space* consists of a number of sample points, and is just a name for the totality or aggregate of them all. Any one of the examples (a)–(f) above can be taken to be a sample space, but so also may any one of the smaller sets in (a′)–(f′). What we choose to call a space [a *universe*] is a relative matter.

Let us then fix a sample space to be denoted by Ω, the capital Greek letter *omega*. It may contain any number of points, possibly infinite but at least one. (As you have probably found out before, mathematics can be very pedantic!) Any of these points may be denoted by ω, the small Greek letter omega, to be distinguished from one another by various devices such as adding subscripts or dashes (as in the case of the two Barbaras if we do not know their family names), thus $\omega_1, \omega_2, \omega', \ldots$. Any partial collection of the points is a *subset* of Ω, and since we have fixed Ω we will just call it a set. In extreme cases a set may be Ω itself or the *empty set* which has no point in it. You may be surprised to hear that the empty set is an important entity and is given a special symbol $\varnothing$. The number of points in a set S will be called its *size* and denoted by $|S|$, thus it is a nonnegative integer or ∞. In particular $|\varnothing| = 0$.

A particular set S is well defined if it is possible to tell whether any given point *belongs to* it or not. These two cases are denoted respectively by

$$\omega \in S; \quad \omega \notin S.$$

Thus a set is determined by a specified rule of membership. For instance, the sets in (a′)–(f′) are well defined up to the limitations of verbal descriptions. One can always quibble about the meaning of words such as "a rotten apple," or attempt to be funny by observing, for instance, that when dice are rolled on a pavement some of them may disappear into the sewer. Some people of a pseudo-philosophical turn of mind get a lot of mileage out of such *caveats*, but we will not indulge in them here. Now, one sure way of specifying a rule to determine a set is to enumerate all its members, namely to make a complete list as the second grader did. But this may be tedious if not impossible. For example, it will be shown in §3.1 that the size of the set in (e) is equal to $6^6 = 46656$. Can you give a quick guess how many pages of a book like this will be needed just to record all these possibilities of a mere throw of six dice? On the other hand it can be described in a systematic and unmistakable way as the set of all ordered 6-tuples of the form below:

$$(s_1, s_2, s_3, s_4, s_5, s_6)$$

where each of the symbols s_j, $1 \leq j \leq 6$, may be any of the numbers 1, 2, 3, 4, 5, 6. This is a good illustration of mathematics being economy of thought (and printing space).

If every point of A belongs to B, then A is *contained* or *included* in B and is a *subset* of B, while B is a *superset* of A. We write this in one of the two ways below:

$$A \subset B, \quad B \supset A.$$

Two sets are *identical* if they contain exactly the same points, and then we write

$$A = B.$$

Another way to say this is: $A = B$ if and only if $A \subset B$ and $B \subset A$. This may sound unnecessarily roundabout to you, but is often the only way to check that two given sets are really identical. It is not always easy to identify two sets defined in different ways. Do you know for example that the set of even integers is identical with the set of all solutions x of the equation $\sin(\pi x/2) = 0$? We shall soon give some examples of showing the identity of sets by the roundabout method.

1.2. Operations with sets

We learn about sets by operating on them, just as we learn about numbers by operating on them. In the latter case we say also that we compute with numbers: add, subtract, multiply, and so on. These operations performed on given numbers produce other numbers which are called their sum, difference, product, etc. In the same way, operations performed on sets produce other sets with new names. We are now going to discuss some of these and the laws governing them.

Complement. The complement of a set A is denoted by A^c and is the set of points which do not belong to A. Remember we are talking only about points in a fixed Ω! We write this symbolically as follows:

$$A^c = \{\omega \mid \omega \notin A\}$$

which reads: "A^c is the set of ω which does not belong to A." In particular $\Omega^c = \varnothing$ and $\varnothing^c = \Omega$. The operation has the property that if it is performed twice in succession on A, we get A back:

$$(A^c)^c = A. \tag{1.2.1}$$

Union. The union $A \cup B$ of two sets A and B is the set of points which belong to at least one of them. In symbols:

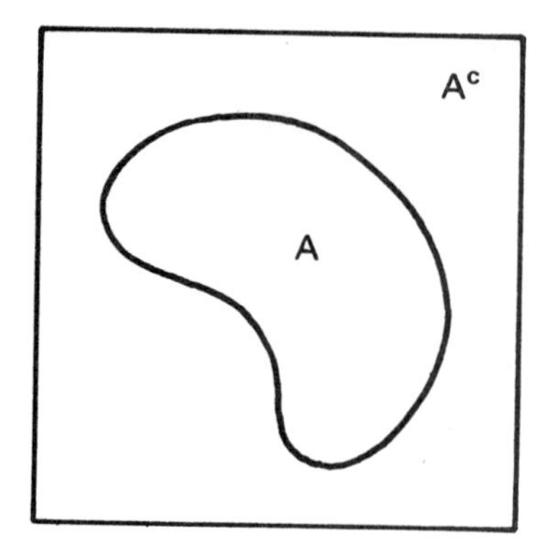

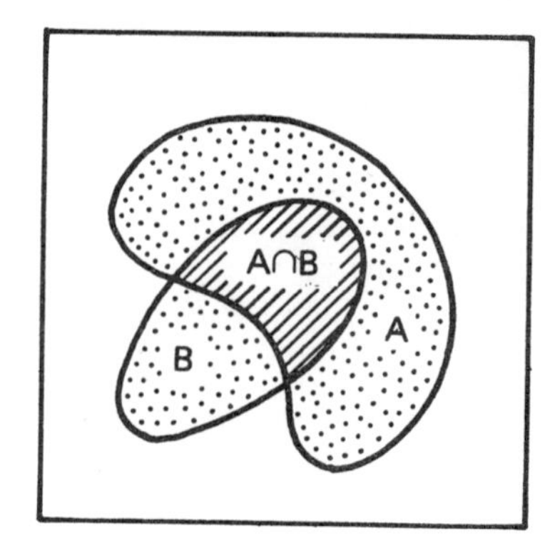

Figure 1

$$A \cup B = \{\omega \mid \omega \in A \text{ or } \omega \in B\}$$

where "or" means "and/or" in pedantic [legal] style, and will always be used in this sense.

Intersection. The intersection $A \cap B$ of two sets A and B is the set of points which belong to both of them. In symbols:

$$A \cap B = \{\omega \mid \omega \in A \text{ and } \omega \in B\}.$$

We hold the truth of the following laws as self-evident:

Commutative Law. $A \cup B = B \cup A$, $A \cap B = B \cap A$.

Associative Law. $(A \cup B) \cup C = A \cup (B \cup C)$,
$(A \cap B) \cap C = A \cap (B \cap C)$.

But observe that these relations are instances of identity of sets mentioned above, and are subject to proof. They should be compared, but not confused, with analogous laws for sum and product of numbers:

$$a + b = b + a, \; a \times b = b \times a$$

$$(a + b) + c = a + (b + c), \; (a \times b) \times c = a \times (b \times c).$$

Brackets are needed to indicate the order in which the operations are to be performed. Because of the associative laws, however, we can write

$$A \cup B \cup C, \quad A \cap B \cap C \cap D,$$

without brackets. But a string of symbols like $A \cup B \cap C$ is ambiguous, therefore not defined; indeed $(A \cup B) \cap C$ is not identical with $A \cup (B \cap C)$. You should be able to settle this easily by a picture.

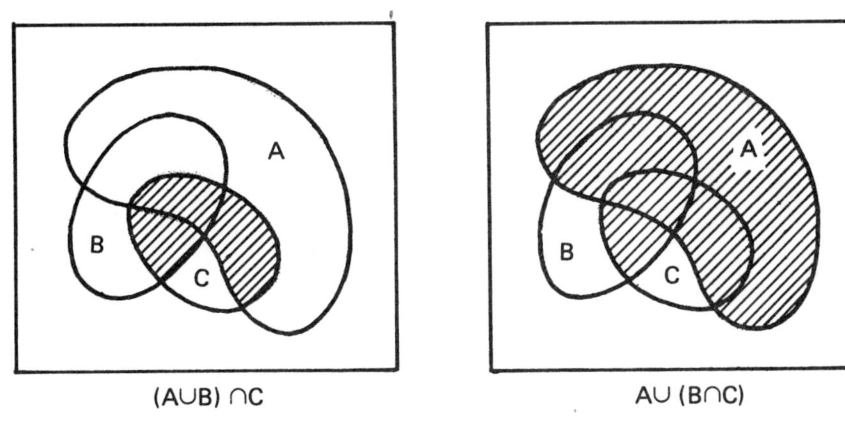

Figure 2

The next pair of *distributive laws* connect the two operations as follows:

(D$_1$) $$(A \cup B) \cap C = (A \cap C) \cup (B \cap C);$$

(D$_2$) $$(A \cap B) \cup C = (A \cup C) \cap (B \cup C).$$

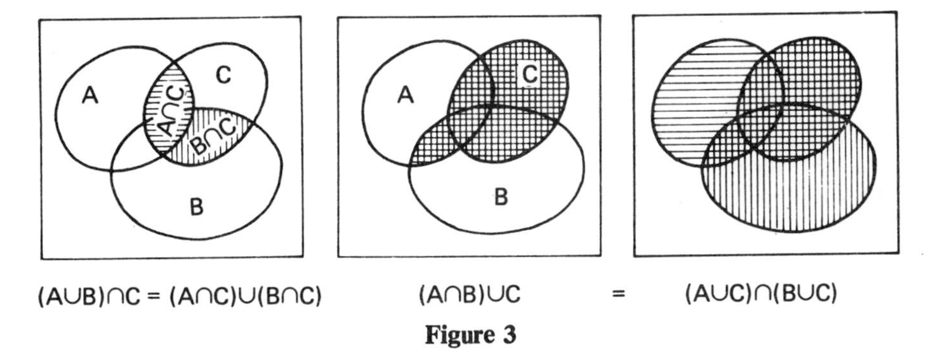

Figure 3

Several remarks are in order. First, the analogy with arithmetic carries over to (D$_1$):

$$(a + b) \times c = (a \times c) + (b \times c);$$

but breaks down in (D$_2$):

$$(a \times b) + c \neq (a + c) \times (b + c).$$

Of course the alert reader will have observed that the analogy breaks down already at an earlier stage, for

$$A = A \cup A = A \cap A;$$

but the only number a satisfying the relation $a + a = a$ is 0; while there are exactly two numbers satisfying $a \times a = a$, namely 0 and 1.

Second, you have probably already discovered the use of diagrams to prove or disprove assertions about sets. It is also a good practice to see the truth of such formulas as (D_1) and (D_2) by well-chosen examples. Suppose then

$$A = \text{inexpensive things}, \; B = \text{really good things},$$
$$C = \text{food [edible things]}.$$

Then $(A \cup B) \cap C$ means "(inexpensive or really good) food," while $(A \cap C) \cup (B \cap C)$ means "(inexpensive food) or (really good food)." So they are the same thing alright. This does not amount to a proof, as one swallow does not make a summer, but if one is convinced that whatever logical structure or thinking process involved above in no way depends on the precise nature of the three things A, B and C, so much so that they can be *anything*, then one has in fact landed a general proof. Now it is interesting that the same example applied to (D_2) somehow does not make it equally obvious (at least to the author). Why? Perhaps because some patterns of logic are in more common use in our everyday experience than others.

This last remark becomes more significant if one notices an obvious duality between the two distributive laws. Each can be obtained from the other by switching the two symbols $\cup$ and $\cap$. Indeed each can be deduced from the other by making use of this duality (Exercise 11).

Finally, since (D_2) comes less naturally to the intuitive mind, we will avail ourselves of this opportunity to demonstrate the roundabout method of identifying sets mentioned above by giving a rigorous proof of the formula. According to this method, we must show: (i) each point on the left side of (D_2) belongs to the right side; (ii) each point on the right side of (D_2) belongs to the left side.

(i) Suppose ω belongs to the left side of (D_2), then it belongs either to $A \cap B$ or to C. If $\omega \in A \cap B$, then $\omega \in A$, hence $\omega \in A \cup C$; similarly $\omega \in B \cup C$. Therefore ω belongs to the right side of (D_2). On the other hand if $\omega \in C$, then $\omega \in A \cup C$ and $\omega \in B \cup C$ and we finish as before.

(ii) Suppose ω belongs to the right side of (D_2), then ω may or may not belong to C, and the trick is to consider these two alternatives. If $\omega \in C$, then it certainly belongs to the left side of (D_2). On the other hand, if $\omega \notin C$, then since it belongs to $A \cup C$, it must belong to A; similarly it must belong to B. Hence it belongs to $A \cap B$, and so to the left side of (D_2). Q.E.D.

1.3. Various relations

The three operations so far defined: complement, union and intersection obey two more laws called *De Morgan's laws:*

$$(C_1) \qquad (A \cup B)^c = A^c \cap B^c;$$

$$(C_2) \qquad (A \cap B)^c = A^c \cup B^c.$$

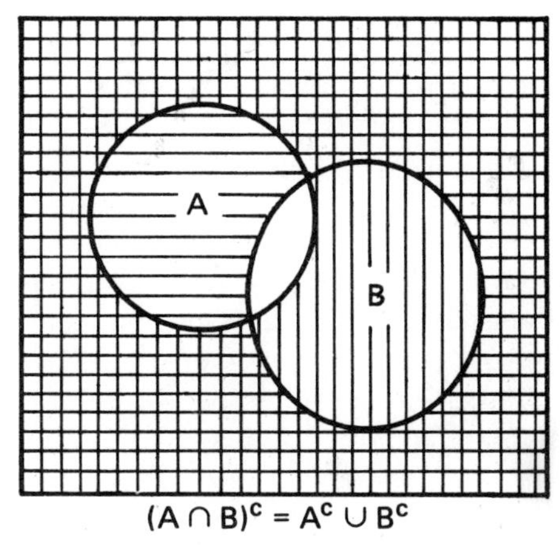

$(A \cap B)^c = A^c \cup B^c$

Figure 4

They are dual in the same sense as (D_1) and (D_2) are. Let us check these by our previous example. If A = inexpensive, and B = really good, then clearly $(A \cup B)^c$ = not inexpensive nor really good, namely high-priced junk, which is the same as $A^c \cap B^c$ = expensive and not really good. Similarly we can check (C_2).

Logically, we can deduce either (C_1) or (C_2) from the other; let us show it one way. Suppose then (C_1) is true, then since A and B are arbitrary sets we can substitute their complements and get

$$(1.3.1) \qquad (A^c \cup B^c)^c = (A^c)^c \cap (B^c)^c = A \cap B$$

where we have also used (1.2.1) for the second equation. Now taking the complements of the first and third sets in (1.3.1) and using (1.2.1) again we get

$$A^c \cup B^c = (A \cap B)^c.$$

This is (C_2). Q.E.D.

It follows from the De Morgan's laws that if we have complementation, then either union or intersection can be expressed in terms of the other. Thus we have

$$A \cap B = (A^c \cup B^c)^c,$$
$$A \cup B = (A^c \cap B^c)^c;$$

and so there is redundancy among the three operations. On the other hand it is impossible to express complementation by means of the other two, although there is a magic symbol from which all three can be derived (Exercise 14). It is convenient to define some other operations, as we now do.

Difference. The set $A\backslash B$ is the set of points which belong to A and (but) not to B. In symbols:

$$A\backslash B = A \cap B^c = \{\omega \mid \omega \in A \text{ and } \omega \notin B\}.$$

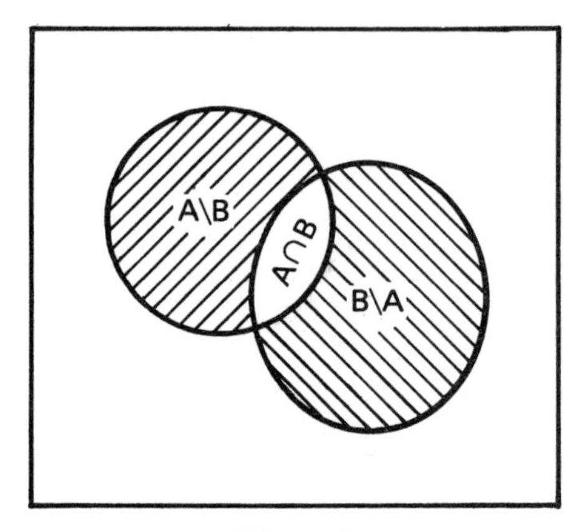

Figure 5

This operation is neither commutative nor associative. Let us find a *counter-example* to the associative law, namely, to find some A, B, C for which

$$(A\backslash B)\backslash C \neq A\backslash(B\backslash C). \tag{1.3.2}$$

Note that in contrast to a proof of identity discussed above, a single instance of falsehood will destroy the identity. In looking for a counter-example one usually begins by specializing the situation to reduce the "unknowns." So try $B = C$. The left side of (1.3.2) becomes $A\backslash B$, while the right side becomes $A\backslash\emptyset = A$. Thus we need only make $A\backslash B \neq A$, and that is easy.

In case $A \supset B$ we write $A - B$ for $A\backslash B$. Using this new symbol we have

$$A\backslash B = A - (A \cap B);$$

and

$$A^c = \Omega - A.$$

The operation "$-$" has some resemblance to the arithmetic operation of subtracting, in particular $A - A = \emptyset$, but the analogy does not go very far. For instance, there is no analogue to $(a + b) - c = a + (b - c)$.

Symmetric Difference. The set $A \triangle B$ is the set of points which belong to exactly one of the two sets A and B. In symbols:

$$A \triangle B = (A \cap B^c) \cup (A^c \cap B) = (A\backslash B) \cup (B\backslash A).$$

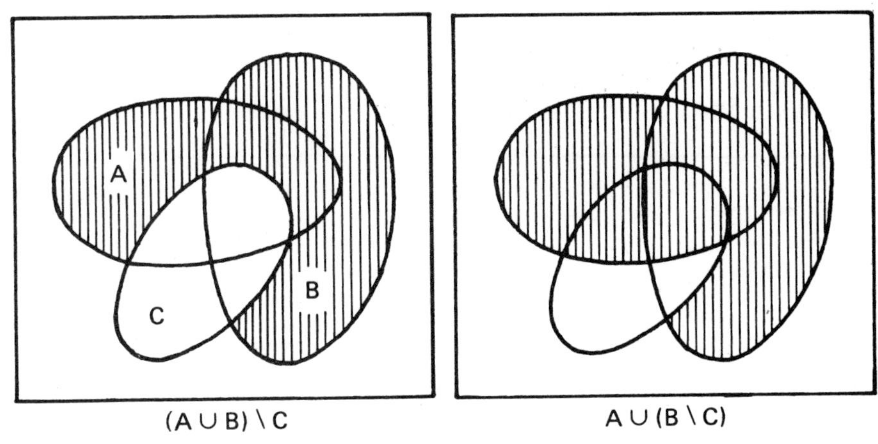

Figure 6

This operation is useful in advanced theory of sets. As its name indicates, it is symmetric with respect to A and B, which is the same as saying that it is commutative. Is it associative? Try some concrete examples or diagrams which have succeeded so well before, and you will probably be as quickly confused as I am. But the question can be neatly resolved by a device to be introduced in §1.4.

Having defined these operations, we should let our fancy run free for a few moments and imagine all kinds of sets that can be obtained by using them in succession in various combinations and permutations, such as

$$[(A\backslash C^c) \cap (B \cup C)^c]^c \cup (A^c \triangle B).$$

But remember we are talking about subsets of a fixed Ω, and if Ω is a finite set the number of distinct subsets is certainly also finite, so there must be a tremendous amount of inter-relationship among these sets that we can build up. The various laws discussed above are just some of the most basic ones, and a few more will be given among the exercises below.

An extremely important relation between sets will now be defined. Two sets A and B are said to be *disjoint* when they do not intersect, namely, have no point in common:

$$A \cap B = \varnothing.$$

This is equivalent to either one of the following inclusion conditions:

$$A \subset B^c; \quad B \subset A^c.$$

Any number of sets are said to be disjoint when every pair of them are disjoint as just defined. Thus, "A, B, C are disjoint" means more than just $A \cap B \cap C = \emptyset$; it means

$$A \cap B = \emptyset, A \cap C = \emptyset, B \cap C = \emptyset.$$

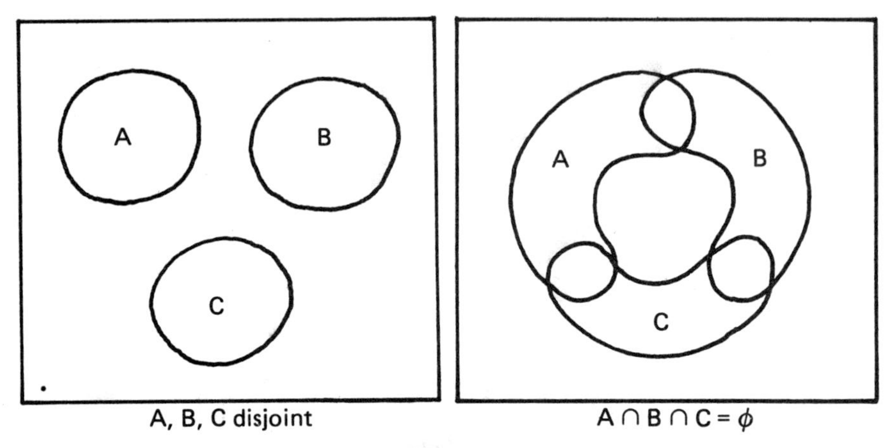

Figure 7

From here on we will omit the intersection symbol and write simply

$$AB \quad \text{for} \quad A \cap B$$

just as we write ab for $a \times b$. When A and B are disjoint we will write sometimes

$$A + B \quad \text{for} \quad A \cup B.$$

But be careful: not only does "+" mean addition for numbers but even when A and B are sets there are other usages of $A + B$ such as their vectorial sum.

For any set A, we have the obvious *decomposition:*

$$\Omega = A + A^c. \tag{1.3.3}$$

The way to think of this is: the set A gives a *classification* of all points ω in Ω according as ω belongs to A or to A^c. A college student may be classified according as he is a mathematics major or not, but he can also be classified according as he is a freshman or not, of voting age or not, has a car or not, . . . , is a girl or not. Each two-way classification divides the sample space into two disjoint sets, and if several of these are superimposed on each other we get, e.g.,

$$\Omega = (A + A^c)(B + B^c) = AB + AB^c + A^cB + A^cB^c, \tag{1.3.4}$$

$$(1.3.5) \quad \Omega = (A + A^c)(B + B^c)(C + C^c) = ABC + ABC^c + AB^cC + AB^cC^c + A^cBC + A^cBC^c + A^cB^cC + A^cB^cC^c.$$

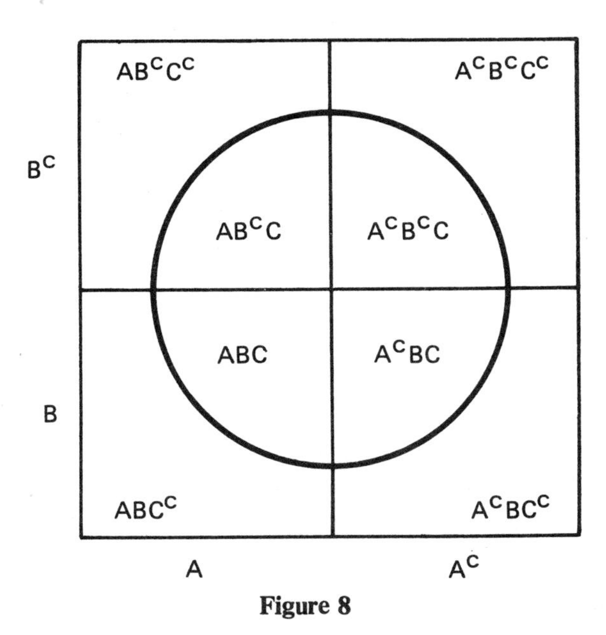

Figure 8

Let us call the pieces of such a decomposition the *atoms*. There are 2, 4, 8 atoms respectively above according as 1, 2, 3 sets are considered. In general there will be 2^n atoms if n sets are considered. Now these atoms have a remarkable property which will be illustrated in the case (1.3.5), as follows: no matter how you operate on the three sets A, B, C, and no matter how many times you do it, the resulting set can always be written as the union of some of the atoms. Here are some examples:

$$A \cup B = ABC + ABC^c + AB^cC + AB^cC^c + A^cBC^c + A^cBC$$

$$(A \backslash B) \backslash C^c = AB^cC$$

$$(A \triangle B)C^c = AB^cC^c + A^cBC^c.$$

Can you see why?

Up to now we have considered only the union or intersection of a finite number of sets. There is no difficulty in extending this to an infinite number of sets. Suppose a finite or infinite sequence of sets A_n, $n = 1, 2, \ldots,$ is given, then we can form their union and intersection as follows:

$$\bigcup_n A_n = \{\omega \mid \omega \in A_n \text{ for at least one value of } n\},$$

$$\bigcap_n A_n = \{\omega \mid \omega \in A_n \text{ for all values of } n\}.$$

When the sequence is infinite these may be regarded as obvious "set limits" of finite unions or intersections, thus:

$$\bigcup_{n=1}^{\infty} A_n = \lim_{m\to\infty} \bigcup_{n=1}^{m} A_n; \quad \bigcap_{n=1}^{\infty} A_n = \lim_{m\to\infty} \bigcap_{n=1}^{m} A_n .$$

Observe that as m increases, $\bigcup_{n=1}^{m} A_n$ does not decrease while $\bigcap_{n=1}^{m} A_n$ does not increase, and we may say that the former *swells up* to $\bigcup_{n=1}^{\infty} A_n$, the latter *shrinks down* to $\bigcap_{n=1}^{\infty} A_n$.

The distributive laws and De Morgan's laws have obvious extensions to a finite or infinite sequence of sets. For instance

$$\left(\bigcup_{n} A_n\right) \cap B = \bigcup_{n} (A_n \cap B) \tag{1.3.6}$$

$$\left(\bigcap_{n} A_n\right)^c = \bigcup_{n} A_n^c . \tag{1.3.7}$$

Really interesting new sets are produced by using both union and intersection an infinite number of times, and in succession. Here are the two most prominent ones:

$$\bigcap_{m=1}^{\infty}\left(\bigcup_{n=m}^{\infty} A_n\right); \quad \bigcup_{m=1}^{\infty}\left(\bigcap_{n=m}^{\infty} A_n\right).$$

These belong to a more advanced course (see [Chung 1; §4.2] of the References). They are shown here as a preview to arouse your curiosity.

1.4.* Indicator

The idea of classifying ω by means of a dichotomy: to be or not to be in A, which we discussed toward the end of §1.3, can be quantified into a useful device. This device will generalize to the fundamental notion of "random variable" in Chapter 4.

Imagine Ω to be a target board and A a certain marked area on the board as in Examples (f) and (f′) above. Imagine that "pick a point ω in Ω" is done by shooting a dart at the target. Suppose a bell rings (or a bulb lights up) when the dart hits within the area A; otherwise it is a dud. This is the intuitive picture expressed below by a mathematical formula:

$$I_A(\omega) = \begin{cases} 1 \text{ if } \omega \in A, \\ 0 \text{ if } \omega \notin A. \end{cases}$$

* This section may be omitted after the first three paragraphs.

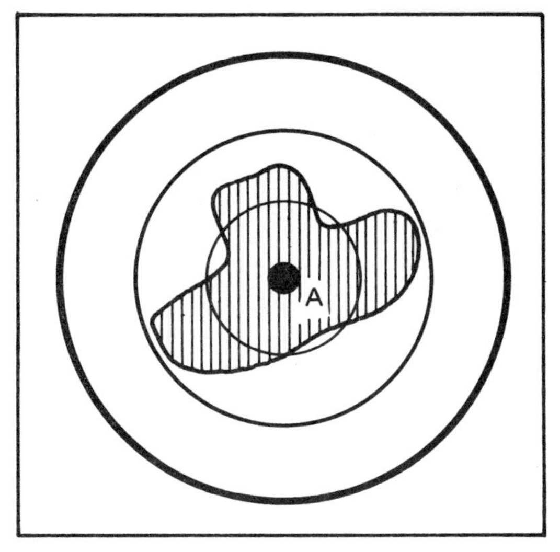

Figure 9

Thus the symbol I_A is a function which is defined on the whole sample space Ω and takes only the two values 0 and 1, corresponding to a dud and a ring. You may have learned in a calculus course the importance of distinguishing between a function (sometimes called a mapping) and one of its values. Here it is the function I_A that indicates the set A, hence it is called the indicator function, or briefly, *indicator* of A. Another set B has *its* indicator I_B. The two functions I_A and I_B are identical (what does *that* mean?) if and only if the two sets are identical.

To see how we can put indicators to work, let us figure out the indicators for some of the sets discussed before. We need two mathematical symbols $\vee$ (cup) and $\wedge$ (cap) which may be new to you. For any two real numbers a and b, they are defined as follows:

$$\begin{aligned} a \vee b &= \text{maximum of } a \text{ and } b, \\ a \wedge b &= \text{minimum of } a \text{ and } b. \end{aligned} \tag{1.4.1}$$

In case $a = b$, either one of them will serve as maximum as well as minimum. Now the salient properties of indicators are given by the formulas below:

$$I_{A \cap B}(\omega) = I_A(\omega) \wedge I_B(\omega) = I_A(\omega) \cdot I_B(\omega); \tag{1.4.2}$$

$$I_{A \cup B}(\omega) = I_A(\omega) \vee I_B(\omega). \tag{1.4.3}$$

You should have no difficulty checking these equations, after all there are only two possible values 0 and 1 for each of these functions. Since the equations are true for every ω, they can be written more simply as equations (identities) between *functions:*

(1.4.4) $$I_{A \cap B} = I_A \wedge I_B = I_A \cdot I_B,$$

(1.4.5) $$I_{A \cup B} = I_A \vee I_B.$$

Here for example the function $I_A \wedge I_B$ is that mapping which assigns to each ω the value $I_A(\omega) \wedge I_B(\omega)$, just as in calculus the function $f + g$ is that mapping which assigns to each x the number $f(x) + g(x)$.

After observing the product $I_A(\omega) \cdot I_B(\omega)$ at the end of (1.4.2) you may be wondering why we do not have the sum $I_A(\omega) + I_B(\omega)$ in (1.4.3). But if this were so we could get the value 2 there, which is impossible since the first member $I_{A \cup B}(\omega)$ cannot take this value. Nevertheless, shouldn't $I_A + I_B$ mean something? Consider target shooting again but this time mark out two overlapping areas A and B. Instead of bell-ringing, you get 1 penny if you hit within A, and also if you hit within B. What happens if you hit the intersection AB? That depends on the rule of the game. Perhaps you still get 1 penny, perhaps you get 2 pennies. Both rules are legitimate. In formula (1.4.3) it is the first rule that applies. If you want to apply the second rule, then you are no longer dealing with the set $A \cup B$ alone as in Figure 10a, but something like Figure 10b:

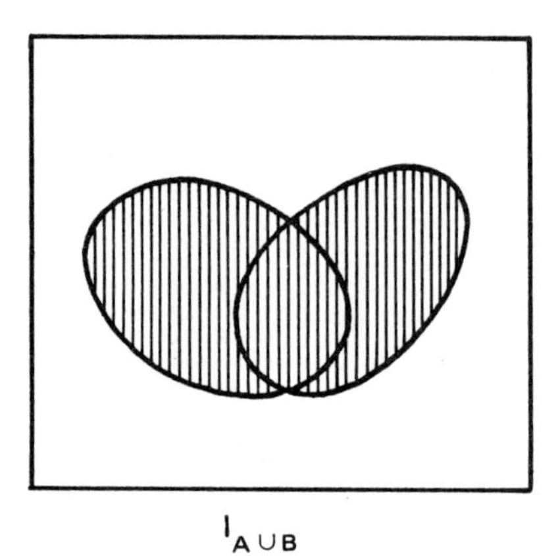

$I_{A \cup B}$

Figure 10a

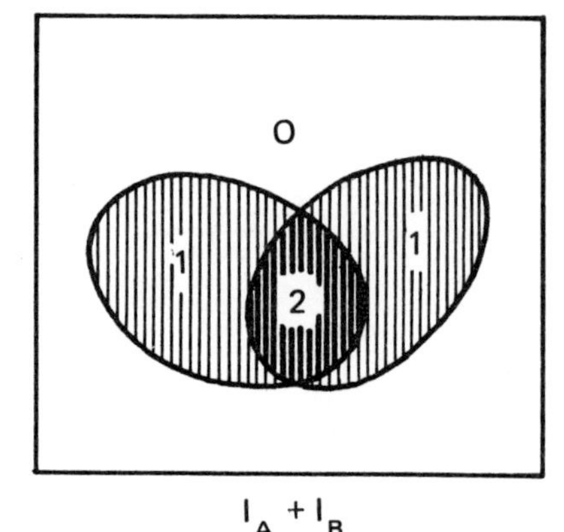

$I_A + I_B$

Figure 10b

This situation can be realized electrically by laying first a uniform charge over the area A, and then on top of this, another charge over the area B, so that the resulting total charge is distributed as shown in Figure 10b. In this case the variable charge will be represented by the function $I_A + I_B$. Such a sum of indicators is a very special case of sum of random variables which will occupy us in later chapters.

For the present let us return to formula (1.4.5) and note that if the two sets A and B are disjoint, then it indeed reduces to the sum of the indicators, because then at most one of the two indicators can take the value 1, so that the maximum coincides with the sum, namely

$$0 \vee 0 = 0 + 0, 0 \vee 1 = 0 + 1, 1 \vee 0 = 1 + 0.$$

Thus we have

(1.4.6) $$I_{A+B} = I_A + I_B \quad \text{provided} \quad A \cap B = \varnothing.$$

As a particular case, we have for any set A:

$$I_\Omega = I_A + I_{A^c}.$$

Now I_Ω is the constant function 1 (on Ω), hence we may rewrite the above as

(1.4.7) $$I_{A^c} = 1 - I_A.$$

We can now derive an interesting formula. Since $(A \cup B)^c = A^c B^c$, we get by applying (1.4.7), (1.4.4) and then (1.4.7) again:

$$I_{A \cup B} = 1 - I_{A^c B^c} = 1 - I_{A^c} I_{B^c} = 1 - (1 - I_A)(1 - I_B).$$

Multiplying out the product (we are dealing with numerical functions!) and transposing terms we obtain

(1.4.8) $$I_{A \cup B} + I_{A \cap B} = I_A + I_B.$$

Finally we want to investigate $I_{A \triangle B}$. We need a bit of arithmetic (also called number theory) first. All integers can be classified as even or odd, according as the remainder we get when we divide it by 2 is 0 or 1. Thus each integer may be identified with (or reduced to) 0 or 1, provided we are only interested in its *parity* and not its exact value. When integers are added or subtracted subject to this reduction, we say we are operating *modulo* 2. For instance:

$$5 + 7 + 8 - 1 + 3 = 1 + 1 + 0 - 1 + 1 = 2 = 0, \quad \text{modulo 2.}$$

A famous case of this method of counting occurs when the maiden picks off the petals of some wild flower one by one and murmers: "he loves me," "he loves me not" in turn. Now you should be able to verify the following equation for every ω:

(1.4.9) $$\begin{aligned} I_{A \triangle B}(\omega) &= I_A(\omega) + I_B(\omega) - 2I_{AB}(\omega) \\ &= I_A(\omega) + I_B(\omega), \qquad \text{modulo 2.} \end{aligned}$$

We can now settle a question raised in Sec. 1.3, and establish without pain the identity:

(1.4.10) $$(A \triangle B) \triangle C = A \triangle (B \triangle C).$$

Proof. Using (1.4.9) twice we have

(1.4.11) $$I_{(A \triangle B) \triangle C} = I_{A \triangle B} + I_C = (I_A + I_B) + I_C, \qquad \text{modulo } 2.$$

Now if you have understood the meaning of addition modulo 2 you should see at once that it is an associative operation (what does that mean, "modulo 2"?). Hence the last member of (1.4.11) is equal to

$$I_A + (I_B + I_C) = I_A + I_{B \triangle C} = I_{A \triangle (B \triangle C)}, \qquad \text{modulo } 2.$$

We have therefore shown that the two sets in (1.4.10) have identical indicators, hence they are identical. Q.E.D.

We do not need this result below. We just want to show that a trick is sometimes neater than a picture!

Exercises

1. Why is the sequence of numbers $\{1, 2, 1, 2, 3\}$ not a set?
2. If two sets have the same size, are they then identical?
3. Can a set and a proper subset have the same size? (A *proper* subset is a subset which is not also a superset!)
4. If two sets have identical complements, then they are themselves identical. Show this in two ways: (i) by verbal definition, (ii) by using formula (1.2.1).
5. If A, B, C have the same meanings as in Section 1.2, what do the following sets mean:

$$A \cup (B \cap C); \; (A \backslash B) \backslash C; \; A \backslash (B \backslash C).$$

6. Show that

$$(A \cup B) \cap C \neq A \cup (B \cap C);$$

but also give some special cases where there *is* equality.

7. Using the atoms given in the decomposition (1.3.5), express

$$A \cup B \cup C; \; (A \cup B)(B \cup C); \; A \backslash B; \; A \triangle B;$$

the set of ω which belongs to exactly 1 [exactly 2; at least 2] of the sets A, B, C.

8. Show that $A \subset B$ if and only if $AB = A$; or $A \cup B = B$. (So the relation of inclusion can be defined through identity and the operations.)
9. Show that A and B are disjoint if and only if $A \backslash B = A$; or $A \cup B = A \triangle B$. (After No. 8 is done, this can be shown purely symbolically without going back to the verbal definitions of the sets.)
10. Show that there is a distributive law also for difference:

$$(A \backslash B) \cap C = (A \cap C) \backslash (B \cap C).$$

Is the dual

$$(A \cap B) \backslash C = (A \backslash C) \cap (B \backslash C)$$

also true?

11. Derive (D_2) from (D_1) by using (C_1) and (C_2).
12.* Show that

$$(A \cup B) \backslash (C \cup D) \subset (A \backslash C) \cup (B \backslash D).$$

13.* Let us define a new operation "/" as follows:

$$A/B = A^c \cup B.$$

Show that

(i) $(A/B) \cap (B/C) \subset A/C$;

(ii) $(A/B) \cap (A/C) = A/BC$;

(iii) $(A/B) \cap (B/A) = (A \triangle B)^c$.

In intuitive logic, "A/B" may be read as "A implies B." Use this to interpret the relations above.

14.* If you like a "dirty trick" this one is for you. There is an operation between two sets A and B from which alone all the operations defined above can be derived. [Hint: It is sufficient to derive complement and union from it. Look for some combination which contains these two. It is not unique.]
15. Show that $A \subset B$ if and only if $I_A \leq I_B$; and $A \cap B = \varnothing$ if and only if $I_A I_B = 0$.
16. Think up some concrete schemes which illustrate formula (1.4.8).
17. Give a direct proof of (1.4.8) by checking it for all ω. You may use the atoms in (1.3.4) if you want to be well organized.
18. Show that for any real numbers a and b, we have

$$a + b = (a \vee b) + (a \wedge b).$$

Use this to prove (1.4.8) again.

19. Express $I_{A \backslash B}$ and I_{A-B} in terms of I_A and I_B.
20. Express $I_{A \cup B \cup C}$ as a *polynomial* of I_A, I_B, I_C. [Hint: Consider $1 - I_{A \cup B \cup C}$.]
21.* Show that

$$I_{ABC} = I_A + I_B + I_C - I_{A \cup B} - I_{A \cup C} - I_{B \cup C} + I_{A \cup B \cup C}.$$

You can verify this directly, but it is nicer to derive it from No. 20 by duality.

Chapter 2

Probability

2.1. Examples of probability

We learned something about sets in Chapter 1; now we are going to measure them. The most primitive way of measuring is to count the number, so we will begin with such an example.

Example 1. In Example (a′) of §1.1, suppose that the number of rotten apples is 28. This gives a measure to the set A described in (a′), called its size and denoted by $|A|$. But it does not tell anything about the total number of apples in the bushel, namely the size of the sample space Ω given in Example (a). If we buy a bushel of apples we are more likely to be concerned with the relative *proportion* of rotten ones in it rather than their absolute number. Suppose then the total number is 550. If we now use the letter P provisionarily for "proportion," we can write this as follows:

$$P(A) = \frac{|A|}{|\Omega|} = \frac{28}{550}. \tag{2.1.1}$$

Suppose next that we consider the set B of unripe apples in the same bushel, whose number is 47. Then we have similarly

$$P(B) = \frac{|B|}{|\Omega|} = \frac{47}{550}.$$

It seems reasonable to suppose that an apple cannot be both rotten and unripe (this is really a matter of definition of the two adjectives); then the two sets are disjoint so their members do not overlap. Hence the number of "rotten or unripe apples" is equal to the sum of the number of "rotten apples" and the number of "unripe apples": $28 + 47 = 75$. This may be written in symbols as:

$$|A + B| = |A| + |B|. \tag{2.1.2}$$

If we now divide through by $|\Omega|$, we obtain

$$P(A + B) = P(A) + P(B). \tag{2.1.3}$$

On the other hand, if some apples can be rotten and unripe at the same time, such as when worms got into green ones, then the equation (2.1.2) must be replaced by an inequality:

$$|A \cup B| \leq |A| + |B|$$

which leads to

$$P(A \cup B) \leq P(A) + P(B). \tag{2.1.4}$$

Now what is the excess of $|A| + |B|$ over $|A \cup B|$? It is precisely the number of "rotten and unripe apples," that is, $|A \cap B|$. Thus

$$|A \cup B| + |A \cap B| = |A| + |B|$$

which yields the pretty equation

$$P(A \cup B) + P(A \cap B) = P(A) + P(B). \tag{2.1.5}$$

Example 2. A more sophisticated way of measuring a set is the area of a plane set as in Examples (f) and (f′) of Section 1.1, or the volume of a solid. It is said that the measurement of land areas was the origin of geometry and trigonometry in ancient times. While the nomads were still counting on their fingers and toes as in Example 1, the Chinese and Egyptians, among other peoples, were subdividing their arable lands, measuring them in units and keeping accounts of them on stone tablets or papyrus. This unit varied a great deal from one civilization to another (who knows the conversion rate of an acre into *mou*'s or hectares?). But again it is often the ratio of two areas which concerns us as in the case of a wild shot which hits the target board. The proportion of the area of a subset A to that of Ω may be written, if we denote the area by the symbol $|\quad|$:

$$P(A) = \frac{|A|}{|\Omega|}. \tag{2.1.6}$$

This means also that if we fix the unit so that the total area of Ω is 1 unit, then the area of A is equal to the fraction $P(A)$ on this scale. Formula (2.1.6) looks just like formula (2.1.1) by the deliberate choice of notation in order to underline the similarity of the two situations. Furthermore, for two sets A and B the previous relations (2.1.3) to (2.1.5) hold equally well in their new interpretations.

Example 3. When a die is thrown there are six possible outcomes. If we compare the process of throwing a particular number [face] with that of picking a particular apple in Example 1, we are led to take $\Omega = \{1, 2, 3, 4, 5, 6\}$ and define

$$P(\{k\}) = \frac{1}{6}, \quad k = 1, 2, 3, 4, 5, 6. \tag{2.1.7}$$

Here we are treating the six outcomes as "equally likely," so that the same measure is assigned to all of them, just as we have done tacitly with the apples.

This hypothesis is usually implied by saying that the die is "perfect." In reality of course no such die exists. For instance the mere marking of the faces would destroy the perfect symmetry; and even if the die were a perfect cube, the outcome would still depend on the way it is thrown. Thus we must stipulate that this is done in a perfectly symmetrical way too, and so on. Such conditions can be approximately realized and constitute the basis of an assumption of equal likelihood on grounds of symmetry.

Now common sense demands an empirical interpretation of the "probability" given in (2.1.7). It should give a measure of what is *likely* to happen, and this is associated in the intuitive mind with the observable frequency of occurrence. Namely, if the die is thrown a number of times, how often will a particular face appear? More generally, let A be an event determined by the outcome; e.g. "to throw a number not less than 5 [or an odd number]." Let $N_n(A)$ denote the number of times the event A is observed in n throws, then the *relative frequency* of A in these trials is given by the ratio

$$Q_n(A) = \frac{N_n(A)}{n}. \tag{2.1.8}$$

There is good reason to take this Q_n as a measure of A. Suppose B is another event such that A and B are *incompatible* or *mutually exclusive* in the sense that they cannot occur in the same trial. Clearly we have $N_n(A + B) = N_n(A) + N_n(B)$ and consequently

$$\begin{aligned} Q_n(A + B) &= \frac{N_n(A + B)}{n} \\ &= \frac{N_n(A) + N_n(B)}{n} = \frac{N_n(A)}{n} + \frac{N_n(B)}{n} = Q_n(A) + Q_n(B). \end{aligned} \tag{2.1.9}$$

Similarly for any two events A and B in connection with the same game, not necessarily incompatible, the relations (2.1.4) and (2.1.5) hold with the P's there replaced by our present Q_n. Of course this Q_n depends on n, and will fluctuate, even wildly, as n increases. But if you let n go to infinity, will the sequence of ratios $Q_n(A)$ "settle down to a steady value"? Such a question can never be answered empirically, since by the very nature of a limit we cannot put an end to the trials. So it is a mathematical idealization to assume that such a limit does exist, and then write

$$Q(A) = \lim_{n \to \infty} Q_n(A). \tag{2.1.10}$$

We may call this the empirical *limiting frequency* of the event A. If you know how to operate with limits then you can see easily that the relation (2.1.9) remains true "in the limit." Namely when we let $n \to \infty$ everywhere in that formula and use the definition (2.1.10), we obtain (2.1.3) with P replaced by Q. Similarly (2.1.4) and (2.1.5) also hold in this context.

But the limit Q still depends on the actual sequence of trials which are carried out to determine its value. On the face of it, there is no guarantee whatever that another sequence of trials, even if it is carried out under the same circumstances, will yield the same value. Yet our intuition demands that a measure of the likelihood of an event such as A should tell something more than the mere record of one experiment. A viable theory built on the frequencies will have to assume that the Q defined above is in fact the same for all similar sequences of trials. Even with the hedge implicit in the word "similar," that is assuming a lot to begin with. Such an attempt has been made with limited success, and has a great appeal to common sense, but we will not pursue it here. Rather, we will use the definition in (2.1.7) which implies that if A is any subset of Ω and $|A|$ its size, then

(2.1.11) $$P(A) = \frac{|A|}{|\Omega|} = \frac{|A|}{6}.$$

For example, if A is the event "to throw an odd number," then A is identified with the set $\{1, 3, 5\}$ and $P(A) = 3/6 = 1/2$.

It is a fundamental proposition in the theory of probability that under certain conditions (repeated *independent* trials with *identical* die), the limiting frequency in (2.1.10) will indeed exist and be equal to $P(A)$ defined in (2.1.11), for "practically all" conceivable sequences of trials. This celebrated theorem, called the *Law of Large Numbers*, is considered to be the cornerstone of all empirical sciences. In a sense it justifies the intuitive foundation of probability as frequency discussed above. The precise statement and derivation will be given in Chapter 7. We have made this early announcement to quiet your feelings or misgivings about frequencies and to concentrate for the moment on sets and probabilities in the following sections.

2.2. Definition and illustrations

First of all, a probability is a number associated with or assigned to a set in order to measure it in some sense. Since we want to consider many sets at the same time (that is why we studied Chapter 1), and each of them will have a probability associated with it, this makes probability a "function of sets." You should have already learned in some mathematics course what a function means, in fact this notion has been used a little in Chapter 1. Nevertheless, let us review it in the familiar notation: a function f defined for some or all real numbers is a rule of association, by which we assign the number $f(x)$ to the number x. It is sometimes written as $f(\cdot)$, or more painstakingly as follows:

(2.2.1) $$f: x \rightarrow f(x).$$

So when we say a probability is a function of sets we mean a similar association, except that x is replaced by a set S:

(2.2.2) $$P: S \rightarrow P(S).$$

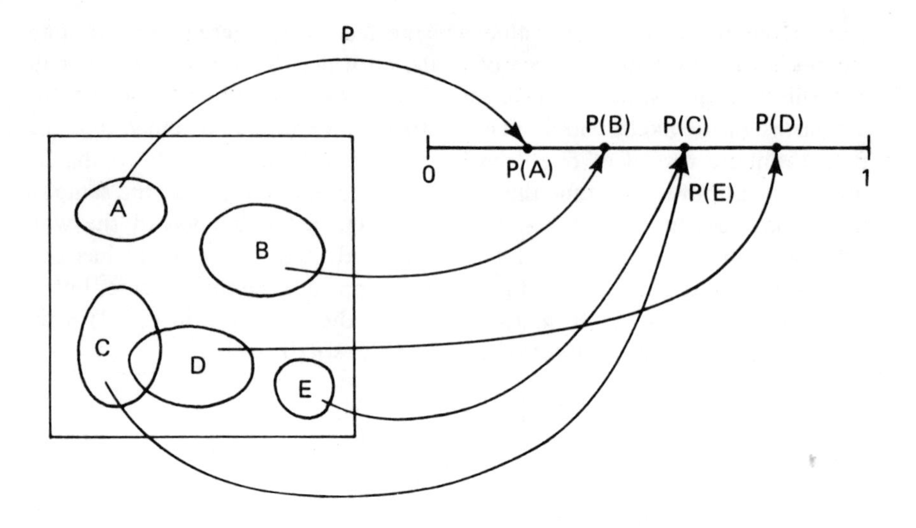

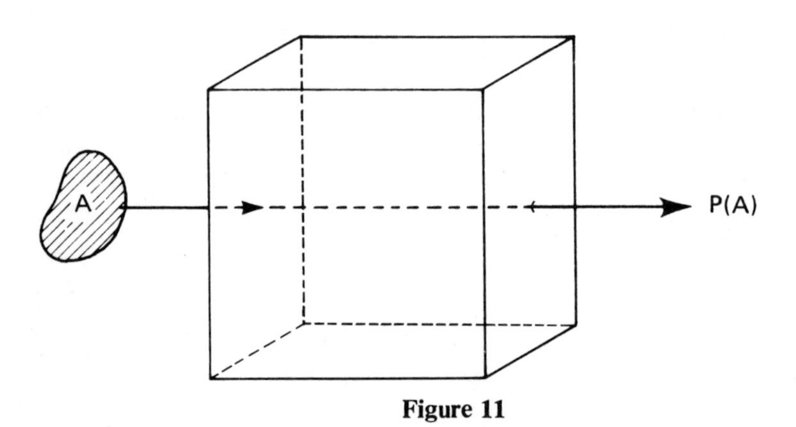

Figure 11

The *value* $P(S)$ is still a number, indeed it will be a number between 0 and 1. We have not been really precise in (2.2.1), because we have not specified the set of x there for which it has a meaning. This set may be the interval (a, b) or the half line $(0, \infty)$ or some more complicated set called the domain of f. Now what is the domain of our probability function P? It must be a *set of sets*, or to avoid the double usage, a *family* (*class*) of sets. As in Chapter 1 we are talking about subsets of a fixed sample space Ω. It would be nice if we could use the family of *all* subsets of Ω, but unexpected difficulties will arise in this case if no restriction is imposed on Ω. We might say that if Ω is too large, namely when it contains uncountably many points, then it has too many subsets, and it becomes impossible to assign a probability to each of them and still satisfy a basic rule (Axiom (ii*) below) governing the assignments. However, if Ω is a finite or countably infinite set then no such trouble can arise and we may indeed assign a probability to each and all of its subsets. This will be shown at the beginning of §2.4. You are supposed to know what a finite set is (although it is by no means easy to give a logical definition, while it is mere tautology to say that "it has only a finite number of points"); let us review what a countably infinite set is. This notion will be of sufficient importance to us, even if it only lurks in the background most of the time.

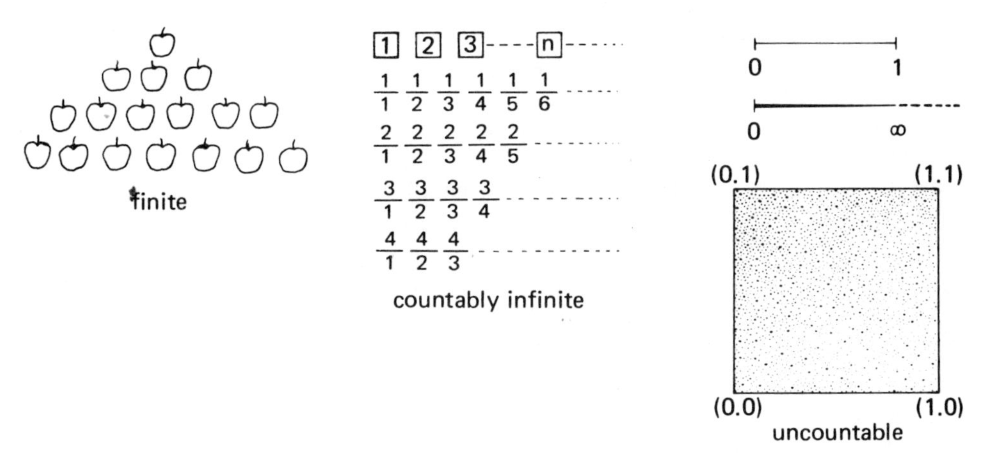

Figure 12

A set is countably infinite when it can be put into 1-to-1 correspondence with the set of positive integers. This correspondence can then be exhibited by labeling the elements as $\{s_1, s_2, \ldots, s_n, \ldots\}$. There are of course many ways of doing this, for instance we can just let some of the elements swap labels (or places if they are thought of being laid out in a row). The set of positive rational numbers is countably infinite, hence they can be labeled in some way as $\{r_1, r_2, \ldots, r_n, \ldots\}$ but don't think for a moment that you can do this by putting them in increasing order as you can with the positive integers $1 < 2 < \cdots < n < \cdots$. From now on we shall call a set *countable* when it is either finite or countably infinite. Otherwise it is called *uncountable*. For example, the set of all real numbers is uncountable. We shall deal with uncountable sets later, and we will review some properties of a countable set when we need them. For the present we will assume the sample space Ω to be countable in order to give the following definition in its simplest form, without a diverting complication. As a matter of fact, we could even assume Ω to be finite as in Examples (a) to (e) of §1.1, without losing the essence of the discussion below.

Definition. A *probability measure* on the sample space Ω is a function of subsets of Ω satisfying three axioms:

(i) For every set $A \subset \Omega$, the value of the function is a non-negative number: $P(A) \geq 0$.

(ii) For any two disjoint sets A and B, the value of the function for their union $A + B$ is equal to the sum of its value for A and its value for B:

$$P(A + B) = P(A) + P(B) \quad \text{provided} \quad AB = \varnothing.$$

(iii) The value of the function for Ω (as a subset) is equal to 1:

$$P(\Omega) = 1.$$

Observe that we have been extremely careful in distinguishing the function $P(\cdot)$ from its values such as $P(A)$, $P(B)$, $P(A + B)$, $P(\Omega)$. Each of these is "a probability," but the function itself should properly be referred to as a "probability measure" as indicated.

Example 1 in §2.1 shows that the proportion P defined there is in fact a probability measure on the sample space, which is a bushel of 550 apples. It assigns a probability to every subset of these apples and this assignment satisfies the three axioms above. In Example 2 if we take Ω to be all the land that belonged to the Pharaoh, it is unfortunately not a countable set. Nevertheless we can define the area for a very large class of subsets which are called "measurable," and if we restrict ourselves to these subsets only, the "area function" is a probability measure as shown in Example 2 where this restriction is ignored. Note that Axiom (iii) reduces to a convention: the decree of a unit. Now how can a land area not be measurable? While this is a sophisticated mathematical question which we will not go into in this book, it is easy to think of practical reasons for the possibility: the piece of land may be too jagged, rough or inaccessible. (See Fig. 13 on page 25)

In Example 3 we have shown that the empirical relative frequency is a probability measure. But we will not use this definition in this book. Instead, we will use the first definition given at the beginning of Example 3, which is historically the earliest of its kind. The general formulation will now be given.

Example 4. A classical enunciation of probability runs as follows. The probability of an event is the ratio of the number of cases *favorable* to that event to the total number of cases, provided that all these are *equally likely*.

To translate this into our language: the sample space is a finite set of possible cases: $\{\omega_1, \omega_2, \ldots, \omega_m\}$, each ω_i being a "case." An event A is a subset $\{\omega_{i_1}, \omega_{i_2}, \ldots, \omega_{i_n}\}$, each ω_{i_j} being a "favorable case." The probability of A is then the ratio

$$P(A) = \frac{|A|}{|\Omega|} = \frac{n}{m}. \tag{2.2.3}$$

As we see from the discussion in Example 1, this defines a probability measure P on Ω anyway, so that the stipulation above that the cases be equally likely

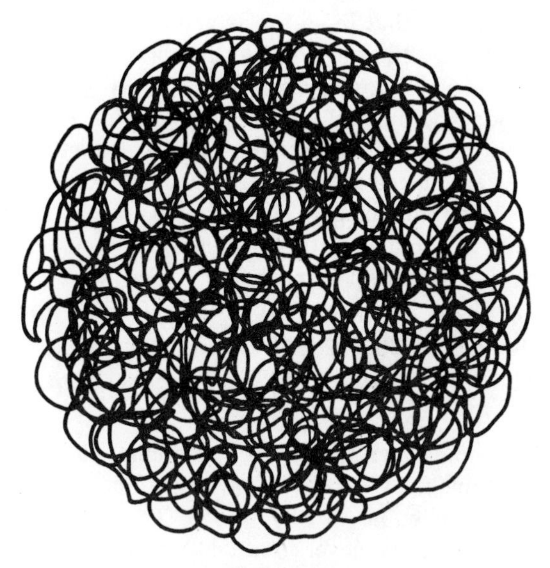

Figure 13

is superfluous from the axiomatic point of view. Besides, what does it really mean? It sounds like a bit of tautology, and how is one going to decide whether the cases are equally likely or not?

A celebrated example will illustrate this. Let two coins be tossed. D'Alembert (mathematician, philosopher and encyclopedist, 1717–83) argued that there are three possible cases, namely:

(i) both heads, (ii) both tails, (iii) a head and a tail.

So he went on to conclude that the probability of "a head and a tail" is equal to $1/3$. If he had figured that this *probability* should have something to do with the experimental *frequency* of the occurrence of the event, he might have changed his mind after tossing two coins more than a few times. (History does not record if he ever did that, but it is said that for centuries people believed that men had more teeth than women because Aristotle had said so, and apparently nobody bothered to look into a few mouths.) For the three cases he considered are not equally likely. Case (iii) should be split into two:

(iiia) first coin shows head and second coin shows tail.

(iiib) first coin shows tail and second coin shows head.

It is the four cases (i), (ii), (iiia) and (iiib) that are equally likely by symmetry and on empirical evidence. This should be obvious if we toss the two coins one after the other rather than simultaneously. However, there is an important point to be made clear here. The two coins may be physically indistinguishable so that in so far as actual observation is concerned, D'Alembert's 3 cases are the only distinct *patterns* to be recognized. In the model of two coins they happen not to be equally likely on the basis of common sense and experimental evidence. But in an analogous model for certain microcosmic particles, called Bose-Einstein statistics (see Exercise 24 of Chapter 3), they are indeed assumed to be equally likely in order to explain some types of physical phenomena. Thus what we regard as "equally likely" is a matter outside of the axiomatic formulation. To put it another way, if we use (2.2.3) as our definition of probability then we are in effect treating the ω's as equally likely, in the sense that we count only their numbers and do not attach different weights to them.

Example 5. If six dice are rolled, what is the probability that all show different faces?

This is just Example (e) and (e′). It is stated elliptically on purpose to get you used to such problems. We have already mentioned that the total number of possible outcomes is equal to $6^6 = 46656$. They are supposed to be all "equally likely" although we never breathed a word about this assumption. Why, nobody can solve the problem as announced without such an assumption. Other data about the dice would have to be given before we could begin—which is precisely the difficulty when similar problems arise in practice. Now if the dice are all perfect, and the mechanism by which they are rolled is also perfect, which excludes any collusion between the movements of the several dice, then our hypothesis of equal likelihood may be justified. Such conditions are taken for granted in a problem like this when nothing is said about the dice. The solution is then given by (2.2.3) with $n = 6^6$ and $m = 6!$ (see Example 2 in §3.1 for these computations):

$$\frac{6!}{6^6} = \frac{720}{46656} = .015432$$

approximately.

Let us note that if the dice are not distinguishable from each other, then to the observer there is exactly one *pattern* in which the six dice show different faces. Similarly, the total number of different patterns when six dice are rolled is much smaller than 6^6 (see Example 3 of §3.2). Yet when we count the possible outcomes we must think of the dice as distinguishable, as if they were painted in different colors. This is one of the vital points to grasp in the counting of cases; see Chapter 3.

In some situations the equally likely cases must be searched out. This point will be illustrated by a famous historical problem called the "problem of points."

Example 6. Two players A and B play a series of games in which the probability of each winning a single game is equal to 1/2, irrespective [independent] of the outcomes of other games. For instance, they may play tennis in which they are equally matched, or simply play "heads or tails" by tossing an unbiased coin. Each player gains a "point" when he wins a game, and nothing when he loses. Suppose that they stop playing when A needs 2 more points, and B needs 3 more points to win the stake. How should they divide it fairly?

It is clear that the winner will be decided in 4 more games. For in those 4 games either A will have won ≥ 2 points or B will have won ≥ 3 points, but not both. Let us enumerate all the possible outcomes of these 4 games using the letter A or B to denote the winner of each game:

AAAA	*AAAB*	*AABB*	*ABBB*	*BBBB*
	AABA	*ABAB*	*BABB*	
	ABAA	*ABBA*	*BBAB*	
	BAAA	*BAAB*	*BBBA*	
		BABA		
		BBAA		

These are equally likely cases on grounds of symmetry. There are* $\binom{4}{4} + \binom{4}{3} + \binom{4}{2} = 11$ cases in which A wins the stake; and $\binom{4}{3} + \binom{4}{4} = 5$ cases in which B wins the stake. Hence the stake should be divided in the ratio 11:5. Suppose it is \$64000; then A gets \$44000, B gets \$20000. [We are taking the liberty of using the dollar as currency; the U.S.A. did not exist at the time when the problem was posed.]

This is Pascal's solution in a letter to Fermat dated August 24, 1654. [Blaise Pascal (1623–62); Pierre de Fermat (1601–65); both among the greatest mathematicians of all time.] Objection was raised by a learned contemporary (and repeated through the ages) that the enumeration above was not reasonable, because the series would have stopped as soon as the winner was decided and not have gone on through all 4 games in some cases. Thus the real possibilities are as follows:

AA	*ABBB*
ABA	*BABB*
ABBA	*BBAB*
BAA	*BBB*
BABA	
BBAA	

But these are not equally likely cases. In modern terminology, if these 10 cases are regarded as constituting the sample space, then

* See (3.2.3) for notation used below.

$$P(AA) = \frac{1}{4}, \quad P(ABA) = P(BAA) = P(BBB) = \frac{1}{8},$$

$$P(ABBA) = P(BABA) = P(BBAA) = P(ABBB) = P(BABB) = P(BBAB) = \frac{1}{16}$$

since A and B are independent events with probability 1/2 each (see §2.4). If we add up these probabilities we get of course

$$P\ (A \text{ wins the stake}) = \frac{1}{4} + \frac{1}{8} + \frac{1}{16} + \frac{1}{8} + \frac{1}{16} + \frac{1}{16} = \frac{11}{16};$$

$$P\ (B \text{ wins the stake}) = \frac{1}{16} + \frac{1}{16} + \frac{1}{16} + \frac{1}{8} = \frac{5}{16}.$$

Pascal did not quite explain his method this way, saying merely that "it is absolutely equal and indifferent to each whether they play in the natural way of the game, which is to finish as soon as one has his score, or whether they play the entire four games." A later letter by him seems to indicate that he fumbled on the same point in a similar problem with three players. The student should take heart that this kind of reasoning was not easy even for past masters.

2.3. Deductions from the axioms

In this section we will do some simple "axiomatics." That is to say, we shall deduce some properties of the probability measure from its definition, using of course the axioms but nothing else. In this respect the axioms of a mathematical theory are like the constitution of a government. Unless and until it is changed or amended, every *law* must be made to follow from it. In mathematics we have the added assurance that there are no divergent views as to how the constitution should be construed.

We record some consequences of the axioms in (iv) to (viii) below. First of all, let us show that a probability is indeed a number between 0 and 1.

(iv) for any set A, we have

$$P(A) \leq 1.$$

This is easy but you will see that in the course of deducing it we shall use all three axioms. Consider the complement A^c as well as A. These two sets are disjoint and their union is Ω:

$$A + A^c = \Omega. \tag{2.3.1}$$

So far, this is just set theory, no probability theory yet. Now use Axiom (ii) on the left side of (2.3.1) and Axiom (iii) on the right:

$$P(A) + P(A^c) = P(\Omega) = 1. \tag{2.3.2}$$

Finally use Axiom (i) for A^c to get

$$P(A) = 1 - P(A^c) \leq 1.$$

Of course the first inequality above is just Axiom (i). You might object to our slow pace above by pointing out that since A is *contained in* Ω, it is obvious that $P(A) \leq P(\Omega) = 1$. This reasoning is certainly correct but we still have to pluck it from the axioms, and that is the point of the little proof above. We can also get it from the following more general proposition.

(v) For any two sets such that $A \subset B$, we have

$$P(A) \leq P(B), \text{ and } P(B - A) = P(B) - P(A).$$

The proof is an imitation of the preceding one with B playing the role of Ω. We have

$$B = A + (B - A)$$
$$P(B) = P(A) + P(B - A) \geq P(A).$$

The next proposition is such an immediate extension of Axiom (ii) that we could have adopted it instead as an axiom.

(vi) For any finite number of disjoint sets $A_1, \ldots, A_n$, we have

$$P(A_1 + \cdots + A_n) = P(A_1) + \cdots + P(A_n). \tag{2.3.3}$$

This property of the probability measure is called *finite additivity*. It is trivial if we recall what "disjoint" means and use (ii) a few times; or we may proceed by induction if we are meticulous. There is an important extension of (2.3.3) to a countable number of sets later, *not* obtainable by induction!

As already checked in several special cases, there is a generalization of Axiom (ii), hence also of (2.3.3), to sets which are not necessarily disjoint. You may find it trite, but it has the dignified name of *Boole's inequality*. Boole (1815–1864) was a pioneer in the "laws of thought" and author of *Theories of Logic and Probabilities*.

(vii) For any finite number of arbitrary sets $A_1, \ldots, A_n$, we have

$$P(A_1 \cup \cdots \cup A_n) \leq P(A_1) + \cdots + P(A_n). \tag{2.3.4}$$

Let us first show this when $n = 2$. For any two sets A and B, we can write their union as the sum of disjoint sets as follows:

$$A \cup B = A + A^cB. \tag{2.3.5}$$

Now we can apply axiom (ii) to get

$$P(A \cup B) = P(A) + P(A^cB). \tag{2.3.6}$$

Since $A^cB \subset B$ we can apply (v) to get (2.3.4).

The general case follows easily by mathematical induction, and you should write it out as a good exercise on this method. You will find that you need the associative law for union of sets as well as that for the addition of numbers.

The next question is the difference between the two sides of the inequality (2.3.4). The question is somewhat moot since it depends on what we want to use to express the difference. However, when $n = 2$ there is a clear answer.

(viii) For any two sets A and B we have

$$P(A \cup B) + P(A \cap B) = P(A) + P(B). \tag{2.3.7}$$

This can be gotten from (2.3.6) by observing that $A^cB = B - AB$, so that we have by virtue of (v):

$$P(A \cup B) = P(A) + P(B - AB) = P(A) + P(B) - P(AB).$$

which is equivalent to (2.3.7). Another neat proof is given in Exercise 12.

We shall postpone a discussion of the general case until Section 6.2. In practice, the inequality is often more useful than the corresponding identity which is rather complicated.

We will not quit formula (2.3.7) without remarking on its striking resemblance to formula (1.4.8) of §1.4, which is repeated below for the sake of comparison:

$$I_{A \cup B} + I_{A \cap B} = I_A + I_B. \tag{2.3.8}$$

There is indeed a deep connection between the pair, as follows. The probability $P(S)$ of each set S can be obtained from its indicator function I_S by a procedure (operation) called "taking expectation" or "integration." If we perform this on (2.3.8) term-by-term their result is (2.3.7). This procedure is an essential part of probability theory and will be thoroughly discussed in Chapter 6. See Exercise 19 for a special case.

To conclude our axiomatics, we will now strengthen Axiom (ii) or its immediate consequence (vi), namely the finite additivity of P, into a new axiom.

(ii*) Axiom of countable additivity. For a countably infinite collection of disjoint sets A_k, $k = 1, 2, \ldots$, we have

$$P\left(\sum_{k=1}^{\infty} A_k\right) = \sum_{k=1}^{\infty} P(A_k). \tag{2.3.9}$$

This axiom includes (vi) as a particular case, for we need only put $A_k = \varnothing$ for $k > n$ in (2.3.9) to obtain (2.3.3). The empty set is disjoint from any other set including itself, and has probability zero (why?). If Ω is a finite set, then the new axiom reduces to the old one. But it is important to see why (2.3.9) *cannot* be deduced from (2.3.3) by letting $n \to \infty$. Let us try this by rewriting (2.3.3) as follows:

$$P\left(\sum_{k=1}^{n} A_k\right) = \sum_{k=1}^{n} P(A_k). \tag{2.3.10}$$

Since the left side above cannot exceed 1 for all n, the series on the right side must converge and we obtain

$$\lim_{n\to\infty} P\left(\sum_{k=1}^{n} A_k\right) = \lim_{n\to\infty} \sum_{k=1}^{n} P(A_k) = \sum_{k=1}^{\infty} P(A_k). \tag{2.3.11}$$

Comparing this established result with the desired result (2.3.9), we see that the question boils down to:

$$\lim_{n\to\infty} P\left(\sum_{k=1}^{n} A_k\right) = P\left(\sum_{k=1}^{\infty} A_k\right)?$$

which can be exhibited more suggestively as

$$\lim_{n\to\infty} P\left(\sum_{k=1}^{n} A_k\right) = P\left(\lim_{n\to\infty} \sum_{k=1}^{n} A_k\right). \tag{2.3.12}$$

See end of §1.3. (See Fig. 14 on page 32)
Thus it is a matter of interchanging the two operations "lim" and "P" in (2.3.12), or you may say, "taking the limit inside the probability relation." If you have had enough calculus you know this kind of interchange is often hard to justify and may be illegitimate or even invalid. The new axiom is created to secure it in the present case and has fundamental consequences in the theory of probability.

2.4. Independent events

From now on, a "probability measure" will satisfy Axioms (i), (ii*) and (iii). The subsets of Ω to which such a probability has been assigned will also be called an *event*.

We shall show how easy it is to *construct* probability measures for any countable space $\Omega = \{\omega_1, \omega_2, \ldots, \omega_n, \ldots\}$. To each sample point ω_n let us attach an arbitrary "weight" p_n subject only to the conditions:

$$\forall n: \quad p_n \geq 0; \sum_n p_n = 1. \tag{2.4.1}$$

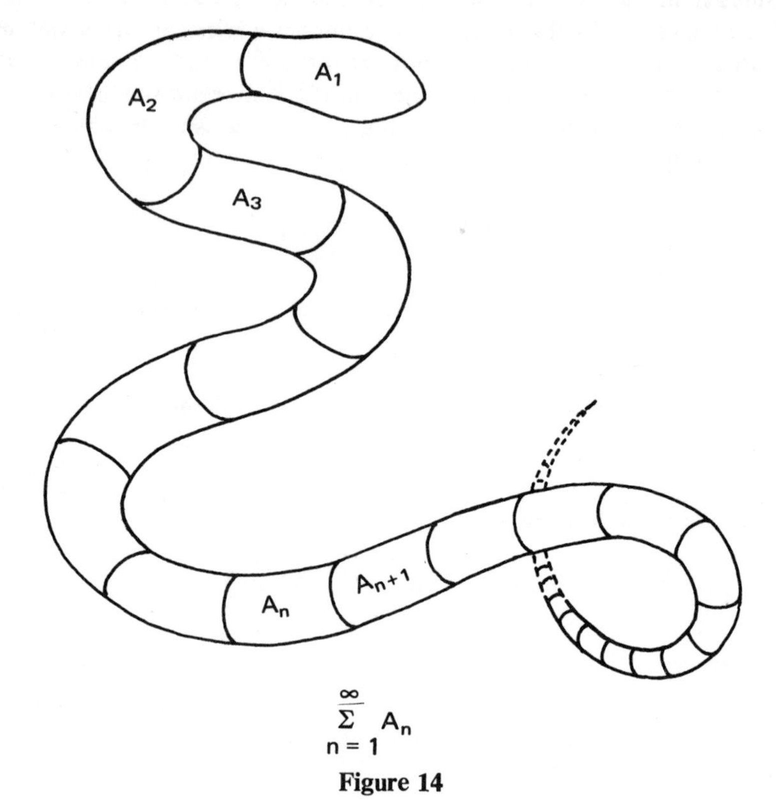

Figure 14

This means that the weights are positive or zero, and add up to 1 altogether. Now for any subset A of Ω, we define its probability to be the *sum of the weights of all the points in it*. In symbols, we put first

$$\forall n:\ P(\{\omega_n\}) = p_n; \tag{2.4.2}$$

and then for every $A \subset \Omega$:

$$P(A) = \sum_{\omega_n \in A} p_n = \sum_{\omega_n \in A} P(\{\omega_n\}).$$

We may write the last term above more neatly as

$$P(A) = \sum_{\omega \in A} P(\{\omega\}). \tag{2.4.3}$$

Thus P is a function defined for all subsets of Ω and it remains to check that it satisfies axioms (i), (ii*) and (iii). This requires nothing but a bit of clear-headed thinking and is best done by yourself. Since the weights are quite arbitrary apart from the easy conditions in (2.4.1), you see that probability measures come "a dime a dozen" in a countable sample space. In fact, we can get them all by the above method of construction. For if any probability measure P is given, never mind how, we can define p_n to be $P(\{\omega_n\})$ as a (2.4.2), and then $P(A)$ must be given as in (2.4.3), *because of Axiom* (ii*). Furthermore the p_n's will satisfy (2.4.1) as a simple consequence of the axioms. In other words, any given P is necessarily of the type described by our construction.

In the very special case that Ω is finite and contains exactly m points, we may attach equal weights to all of them, so that

$$p_n = \frac{1}{m}, \quad n = 1, 2, \ldots, m.$$

Then we are back to the "equally likely" situation in Example 4 of §2.2. But in general the p_n's need not be equal, and when Ω is countably infinite they cannot be all equal (why?). The preceding discussion shows the degree of arbitrariness involved in the general concept of a probability measure.

An important model of probability space is that of *repeated independent trials:* This is the model used when a coin is tossed, a die thrown, a card drawn from a deck (with replacement) several times. Alternately, we may toss several coins or throw several dice at the same time. Let us begin with an example.

Example 7. First toss a coin, then throw a die, finally draw a card from a deck of poker cards. Each trial produces an event; let

$$\begin{aligned} A &= \text{coin falls heads;} \\ B &= \text{die shows number 5 or 6;} \\ C &= \text{card drawn is a spade.} \end{aligned}$$

Assuming that the coin is fair, the die is perfect and the deck thoroughly shuffled. Furthermore assume that these three trials are carried out "independently" of each other, which means intuitively that the outcome of each trial does not influence that of the others. For instance this condition is approximately fulfilled if they are done by different people in different places, or by the same person in different months! Then all possible joint outcomes may be regarded as equally likely. There are respectively 2, 6 and 52 possible cases for the individual trials, and the total number of cases for the whole set of trials is obtained by multiplying these numbers together: $2 \cdot 6 \cdot 52$ (as you will soon see it is better not to compute this product). This follows from a fundamental rule of counting which is fully discussed in §3.1, and which you should read now if need be. [In general, many parts of this book may be read in different orders, back and forth.] The same rule yields the numbers of favorable cases to the events A, B, C, AB, AC, BC, ABC given below, where the symbol $|\ldots|$ for size is used:

$$|A| = 1\cdot 6\cdot 52, \quad |B| = 2\cdot 2\cdot 52, \quad |C| = 2\cdot 6\cdot 13,$$
$$|AB| = 1\cdot 2\cdot 52, \ |AC| = 1\cdot 6\cdot 13, \ |BC| = 2\cdot 2\cdot 13,$$
$$|ABC| = 1\cdot 2\cdot 13.$$

Dividing these numbers by $|\Omega| = 2\cdot 6\cdot 52$, we obtain after quick cancellation of factors:

$$P(A) = \frac{1}{2}, \quad P(B) = \frac{1}{3}, \quad P(C) = \frac{1}{4},$$
$$P(AB) = \frac{1}{6}, \ P(AC) = \frac{1}{8}, \ P(BC) = \frac{1}{12},$$
$$P(ABC) = \frac{1}{24}.$$

We see at a glance that the following set of equations hold:

$$(2.4.4) \quad P(AB) = P(A)P(B), P(AC) = P(A)P(C), P(BC) = P(B)P(C)$$
$$P(ABC) = P(A)P(B)P(C).$$

The reader is now asked to convince himself that this set of relations will also hold for any three events A, B, C such that A is determined by the coin, B by the die and C by the card drawn *alone*. When this is the case we say that these trials are *stochastically independent* as well as the events so produced. The adverb "stochastically" is usually omitted for brevity.

The astute reader may observe that we have not formally defined the word "trial," and yet we are talking about independent trials! A logical construction of such objects is quite simple but perhaps a bit too abstract for casual introduction. It is known as "product space"; see Exercise 29. However, it takes less fuss to define "independent events" and we shall do so at once.

Two events A and B are said to be independent if we have $P(AB) = P(A)P(B)$. Three events A, B and C are said to be independent if the relations in (2.4.4) hold. Thus independence is a notion relative to a given probability measure (by contrast, the notion of disjointness e.g. does not depend on any probability). More generally, the n events $A_1, A_2, \ldots, A_n$ are independent if the intersection [joint occurrence] of any subset of them has as its probability the product of probabilities of the individual events. If you find this sentence too long and involved, you may prefer the following symbolism. For any subset $(i_1, i_2, \ldots, i_k)$ of $(1, 2, \ldots, n)$, we have

$$(2.4.5) \qquad P(A_{i_1} \cap A_{i_2} \cap \cdots \cap A_{i_k}) = P(A_{i_1})P(A_{i_2}) \cdots P(A_{i_k}).$$

Of course here the indices $i_1, \ldots, i_k$ are distinct and $1 \leq k \leq n$.

Further elaboration of the notion of independence is postponed to §5.5, because it will be better explained in terms of random variables. But we shall describe briefly a classical scheme—the grand daddy of repeated trials, and subject of intensive and extensive research by J. Bernoulli, De Moivre, Laplace, . . . , Borel, ...

Example 8. (The coin-tossing scheme). A coin is tossed repeatedly n times. The joint outcome may be recorded as a sequence of H's and T's, where $H =$ "head," $T =$ "tail." It is often convenient to *quantify* by putting $H = 1$, $T = 0$; or $H = 1$, $T = -1$; we shall adopt the first usage here. Then the result is a sequence of 0's and 1's consisting of n terms such as 110010110 with $n = 9$. Since there are 2 outcomes for each trial, there are 2^n possible joint outcomes. This is another application of the Fundamental Rule in §3.1. If all of these are assumed to be equally likely so that each particular joint outcome has probability $1/2^n$ then we can proceed as in Example 7 to verify that the trials are independent and the coin is fair. You will find this a dull exercise but it is recommended that you go through it in your head if not on paper. However, we will turn the table around here by *assuming at the outset* that the successive tosses do form independent trials. On the other hand, we do not assume the coin to be "fair," but only that the probabilities for head (H) and tail (T) remain constant throughout the trials. Empirically speaking, this is only approximately true since things do not really remain unchanged over long periods of time. Now we need a precise notation to record complicated statements, ordinary words being often awkward or ambiguous. Let then X_i denote the outcome of the ith trial and let ϵ_i denote 0 or 1 for each i, but of course varying with the subscript. Then our hypothesis above may be written as follows:

$$P(X_i = 1) = p, \quad P(X_i = 0) = 1 - p, \quad i = 1, 2, \ldots, n; \tag{2.4.6}$$

where p is the probability of head for each trial. For any particular, namely completely specified, sequence $(\epsilon_1, \epsilon_2, \ldots, \epsilon_n)$ of 0's and 1's, the probability of the corresponding sequence of outcomes is equal to

$$\begin{aligned} &P(X_1 = \epsilon_1, X_2 = \epsilon_2, \ldots, X_n = \epsilon_n) \\ &\quad = P(X_1 = \epsilon_1)P(X_2 = \epsilon_2) \ldots P(X_n = \epsilon_n) \end{aligned} \tag{2.4.7}$$

as a consequence of independence. Now each factor on the right side above is equal to p or $1 - p$ according as the corresponding ϵ_i is 1 or 0. Suppose j of these are 1's and $n - j$ are 0's; then the quantity in (2.4.7) is equal to

$$p^j(1 - p)^{n-j} \tag{2.4.8}$$

Observe that for each sequence of trials, the number of heads is given by the sum $\sum_{i=1}^{n} X_i$. It is important to understand that the number in (2.4.8) is not the probability of obtaining j heads in n tosses, but rather that of obtaining a specific sequence of heads and tails in which there are j heads. In order to compute the former probability, we must count the total number of the latter sequences since all of them have the same probability given in (2.4.8). This number is equal to the binomial coefficient $\binom{n}{j}$; see §3.2 for a full discussion. Each one of these $\binom{n}{j}$ sequences corresponds to one possibility of obtaining j heads in n trials, and these possibilities are disjoint. Hence it follows from the additivity of P that we have

$$
\begin{aligned}
P\left(\sum_{i=1}^{n} X_i = j\right) &= P(\text{exactly } j \text{ heads in } n \text{ trials}) \\
&= \binom{n}{j} P(\text{any specified sequence of } n \text{ trials with exactly } j \text{ heads}) \\
&= \binom{n}{j} p^j (1-p)^{n-j}.
\end{aligned}
$$

This famous result is known as *Bernoulli's formula*. We shall return to it many times in the book.

2.5.* Arithmetical density

We study in this section a very instructive example taken from arithmetic.

Example 9. Let Ω be the first 120 *natural numbers* $\{1, 2, \ldots, 120\}$. For the probability measure P we use the proportion as in Example 1 of §2.1. Now consider the sets

$$
\begin{aligned}
A &= \{\omega \mid \omega \text{ is a multiple of } 3\} \\
B &= \{\omega \mid \omega \text{ is a multiple of } 4\}.
\end{aligned}
$$

Then every third number of Ω belongs to A, and every fourth to B. Hence we get the proportions:

$$P(A) = 1/3, \; P(B) = 1/4.$$

What does the set AB represent? It is the set of integers which are divisible both by 3 and by 4. If you have not forgotten entirely your school arithmetic, you know this is just the set of multiples of $3 \cdot 4 = 12$. Hence $P(AB) = 1/12$. Now we can use (viii) to get $P(A \cup B)$:

$$P(A \cup B) = P(A) + P(B) - P(AB) = 1/3 + 1/4 - 1/12 = 1/2. \tag{2.5.1}$$

What does this mean? $A \cup B$ is the set of those integers in Ω which are divisible by 3 or by 4 (or by both). We can count them one by one, but if

* This section may be omitted.

you are smart you see that you don't have to do this drudgery. All you have to do is to count up to 12 (which is ten percent of the whole population Ω), and check them off as shown:

$$\begin{array}{cccccccccccc} 1, & 2, & 3, & 4, & 5, & 6, & 7, & 8, & 9, & 10, & 11, & 12. \\ & & \checkmark & \checkmark & & \checkmark & & \checkmark & \checkmark & & & \checkmark \\ & & & & & & & & & & & \checkmark \end{array}$$

There are 6 checked (one checked twice), hence the proportion of $A \cup B$ among these 12 is equal to $6/12 = 1/2$ as given by (2.5.1).

An observant reader will have noticed that in the case above we have also

$$P(AB) = 1/12 = 1/3 \cdot 1/4 = P(A) \cdot P(B).$$

This is true because the two numbers 3 and 4 happen to be *relatively prime*, namely they have no common divisor except 1. Suppose we consider another set:

$$C = \{\omega \mid \omega \text{ is a multiple of } 6\}.$$

Then $P(C) = 1/6$ but what is $P(BC)$ now? The set BC consists of those integers which are divisible by both 4 and 6, namely divisible by their *least common multiple* (remember that?) which is 12 and not the product $4 \cdot 6 = 24$. Thus $P(BC) = 1/12$. Furthermore, because 12 is the least common multiple we can again stop counting at 12 in computing the proportion of the set $B \cup C$. An actual counting gives the answer $4/12 = 1/3$, which may also be obtained from the formula (2.3.7):

$$P(B \cup C) = P(B) + P(C) - P(BC) = 1/4 + 1/6 - 1/12 = 1/3. \tag{2.5.2}$$

This example illustrates a point which arose in the discussion in Example 3 of §2.1. Instead of talking about the proportion of the multiples of 3, say, we can talk about its frequency. Here no rolling of any fortuitous dice is needed. God has given us those natural numbers (a great mathematician Kronecker said so), and the multiples of 3 occur at perfectly regular periods with the frequency 1/3. In fact, if we use $N_n(A)$ to denote the number of natural numbers up to and including n which belong to the set A, it is a simple matter to show that

$$\lim_{n \to \infty} \frac{N_n(A)}{n} = \frac{1}{3}.$$

Let us call this $P(A)$, the limiting frequency of A. Intuitively, it should represent the chance of picking a number divisible by 3, if we can reach into the whole bag of natural numbers as if they were so many indistinguishable balls in an urn. Of course similar limits exist for the sets B, C, AB, BC, etc. and have the values computed above. But now with this infinite sample space of "all natural numbers," call it Ω^*, we can treat by the same method any set of the form

$$A_m = \{\omega \mid \omega \text{ is divisible by } m\} \tag{2.5.3}$$

where m is an arbitrary natural number. Why then did we not use this more natural and comprehensive model?

The answer may be a surprise for you. By our definition of probability measure given in §2.2, we should have required that every subset of Ω^* has a probability, provided that Ω^* is countable which is the case here. Now take for instance the set which consists of the single number $\{1971\}$ or if you prefer the set $Z = \{\text{all numbers from 1 to 1971}\}$. Its probability is given by $\lim_{n\to\infty} N_n(Z)/n$ according to the same rule that was applied to the set A. But $N_n(Z)$ is equal to 1971 for all values of $n \geq 1971$, hence the limit above is equal to 0 and we conclude that every finite set has probability 0 by this rule. If P were to be countably additive as required by Axiom (ii*) in §2.3, then $P(\Omega^*)$ would be 0 rather than 1. This contradiction shows that P cannot be a probability measure on Ω^*. Yet it works perfectly well for sets such as A_m.

There is a way out of this paradoxical situation. We must abandon our previous requirement that the measure be defined for all subsets (of natural numbers). Let a finite number of the sets A_m be given, and let us consider the *composite sets* which can be obtained from these by the operations: complementation, union and intersection. Call this class of sets the class *generated by* the original sets. Then it is indeed possible to define P in the manner prescribed above for all sets in *this* class. A set which is not in the class has no probability at all. For example, the set Z does not belong to the class generated by A, B, C. Hence its probability is *not* defined, rather than zero. We may also say that the set Z is *nonmeasurable* in the context of Example 2 of §2.1. This saves the situation but we will not pursue it further here except to give another example.

Example 10. What is the probability of the set of numbers divisible by 3, not divisible by 5, and divisible by 4 or 6?

Using the preceding notation, the set in question is $AD^c(B \cup C)$, where $D = A_5$. Using distributive law, we can write this as $AD^cB \cup AD^cC$. We have also

$$(AD^cB)(AD^cC) = AD^cBC = ABC - ABCD.$$

Hence by (v),

$$P(AD^cBC) = P(ABC) - P(ABCD) = \frac{1}{12} - \frac{1}{60} = \frac{1}{15}.$$

Similarly, we have

$$P(AD^cB) = P(AB) - P(ABD) = \frac{1}{12} - \frac{1}{60} = \frac{4}{60} = \frac{1}{15};$$

$$P(AD^cC) = P(AC) - P(ACD) = \frac{1}{6} - \frac{1}{30} = \frac{4}{30} = \frac{2}{15}.$$

Finally we obtain by (viii):

$$P(AD^cB \cup AD^cC) = P(AD^cB) + P(AD^cC) - P(AD^cBC)$$

$$= \frac{1}{15} + \frac{2}{15} - \frac{1}{15} = \frac{2}{15}.$$

You should check this using the space Ω in Example 9.

The problem can be simplified by a little initial arithmetic, because the set in question is seen to be that of numbers divisible by 2 or 3 and not by 5. Now our method will yield the answer more quickly.

Exercises

1. Consider Example 1 in Section 2.1. Suppose that each good apple costs 1¢ while a rotten one costs nothing. Denote the rotten ones by R, an arbitrary bunch from the bushel by S, and define

$$Q(S) = |S \backslash R| / |\Omega - R|.$$

Q is the relative value of S, with respect to that of the bushel. Show that it is a probability measure.

2. Suppose that the land of a square kingdom is divided into three strips A, B, C of equal area and suppose the value per unit is in the ratio of 1:3:2. For any piece of (measurable) land S in this kingdom the relative value with respect to that of the kingdom is then given by the formula:

$$V(S) = \frac{P(SA) + 3P(SB) + 2P(SC)}{2}$$

where P is as in Example 2 of §2.1. Show that V is a probability measure.

3.* Generalizing No. 2, let $a_1, \ldots, a_n$ be arbitrary positive numbers and let $A_1 + \cdots + A_n = \Omega$ be an arbitrary partition. Let P be a probability measure on Ω and

$$Q(S) = [a_1P(SA_1) + \cdots + a_nP(SA_n)]/[a_1P(A_1) + \cdots + a_nP(A_n)]$$

for any subset of Ω. Show that P is a probability measure.

4. Let A and B denote two cups of coffee you drank at a lunch counter. Suppose the first cup of coffee costs 15¢, and a second cup costs 10¢. Using P to denote "price," write down a formula like Axiom (ii) but with an inequality (P is "subadditive").

5. Suppose that on a shirt sale each customer can buy two shirts at \$4 each, but the regular price is \$5. A customer bought 4 shirts $S_1, \ldots, S_4$. Write down a formula like Axiom (ii) and contrast with Exercise 3. Forget about sales tax! (P is "superadditive.")
6. Show that if P and Q are two probability measures defined on the same (countable) sample space, then $aP + bQ$ is also a probability measure for any two nonnegative numbers a and b satisfying $a + b = 1$. Give a concrete illustration of such a *mixture*.
7.* If P is a probability measure, show that the function $P/2$ satisfies Axioms (i) and (ii) but not (iii). The function P^2 satisfies (i) and (iii) but not necessarily (ii); give a counterexample to (ii) by using Example 1.
8.* If A, B, C are arbitrary sets, show that

(a) $$P(A \cap B \cap C) \leq P(A) \wedge P(B) \wedge P(C);$$

(b) $$P(A \cup B \cup C) \geq P(A) \vee P(B) \vee P(C).$$

9.* Prove that for any two sets A and B, we have

$$P(AB) \geq P(A) + P(B) - 1.$$

Give a concrete example of this inequality. [Hint: Use (2.3.4) with $n = 2$ and DeMorgan's laws.]
10. We have $A \cap A = A$ but when is $P(A) \cdot P(A) = P(A)$? Can $P(A) = 0$ but $A \neq \varnothing$?
11. Find an example where $P(AB) < P(A)P(B)$.
12. Prove (2.3.7) by first showing that

$$(A \cup B) - A = B - (A \cap B).$$

13. Two groups share some members. Suppose that Group A has 123, Group B has 78 members, and the total membership in both groups is 184. How many members belong to both?
14. Groups A, B, C have respectively 57, 49, 43 members. A and B have 13, A and C have 7, B and C have 4 members in common; and there is a lone guy who belongs to all three groups. Find the total number of people in all three groups.
15.* Generalize Exercise 14 when the various numbers are arbitrary, but of course subject to certain obvious inequalities. The resulting formula, divided by the total population (there may be many non-joiners!) is the extension of (2.3.7) to $n = 3$.
16. Compute $P(A \triangle B)$, in terms of $P(A)$, $P(B)$ and $P(AB)$; also in terms of $P(A)$, $P(B)$ and $P(A \cup B)$.
17.* Using the notation (2.5.3) and the probability defined in that context, show that for any two m and n we have

$$P(A_mA_n) \geq P(A_m)P(A_n).$$

When is there equality above?

18.* Recall the computation of plane areas by double integration in calculus; for a nice figure such as a parallelogram, trapezoid or circle we have

$$\text{Area of } S = \iint_S 1\, dx\, dy.$$

Show that this can be written in terms of the indicator I_S as

$$A(S) = \iint_\Omega I_S(x, y)\, dx\, dy,$$

where Ω is the whole plane and $I_S(x, y)$ is the value of the function I_S for (at) the point (x, y) (denoted by ω in §1.4). Show also that for two such figures S_1 and S_2, we have

$$A(S_1) + A(S_2) = \iint (I_{S_1} + I_{S_2}),$$

where we have omitted some unnecessary symbols.

19.* Now you can demonstrate the connection between (2.3.7) and (2.3.8) mentioned there, in the case of plane areas.

20. Find several examples of $\{p_n\}$ satisfying the conditions in (2.4.1); give at least two in which all $p_n > 0$.

21.* Deduce from Axiom (ii*) the following two results. (a) If the sets A_n are nondecreasing, namely $A_n \subset A_{n+1}$ for all $n \geq 1$, and $A_\infty = \bigcup_n A_n$, then $P(A_\infty) = \lim_{n\to\infty} P(A_n)$. (b) If the sets A_n are nonincreasing, namely $A_n \supset A_{n+1}$ for all $n \geq 1$, and $A_\infty = \bigcap_n A_n$, then $P(A_\infty) = \lim_{n\to\infty} P(A_n)$. [Hint: For (a), consider $A_1 + (A_2 - A_1) + (A_3 - A_2) + \cdots$; for (b), dualize by complementation.]

22. What is the probability (in the sense of Example 10) that a natural number picked at random is not divisible by any of the numbers 3, 4, 6 but is divisible by 2 or 5?

23.* Show that if $(m_1, \ldots, m_n)$ are co-prime positive integers, then the events $(A_{m_1}, \ldots, A_{m_n})$ defined in §2.5 are independent.

24. What can you say about the event A if it is independent of itself? If the events A and B are disjoint and independent, what can you say of them?

25. Show that if the two events (A, B) are independent, then so are (A, B^c), (A^c, B) and (A^c, B^c). Generalize this result to three independent events.

26. Show that if A, B, C are independent events, then A and $B \cup C$ are independent, also $A \backslash B$ and C are independent.

27. Prove that

$$P(A \cup B \cup C) = P(A) + P(B) + P(C) - P(AB) - P(AC) - P(BC) + P(ABC)$$

when A, B, C are independent by considering $P(A^c B^c C^c)$. [The formula remains true without the assumption of independence; see §6.2.]

28. Suppose 5 coins are tossed; the outcomes are independent but the probability of head may be different for different coins. Write the probability of the specific sequence $HHTHT$, and the probability of exactly 3 heads.
29.* How would you build a mathematical model for arbitrary repeated trials, namely without the constraint of independence? In other words, describe a sample space suitable for recording such trials. What is the mathematical definition of an event which is determined by one of the trials alone, two of them, etc.? You do not need a probability measure. Now think how you would cleverly construct such a measure over the space in order to make the trials independent. The answer is given in e.g. [Feller 1, §V.4], but you will understand it better if you first give it a try yourself.

Chapter 3

Counting

3.1. Fundamental rule

The calculation of probabilities often leads to the counting of various possible cases. This has been indicated in Examples 4 and 5 of §2.2 and forms the backbone of the classical theory with its stock in trade the games of chance. But combinatorial techniques are also needed in all kinds of applications arising from sampling, ranking, partitioning, allocating, programming and model building, to mention a few. In this chapter we shall treat the most elementary and basic types of problems and the methods of solving them.

The author has sometimes begun a discussion of "permutations and combinations" by asking in class the following question. If a man has three shirts and two ties, in how many ways can he dress up [put on one of each]? There are only two numbers 2 and 3 involved, and it's anybody's guess that one must combine them in some way. Does one add: $2 + 3$? or multiply: 2×3? (or perhaps make 2^3 or 3^2). The question was meant to be rhetorical but experience revealed an alarming number of wrong answers. So if we dwell on this a little longer than you deem necessary you will know why.

First of all, in a simple example like that, one can simply picture to oneself the various possibilities and count them up mentally:

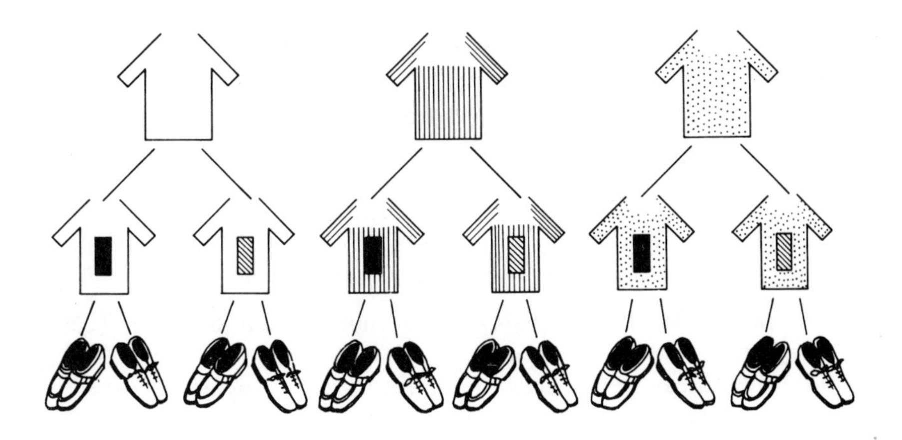

Figure 15

A commonly used tabulation is as follows:

(3.1.1)

T \ S	1	2	3
1	11	21	31
2	12	22	32

As mathematics is economy of thought we can schematize (program) this in a more concise way:

$$(s_1, t_1)\ (s_1, t_2)\ (s_2, t_1)\ (s_2, t_2)\ (s_3, t_1)\ (s_3, t_2)$$

and finally we see that it is enough just to write

(3.1.2) $$(1, 1)\ (1, 2)\ (2, 1)\ (2, 2)\ (3, 1)\ (3, 2)$$

by assigning the first slot to "shirt" and the second to "tie." Thus we have reached the mathematical method of naming the collection in (3.1.2). It is the set of all ordered couples (a, b) such that $a = 1, 2, 3$; $b = 1, 2$; and you see that the answer to my question is $3 \times 2 = 6$.

In general we can talk about ordered k-tuples $(a_1, \ldots, a_k)$ where for each j from 1 to k, the symbol a_j indicates the assignment (choice) for the jth slot, and it may be denoted by a numeral between 1 and m_j. In the example above $k = 2, m_1 = 3, m_2 = 2$, and the collection of all (a_1, a_2) is what is enumerated in (3.1.2).

This symbolic way of doing things is extremely convenient. For instance, if the man has also two pairs of shoes, we simply extend each 2-tuple to a 3-tuple by adding a third slot into which we can put either "1" or "2". Thus each of the original 2-tuples in (3.1.2) splits into two 3-tuples, and so the total of 3-tuples will be $3 \times 2 \times 2 = 12$. This is the number of ways the man can choose a shirt, a tie and a pair of shoes. You see it is all automated as on a computing machine. As a matter of fact, it is mathematical symbolism that taught the machines, not the other way around (at least, not yet).

The idea of splitting mentioned above lends well to visual imagination. It shows why 3 "shirts" *multiply* into 6 "shirt-ties" and 12 "shirt-tie-shoes." Take a good look at it. Here is the general proposition:

Fundamental Rule. A number of multiple choices are to be made. There are m_1 possibilities for the first choice, m_2 for the second, m_3 for the third, etc. If these choices can be combined freely, then the total number of possibilities for the whole set of choices is equal to

$$m_1 \times m_2 \times m_3 \times \cdots$$

A formal proof would amount to repeating what is described above in more cut-and-dried terms, and is left to your own discretion. Let me point out however that "free combination" means in the example above that no matching of shirt and ties is required, etc.

Example 1. A menu in a restaurant reads like this:

Choice of one:
Soup, Juice, Fruit Cocktail

Choice of one:
Beef Hash
Roast Ham
Fried Chicken
Spaghetti with Meat Balls

Choice of one:
Mashed Potatoes, Broccoli, Lima Beans

Choice of one:
Ice Cream, Apple Pie

Choice of one:
Coffee, Tea, Milk

Suppose you take one of each "course" without substituting or skipping, how many options do you have? Or if you like the language nowadays employed in more momentous decisions of this sort, how many *scenarios* of a "complete 5-course dinner" (as advertised) can you make out of this menu? The total number of items you see on the menu is

$$3 + 4 + 3 + 2 + 3 = 15.$$

But you don't eat them all. On the other hand, the number of different dinners available is equal to

$$3 \times 4 \times 3 \times 2 \times 3 = 216,$$

according to the Fundamental Rule. True, you eat only one dinner at a time, but it is quite possible for you to try all these 216 dinners if you have catholic taste in food and patronize that restaurant often enough. More realistically and statistically significant: all these 216 dinners may be actually served to different customers over a period of time and perhaps even on a single day. This possibility forms the empirical basis of combinatorial counting and its relevance to computing probabilities.

Example 2. We can now solve the problem about Example (e) and (e') in Section 1.1: in how many ways can six dice appear when they are rolled? and in how many ways can they show all different faces?

Each die here represents a multiple choice of six possibilities. For the first problem these 6 choices can be freely combined so the rule applies directly to give the answer $6^6 = 46656$. For the second problem the choices cannot be freely combined since they are required to be all different. Offhand the rule does not apply, but the reasoning behind it does. This is what counts in mathematics: not a blind reliance on a rule but a true understanding of its meaning. (Perhaps that is why "permutation and combination" is for many students harder stuff than algebra or calculus.) Look at the following splitting diagram (see Fig. 16).

The first die can show any face, but the second must show a different one. Hence, after the first choice has been made, there are 5 possibilities for the second choice. Which five depends on the first choice *but their number does not*. So there are 6×5 possibilities for the first and second choices together. After these have been made, there are 4 possibilities left for the third, and so on. For the complete *sequence* of six choices we have therefore $6.5.4.3.2.1 = 6! = 720$ possibilities. By the way, make sure by analyzing the diagram that the first die hasn't got a preferential treatment. Besides, which is "first"?

Of course, we can re-enunciate a more general rule to cover the situation just discussed, but is it necessary once the principles are understood?

3.2. Diverse ways of sampling

Let us proceed to several standard methods of counting which constitute the essential elements in the majority of combinatorial problems. These can be conveniently studied either as sampling or as allocating problems. We begin with the former.

An urn contains m distinguishable balls marked 1 to m, from which n balls will be drawn under various specified conditions, and the number of all possible outcomes will be counted in each case.

I. Sampling with replacement and with ordering.

We draw n balls sequentially, each ball drawn being put back into the urn before the next drawing is made, and we record the numbers on the balls together with their order of appearance. Thus we are dealing with ordered n-tuples $(a_1 \ldots a_n)$ in which each a_j can be any number from 1 to m. The Fundamental Rule applies directly and yields the answer m^n. This corresponds to the case of rolling six dice without restriction, but the analogy may be clearer if we think of the same dice being rolled six times in succession, so that each rolling corresponds to a drawing.

II. Sampling without replacement and with ordering.

We sample as in Case I but after each ball is drawn it is left out of the urn. We are dealing with ordered n-tuples $(a_1, \ldots, a_n)$ as above with the restriction that the a_j be all different. Clearly we must have $n \leq m$. The Fundamental Rule does not apply directly but the splitting argument works as in Example 2 and yields the answer

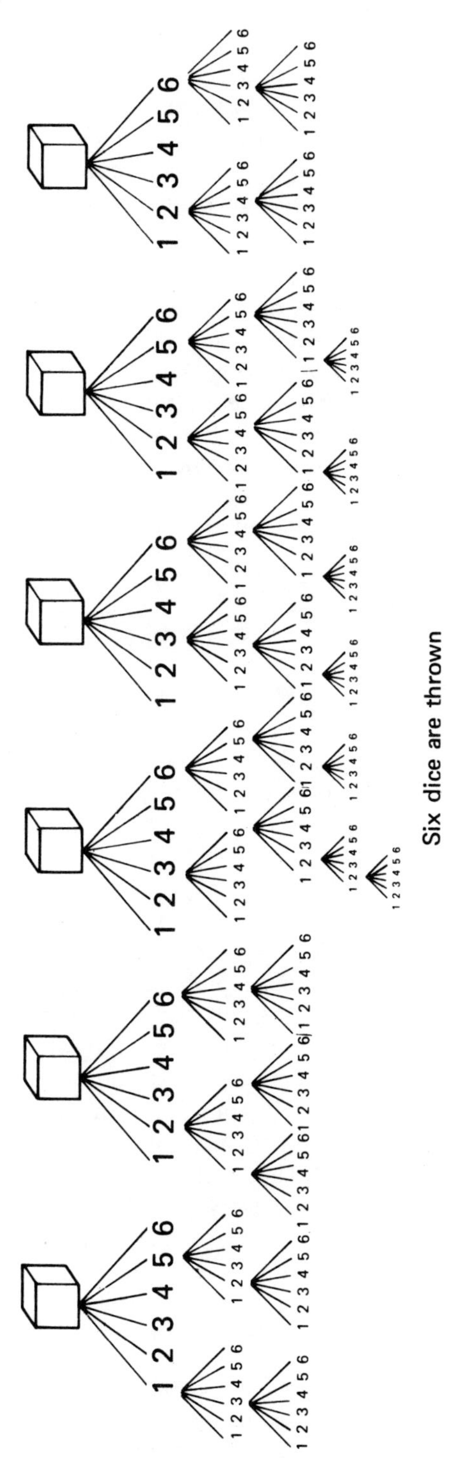

Six dice are thrown

Figure 16

(3.2.1) $$m \cdot (m-1) \cdot (m-2) \cdots (m-n+1) = (m)_n.$$

Observe that there are n factors on the left side of (3.2.1) and that the last factor is $m - (n-1)$ rather than $m - n$, why? We have introduced the symbol $(m)_n$ to denote this "continued product" on the left side of (3.2.1).

Case II has a very important subcase which can be posed as a "permutation" problem.

IIa. Permutation of m distinguishable balls.

This is the case of II when $m = n$. Thus all m balls are drawn out one after another without being put back. The result is therefore just the m numbered balls appearing in some order, and the total number of such possibilities is the same as that of all possible arrangements (ordering, ranking, permuation) of the set $\{1, 2, \ldots, m\}$. This number is called the *factorial* of m and denoted by

$$m! = (m)_m = m(m-1) \cdots 2 \cdot 1$$

n	$n!$
1	1
2	2
3	6
4	24
5	120
6	720
7	5040
8	40320
9	362880
10	3628800

III. Sampling without replacement and without ordering.

Here the balls are not put back and their order of appearance is not recorded; hence we might as well draw all n balls at one grab. We are dealing therefore with subsets of size n from a set (*population*) of size m. To count their number, we will compare with Case II where the balls are ranged in order. Now a bunch of n balls, if drawn one by one, can appear in $n!$ different ways by Case IIa. Thus each unordered sample of size n produces $n!$ ordered ones, and conversely every ordered sample of size n can be produced in this manner. For instance if $m = 5$, $n = 3$, the subset $\{3, 5, 2\}$ can be drawn in $3! = 6$ ways as follows:

$$(2, 3, 5)\ (2, 5, 3)\ (3, 2, 5)\ (3, 5, 2)\ (5, 2, 3)\ (5, 3, 2).$$

In general we know from Case II that the total number of ordered samples of size n is $(m)_n$. Let us denote for one moment the unknown number of unordered samples of size n by x, then the argument above shows that

$$n!x = (m)_n.$$

Solving for x, we get the desired answer which will be denoted by

$$\binom{m}{n} = \frac{(m)_n}{n!}. \tag{3.2.2}$$

If we multiply both numerator and denominator by $(m - n)!$, we see from (3.2.1) that

$$\begin{aligned}\binom{m}{n} &= \frac{(m)_n(m-n)!}{n!(m-n)!} \\ &= \frac{m(m-1)\cdots(m-n+1)(m-n)\cdots 2.1}{n!(m-n)!} = \frac{m!}{n!(m-n)!}.\end{aligned} \tag{3.2.3}$$

When $n = m$, there is exactly one subset of size n, namely the whole set, hence the number in (3.2.3) must reduce to 1 if it is to maintain its significance. So we are obliged to set $0! = 1$. Under this convention, formula (3.2.3) holds for $0 \leq n \leq m$. The number $\binom{m}{n}$ is called a *binomial coefficient* and plays an important role in probability theory. Note that

$$\binom{m}{n} = \binom{m}{m-n} \tag{3.2.4}$$

which is immediate from (3.2.3). It is also obvious without this explicit evaluation from the interpretation of both sides as counting formulas (why?).

The argument used in Case III leads to a generalization of IIa:

IIIa. Permutation of m balls which are distinguishable by groups.

Suppose that there are m_1 balls of color no. 1, m_2 balls of color no. 2, . . . , m_r balls of color no. r. Their colors are distinguishable but balls of the same color are not. Of course $m_1 + m_2 + \cdots + m_r = m$. How many distinguishable arrangements of these m balls are there?

For instance, if $m_1 = m_2 = 2$, $m = 4$ and the colors are black and white, there are 6 distinguishable arrangments as follows:

To answer the question in general, we compare with Case IIa where all balls are distinguishable. Suppose we mark the balls of color no. 1 from 1 to m_1, the balls of color no. 2 from 1 to m_2, and so forth. Then they become all distinguishable and so the total number of arrangements after the markings will be $m!$ by Case IIa. Now the m_1 balls of color no. 1 can be arranged in $m_1!$ ways by their new marks, the m_2 balls of color no. 2 can be arranged in $m_2!$ ways by their new marks, etc. Each arrangement for one color can be freely combined with any arrangement for another color. Hence according to the Fundamental Rule, there are altogether

$$m_1!m_2! \dots m_r!$$

new arrangements produced by the various markings, for each original unmarked arrangement. It follows as in the discussion of Case III that the total number of distinguishable unmarked arrangements is equal to the quotient

$$\frac{m!}{m_1!m_2! \dots m_r!}$$

This is called a *multinomial coefficient*. When $r = 2$ it reduces to the *binomial coefficient* $\binom{m}{m_1} = \binom{m}{m_2}$.

IV. Sampling with replacement and without ordering.

We draw n balls one after another, each ball being put back into the urn before the next drawing is made, but we record the numbers drawn with possible repetitions as a lot without paying attention to their order of appearance. This is a slightly more tricky situation so we will begin by a numerical illustration. Take $m = n = 3$; all the possibilities in this case are listed in the first column below:

(3.2.5)

<table>
<tr><td>111</td><td>✓✓✓| |</td><td>✓✓✓||</td></tr>
<tr><td>112</td><td>✓✓| ✓|</td><td>✓✓|✓|</td></tr>
<tr><td>113</td><td>✓✓| |✓</td><td>✓✓||✓</td></tr>
<tr><td>122</td><td>✓| ✓✓|</td><td>✓|✓✓|</td></tr>
<tr><td>123</td><td>✓| ✓|✓</td><td>✓|✓|✓</td></tr>
<tr><td>133</td><td>✓| |✓✓</td><td>✓||✓✓</td></tr>
<tr><td>222</td><td>|✓✓✓|</td><td>|✓✓✓|</td></tr>
<tr><td>223</td><td>| ✓✓|✓</td><td>|✓✓|✓</td></tr>
<tr><td>233</td><td>| ✓|✓✓</td><td>|✓|✓✓</td></tr>
<tr><td>333</td><td>| |✓✓✓</td><td>||✓✓✓</td></tr>
</table>

Do you see the organization principle used in making the list?

In general think of a "tally sheet" with numbers indicating the balls in the top line:

1	2	3	4		m
✓✓	✓		✓✓✓		

After each drawing we place a check under the number (of the ball) which is drawn. Thus at the end of all the drawings the total number of checks on the sheet will be n (which may be greater than m); there may be as many as that in an entry, and there may be blanks in some entries. Now economize by removing all dispensable parts of the tally sheet, so that column 1 in (3.2.5) becomes the skeleton in column 2. Check this over carefully to see that no information is lost in the simplified method of accounting. Finally align the symbols $\checkmark$ and $|$ in column 2 to get column 3. Now forget about columns 1 and 2 and concentrate on column 3 for a while. Do you see how to reconstruct from each little cryptogram of "checks and bars" the original tally? Do you see that all possible ways of arranging 3 checks and 2 bars are listed there? Thus the total number is (by Case IIIa with $m = 5$, $m_1 = 3$, $m_2 = 2$, or equivalently Case III with $m = 5$, $n = 3$) equal to $5!/3!2! = 10$ as shown. This must therefore also be the number of all possible tally results.

In general each possible record of sampling under Case IV can be transformed by the same method into the problem of arranging n checks and $m - 1$ bars (since m slots have $m - 1$ lines dividing them) in all possible ways. You will have to draw some mental pictures to convince yourself that there is one-to-one correspondence between the two problems as in the particular case illustrated above. From IIIa we know the solution to the second problem is

$$\binom{m+n-1}{n} = \binom{m+n-1}{m-1}. \tag{3.2.6}$$

Hence this is also the total number of outcomes when we sample under Case IV.

Example 3. D'Alembert's way of counting discussed in Example 4 of Section 2.2 is equivalent to sampling under Case IV with $m = n = 2$. Tossing each coin corresponds to drawing a head or a tail, if the results for two coins are tallied without regard to "which shows which" (or without ordering when the coins are tossed one after the other), then there are three possible outcomes:

$\checkmark\checkmark|$	$\checkmark|\checkmark$	$|\checkmark\checkmark$
HH	HT = TH	TT

Similarly, if six dice are rolled and the dice are not distinguishable, then the total number of recognizably distinct patterns is given by (3.2.6) with $m = n = 6$, namely

$$\binom{6+6-1}{6} = \binom{11}{6} = 462.$$

This is less than 1% of the number 46656 under Case I.

We will now illustrate in a simple numerical case the different ways of counting in the four sampling procedures: $m = 4$, $n = 2$.

Case (I)

(1, 1)	(1, 2)	(1, 3)	(1, 4)
(2, 1)	(2, 2)	(2, 3)	(2, 4)
(3, 1)	(3, 2)	(3, 3)	(3, 4)
(4, 1)	(4, 2)	(4, 3)	(4, 4)

Case (II)

	(1, 2)	(1, 3)	(1, 4)
(2, 1)		(2, 3)	(2, 4)
(3, 1)	(3, 2)		(3, 4)
(4, 1)	(4, 2)	(4, 3)	

Case (IV)

(1, 1)	(1, 2)	(1, 3)	(1, 4)
	(2, 2)	(2, 3)	(2, 4)
		(3, 3)	(3, 4)
			(4, 4)

Case (III)

(1, 2)	(1, 3)	(1, 4)
	(2, 3)	(2, 4)
		(3, 4)

3.3. Allocation models; binomial coefficients

A source of inspiration as well as frustration in combinatorics is that the same problem may appear in different guises and it may take an effort to recognize their true identity. Sampling under Case (IV) is a case in point, another will be discussed under (IIIb) below. People have different thinking habits and often prefer one way to another. But it may be worthwhile to learn some other ways as we learn foreign languages. In the above we have treated several basic counting methods as sampling problems. Another formulation, prefered by physicists and engineers, is "putting balls into boxes." Since these balls play a different role from those used above, we will call them *tokens* instead to simplify the translation later.

There are m boxes labeled from 1 to m and there are n tokens which are numbered from 1 to n. The tokens are put into the boxes with or without the condition that no box can contain more than one token. We record the outcome of the allocation (occupancy) by noting the number of tokens in each box, with or without noting the labels on the tokens. The four cases below then correspond respectively with the four cases of sampling discussed above.

I′. Each box may contain any number of tokens and the labels on the tokens are observed.

II′. No box may contain more than one token and the labels on the tokens are observed.

III′. No box may contain more than one token and the labels on the tokens are not observed.

IV′. Each box may contain any number of tokens and the labels on the tokens are not observed.

It will serve no useful purpose to *prove* that the corresponding problems are really identical, for this is the kind of mental exercise one must go through by oneself to be convinced. (Some teachers even go so far as to say that combinatorial thinking cannot be taught.) However, here are the key words in the translation from one to the other description.

Sampling	Allocating
Ball	Box
Number of drawing	Number on token
jth drawing gets ball no. k	jth token put into box no. k

In some way the new formulation is more adaptable in that further conditions on the allocation can be imposed easily. For instance, one may require that no box be left empty when $n \geq m$, or specify "loads" in some boxes. Of course these conditions can be translated into the other language, but they may then become less natural. Here is one important case of this sort which is just Case IIIa in another guise.

IIIb. Partition into numbered groups.

Let a population of m objects be subdivided into r *subpopulations* or just "groups": m_1 into group no. 1, m_2 into group no. 2, . . . , m_r into group no. r; where $m_1 + \cdots + m_r = m$ and all $m_j \geq 1$. This is a trivial paraphrasing of putting m tokens into r boxes so that m_j tokens are put into box no. j. It is important to observe that it is not the same as subdividing into r groups of sizes $m_1, \ldots, m_r$; for the groups are numbered. A simple example will make this clear.

Example 4. In how many ways can 4 people split into two pairs?

The English language certainly does not make this question unambiguous, but offhand one would have to consider the following 3 ways as the answer:

$$(12)(34) \qquad (13)(24) \qquad (14)(23). \tag{3.3.1}$$

This is the correct interpretation if the two pairs are going to play chess or pingpong games and two equally good tables are available to both pairs. But now suppose the two pairs are going to play double tennis together and the "first" pair has the choice of side of the court, or will be the first to serve. It will then make a difference whether (12) precedes (34) or vice versa. So each case in (3.3.1) must be permuted *by pairs* into two orders and the answer is then the following 6 ways:

$$(12)(34) \qquad (34)(12) \qquad (13)(24) \qquad (24)(13) \qquad (14)(23) \qquad (23)(14).$$

This is the situation covered by the general problem of partition under IIIb which will now be solved.

Think of putting tokens (people) into boxes (groups). According to sampling Case III, there are $\binom{m}{m_1}$ ways of choosing m_1 tokens to be put into box 1; after that, there are $\binom{m - m_1}{m_2}$ ways of choosing m_2 tokens from the remaining $m - m_1$ to be put into box 2; and so on. The Fundamental Rule does not appiy but its modification used in sampling Case II does, and so the answer is

$$\binom{m}{m_1}\binom{m - m_1}{m_2}\binom{m - m_1 - m_2}{m_3}\cdots\binom{m - m_1 - m_2 - \cdots - m_{r-1}}{m_r}$$

$$(3.3.2) \qquad = \frac{m!}{m_1!(m - m_1)!}\,\frac{(m - m_1)!}{m_2!(m - m_1 - m_2)!}\,\frac{(m - m_1 - m_2)!}{m_3!(m - m_1 - m_2 - m_3)!}\cdots\frac{(m - m_1 - \cdots - m_{r-1})!}{m_r!0!}$$

$$= \frac{m!}{m_1!m_2!\cdots m_r!}.$$

Observe that there is no duplicate counting involved in the argument, even if some of the groups have the same size as in the tennis player example above. This is because we have given numbers to the boxes (groups). On the other hand, we are not arranging the numbered groups in order (as the words "ordered groups" employed by some authors would seem to imply). To clarify this essentially linguistic confusion let us consider another simple example.

Example 5. Six mountain climbers decide to divide into three groups for the final assault on the peak. The groups will be of size 1, 2, 3 respectively and all manners of deployment are considered. What is the total number of possible grouping and deploying?

The number of ways of splitting in G_1, G_2, G_3 where the subscript denotes the size of group, is given by (3.3.2):

$$\frac{6!}{1!2!3!} = 60.$$

Having formed these three groups, there remains the decision which group leads, which in the middle, and which backs up. This is solved by Case IIa: $3! = 6$. Now each grouping can be combined freely with each deploying, hence the Fundamental Rule gives the final answer $60 \cdot 6 = 360$.

What happens when some of the groups have the same size? Think about the tennis players again.

Returning to (3.3.2), this is the same multinomial coefficient obtained as solution to the permutation problem IIIa. Here it appears as a combination type of problem since we have used sampling Case III repeatedly in its derivation. Thus it is futile to try to pin a label on a problem as permutation or combination. The majority of combinatorial problems involves a mixture of various ways of counting discussed above. We will now illustrate this with several worked problems.

In the remainder of this section we will establish some useful formulas connecting binomial coefficients. First of all, let us lay down the convention:

(3.3.3) $$\binom{m}{n} = 0 \qquad \text{if } m < n, \text{ or if } n < 0.$$

Next we show that

(3.3.4) $$\binom{m}{n} = \binom{m-1}{n-1} + \binom{m-1}{n}, \qquad 0 \leq n \leq m.$$

Since we have the explicit evaluation of $\binom{m}{n}$ from (3.2.3), this can of course be verified at once. But here is a combinatorial argument without computation. Recall that $\binom{m}{n}$ is the number of different ways of choosing n objects out of m objects, which may be thought of as being done at one stroke. Now think of one of the objects as "special." This special one may or may not be included in the choice. If it is included, then the number of ways of choosing $n - 1$ more objects out of the other $m - 1$ objects is equal to $\binom{m-1}{n-1}$. If it is not included, then the number of ways of choosing all n objects from the other $m - 1$ objects is equal to $\binom{m-1}{n}$. The sum of the two alternatives must then give the total number of choices, and this is what (3.3.4) says. Isn't this neat?

As a consequence of (3.3.4), we can obtain $\binom{m}{n}$, $0 \leq n \leq m$, step by step as m increases, as follows:

(3.3.5)
$$\begin{array}{ccccccccccccccc}
 & & & & & & & 1 & & & & & & & \\
 & & & & & & 1 & & 1 & & & & & & \\
 & & & & & 1 & & 2 & & 1 & & & & & \\
 & & & & 1 & & 3 & & 3 & & 1 & & & & \\
 & & & 1 & & 4 & & 6 & & 4 & & 1 & & & \\
 & & 1 & & 5 & & 10 & & 10 & & 5 & & 1 & & \\
 & 1 & & 6 & & 15 & & 20 & & 15 & & 6 & & 1 & \\
1 & & 7 & & 21 & & 35 & & 35 & & 21 & & 7 & & 1 \\
\cdot & & \cdot & & \cdot & & \cdot & & \cdot & & \cdot & & \cdot & & \cdot
\end{array}$$

For example, each number in the last row shown above is obtained by adding its two neighbors in the preceding row, where a vacancy may be regarded as zero:

$$1 = 0 + 1,\ 7 = 1 + 6,\ 21 = 6 + 15,\ 35 = 15 + 20,\ 35 = 20 + 15,$$
$$21 = 15 + 6,\ 7 = 6 + 1,\ 1 = 1 + 0.$$

Thus $\binom{7}{n} = \binom{6}{n-1} + \binom{6}{n}$ for $0 \le n \le 7$. The array in (3.3.5) is called *Pascal's triangle*, though apparently he was not the first one to have used it.

* Observe that we can split the last term $\binom{m-1}{n}$ in (3.3.4) as we split the first term $\binom{m}{n}$ by the same formula applied to $m - 1$. Thus we obtain, successively:

$$\binom{m}{n} = \binom{m-1}{n-1} + \binom{m-1}{n} = \binom{m-1}{n-1} + \binom{m-2}{n-1} + \binom{m-2}{n}$$
$$= \binom{m-1}{n-1} + \binom{m-2}{n-1} + \binom{m-3}{n-1} + \binom{m-3}{n} = \cdots.$$

The final result is

$$\begin{aligned}\binom{m}{n} &= \binom{m-1}{n-1} + \binom{m-2}{n-1} + \cdots + \binom{n}{n-1} + \binom{n}{n} \\ &= \sum_{k=n-1}^{m-1} \binom{k}{n-1} = \sum_{k \le m-1} \binom{k}{n-1};\end{aligned} \tag{3.3.6}$$

since the last term in the sum is $\binom{n}{n} = \binom{n-1}{n-1}$, and for $k < n - 1$ the terms are zero by our convention (3.3.3).

Example. $35 = \binom{7}{4} = \binom{6}{3} + \binom{5}{3} + \binom{4}{3} + \binom{3}{3} = 20 + 10 + 4 + 1.$ Look at Pascal's triangle to see where these numbers are located.

As an application, we can now give another solution to the counting problem for sampling under (IV) in §2. By the second formulation (IV′) above, this is the number of ways of putting n indistinguishable [unnumbered] tokens into m labeled boxes without restriction. We know from (3.2.6) that it is equal to $\binom{m+n-1}{m-1}$, but the argument leading to this answer is pretty tricky. Suppose we were not smart enough to have figured it out that way, but have surmised the result by experimenting with small values of m and n. We can still establish the formula in general as follows. [Actually, that tricky

* The rest of the section may be omitted.

argument was probably invented as an after-thought after the result had been surmised.]

We proceed by mathematical induction on the value of m. For $m = 1$ clearly there is just one way of dumping all the tokens, no matter how many, into the one box, which checks with the formula since $\binom{1+n-1}{1-1} = \binom{n}{0} = 1$. Now suppose that the formula holds true for any number of tokens when the number of boxes is equal to $m - 1$. Introduce a new box; we may put any number of tokens into it. If we put j tokens into the new one, then we must put the remaining $n - j$ tokens into the other $m - 1$ boxes. According to the induction hypothesis, there are $\binom{m-2+n-j}{m-2}$ different ways of doing this. Summing over all possible values of j, we have

$$\sum_{j=0}^{n} \binom{m-2+n-j}{m-2} = \sum_{k=m-2}^{m+n-2} \binom{k}{m-2}$$

where we have changed the index of summation by setting $m - 2 + n - j = k$. The second sum above is equal to $\binom{m+n-1}{m-1}$ by (3.3.6), and the induction is complete.

Next, let us show that

$$\binom{n}{0} + \binom{n}{1} + \binom{n}{2} + \cdots + \binom{n}{n} = \sum_{k=0}^{n} \binom{n}{k} = 2^n; \tag{3.3.7}$$

that is, the sum of the nth row in Pascal's triangle is equal to 2^n [the first row shown in (3.3.5) is the 0th]. If you know Newton's Binomial Theorem this can be shown from

$$(a + b)^n = \sum_{k=0}^{n} \binom{n}{k} a^k b^{n-k} \tag{3.3.8}$$

by substituting $a = b = 1$. But here is a combinatorial proof. The terms on the left side of (3.3.7) represent the various numbers of ways of choosing $0, 1, 2, \ldots, n$ objects out of n objects. Hence the sum is the total number of ways of choosing *any* subset [the empty set and the entire set both included] from a set of size n. Now in such a choice each object may or may not be included, and the inclusion or exclusion of each object may be freely combined with that of any other. Hence the Fundamental Rule yields the total number of choices as

$$\underbrace{2 \times 2 \times \cdots \times 2}_{n \text{ times}} = 2^n.$$

This is the number given on the right side of (3.3.7). It is the total number of distinct subsets of a set of size n.

Example. For $n = 2$, all choices from (a, b) are:

$$\varnothing, \{a\}, \{b\}, \{ab\}.$$

For $n = 3$, all choices from (a, b, c) are:

$$\varnothing, \{a\}, \{b\}, \{c\}, \{ab\}, \{ac\}, \{bc\}, \{abc\}.$$

Finally, let $k \leq n$ be two positive integers. We show

$$\binom{m}{n} = \sum_{j=0}^{k} \binom{k}{j}\binom{m-k}{n-j}. \tag{3.3.9}$$

Observe how the indices on top [at bottom] on the right side add up to the index on top [at bottom] on the left side; it is not necessary to indicate the precise range of j in the summation; we may let j range over all integers because the superfluous terms will automatically be zero by our convention (3.3.3).

To see the truth of (3.3.9), we think of the m objects as being separated into two piles, one containing k objects and the other $m - k$. To choose n objects from the entire set, we may choose j objects from the first pile and $n - j$ objects from the second pile, and combine them. By the Fundamental Rule, for each fixed value j the number of such combinations is equal to $\binom{k}{j}\binom{m-k}{n-j}$. So if we allow j to take all possible values and add up the results we obtain the total number of choices which is equal to $\binom{m}{n}$. You need not worry about "impossible" values for j when $j > n$ or $n - j > m - k$, because the corresponding term will be zero by our convention.

Example.

$$\binom{7}{3} = \binom{3}{0}\binom{4}{3} + \binom{3}{1}\binom{4}{2} + \binom{3}{2}\binom{4}{1} + \binom{3}{3}\binom{4}{0}$$

$$\binom{7}{5} = \binom{3}{1}\binom{4}{4} + \binom{3}{2}\binom{4}{3} + \binom{3}{3}\binom{4}{2}.$$

In particular, if $k = 1$ in (3.3.9), we are back to (3.3.4). In this case our argument also reduces to the one used there.

An algebraic derivation of (3.3.9), together with its extension to the case where the upper indices are no longer positive integers, will be given in Chapter 6.

3.4. How to solve it.

This section may be entitled "How to count." Many students find these problems hard, partly because they have been inured in other elementary mathematics courses to the cook book variety of problems such as: "Solve $x^2 - 5x + 10 = 0$," "differentiate xe^{-x}" (maybe twice), etc. One can do such problems by memorizing certain rules without any independent thought. Of course we have this kind of problem in "permutation and combination" too, and you will find some of these among the exercises. For instance, there is a famous formula to do the "round table" problem: "In how many different ways can 8 people be seated at a round table?" If you learned it you could solve the problem without knowing what the word "different" means. But a little variation may get you into deep trouble. The truth is, and that's also a truism: there is no substitute for true understanding. However, it is not easy to understand the principles without concrete applications, so the handful of examples below are selected to be the "test cases." More are given in the Exercises and you should have a lot of practice if you want to become an expert. Before we discuss the examples in detail, a few general tips will be offered to help you to do your own thing. They are necessarily very broad and rather slippery, but they may be of *some* help *sometimes*.

(a) If you don't *see* the problem well, try some particular (but not too particular) case with small numbers so you can see better. This will fix in your mind what is to be counted, and help you especially in spotting duplicates and omissions.
(b) Break up the problem into pieces provided that they are simpler, cleaner, and easier to concentrate on. This can be done sometimes by fixing one of the "variables," and the number of similar pieces may be counted as a subproblem.
(c) Don't try to argue step by step if you can see complications rising rapidly. Of all the negative advice I gave my classes this was the least heeded but probably the most rewarding. Counting step by step may seem easy for the first couple of steps but do you see how to carry it through to the end?
(d) Don't be turned off if there is ambiguity in the statement of the problem. This is a semantical hang-up, not a mathematical one. Try all interpretations if necessary. This may not be the best strategy in a quiz but it's a fine thing to do if you want to learn the stuff. In any case, don't take advantage of the ambiguities of the English language or the oversight of your instructor to turn a reasonable problem into a trivial one. (See Exercise 13.)

Problem 1. (Quality Control). Suppose that in a bushel of 550 apples there are 2 percent rotten ones. What is the probability that a "random sample" of 25 apples contains 2 rotten apples?

This is the principle behind testing the quality of products by random checking. If the probability turns out to be too small on the basis of the claimed percentage, compared with that figured on some other suspected percentage, then the claim is in doubt. This problem can be done just as easily with arbitrary numbers so we will formulate it in the general case. Suppose there are k defective items in a lot of m products. What is the probability that a random sample of size n contain j defective items? The word "random" here signifies that all samples of size n, under Case III in §3.2, are considered equally likely. Hence the total number is $\binom{m}{n}$. How many of these contain exactly j defective items? To get such a sample we must choose any j out of the k defective items, and combine it freely with $n - j$ out of the $m - k$ non-defective items. The first choice can be made in $\binom{k}{j}$ ways, the second in $\binom{m-k}{n-j}$ ways, by sampling under Case III. By the Fundamental Rule, the total number of samples of size n containing j defective items is equal to the product, and consequently the desired probability is the ratio

$$\binom{k}{j}\binom{m-k}{n-j}\bigg/\binom{m}{n}. \tag{3.4.1}$$

In the case of the apples, we have $m = 550$, $k = 11$, $n = 25$, $j = 2$; so the probability is equal to

$$\binom{11}{2}\binom{539}{23}\bigg/\binom{550}{25}.$$

This number is not easy to compute, but we will learn how to get a good approximation later. Numerical tables are also available.

If we sum the probabilities in (3.4.1) for all j, $0 \leq j \leq n$, the result ought to equal one since all possibilities are counted. We have therefore *proved* the formula

$$\sum_{j=0}^{k}\binom{k}{j}\binom{m-k}{n-j} = \binom{m}{n}$$

by a probabilistic argument. This is confirmed by (3.3.9); indeed a little reflection should convince you that the two arguments are really equivalent.

Problem 2. If a deck of poker cards are thoroughly shuffled, what is the probability that the four aces are found in a row?

There are 52 cards among which are 4 aces. A thorough shuffling signifies that all permutations of the cards are equally likely. For the whole deck, there are (52)! outcomes by Case IIa. In how many of these do the 4 aces stick together? Here we use tip (b) to break up the problem according to where the aces are found. Since they are supposed to appear in a row, we need only locate the first ace as we check the cards in the order they appear in the deck. This may be the top card, the next, and so on, until the 49th. Hence there are 49 positions for the 4 aces. After this has been fixed, the 4 aces can still permute among themselves in 4! ways, and so can the 48 non-aces. This may be regarded as a case of IIIa with $r = 2, m_1 = 4, m_2 = 48$. The Fundamental Rule carries the day and we get the answer

$$\frac{49 \cdot 4!(48)!}{(52)!} = \frac{24}{52 \cdot 51 \cdot 50}.$$

This problem is a case where my tip (a) may be helpful. Try 4 cards with 2 aces. The total number of permutations in which the aces stick together is only $3 \cdot 2!2! = 12$ so you can list them all and look.

Problem 3. Fifteen new students are to be evenly distributed among three classes. Suppose that there are three whiz-kids among the fifteen. What is the probability that each class gets one? one class gets them all?

It should be clear that this is the partition problem discussed under Case IIIb, with $m = 15, m_1 = m_2 = m_3 = 5$. Hence the total number of outcomes is given by

$$\frac{15!}{5!5!5!}.$$

To count the number of these assignments in which each class gets one whiz-kid we will first assign these three kids. This can be done in 3! ways by IIa. The other 12 students can be evenly distributed in the 3 classes by Case IIIb with $m = 12, m_1 = m_2 = m_3 = 4$. The Fundamental Rule applies and we get the desired probability

$$3! \frac{12!}{4!4!4!} \Big/ \frac{15!}{5!5!5!} = \frac{6 \cdot 5^3}{15 \cdot 14 \cdot 13}.$$

Next, if one class gets them all, then there are 3 possibilities according to which class it is, and the rest is similar. So we just replace the numerator above by $3 \cdot \dfrac{12!}{5!5!2!}$ and obtain

$$3 \cdot \frac{12!}{5!5!2!} \Big/ \frac{15!}{5!5!5!} = \frac{5 \cdot 4 \cdot 3^2}{15 \cdot 14 \cdot 13}.$$

By the way, we can now get the probability of the remaining possibility, namely that the number of whiz-kids in the three classes be two, one, zero respectively.

Problem 4. Six dice are rolled. What is the probability of getting three pairs?

One can ask at once "which three pairs?" This means a choice of 3 numbers out of the 6 numbers from 1 to 6. The answer is given by sampling under Case III: $\binom{6}{3} = 20$. Now we can concentrate on one of these cases, say {2, 3, 5} and figure out the probability of getting "a pair of 2, a pair of 3, and a pair of 5." This is surely more clear-cut, so my tip (b) should be used here. To count the number of ways 6 dice can show the pattern {2, 2, 3, 3, 5, 5}, one way is to consider this as putting six labeled tokens (the dice as distinguishable) into three boxes marked $\boxed{2}$, $\boxed{3}$, $\boxed{5}$, with two going into each box. So the number is given by IIIb:

$$\frac{6!}{2!2!2!} = 90.$$

Another way is to think of the dice as six distinguishable performers standing in line waiting for cues to do their routine acts, with two each doing acts nos. 2, 3, 5 respectively, but who does which is up to Boss Chance. This then becomes a permutation problem under IIIa and gives of course the same number above. Finally, we multiply this by the number of choices of the 3 numbers to get

$$\binom{6}{3} \frac{6!}{2!2!2!} = 20 \cdot 90.$$

You may regard this multiplication as another application of the ubiquitous Fundamental Rule but it really just means that 20 mutually exclusive categories are *added* together, each containing 90 cases. The desired probability is given by

$$\frac{20 \cdot 90}{6^6} = \frac{25}{648}.$$

This problem is a case where my negative tip (c) may save you some wasted time as I have seen students trying an argument as follows. If we want to end up with three pairs, the first two dice can be anything; the third die must be one of the two if they are different, and a new one if they are the same. The probability of the first two being different is 5/6, in which case the third die has probability 4/6; on the other hand the probability of the first two being the same is 1/6, in which case the third has probability 5/6. Are you still with us? But what about the next step, and the next?

However, this kind of sequential analysis, based on conditional probabilities, will be discussed in Chapter 5. It works very well sometimes, as in the next problem.

Problem 5. (Birthdays) What is the probability that among n people there are at least two who have the same birthday? We are assuming that they "choose" their birthdays *independently* of one another, so that the result is as if they had drawn n balls marked from 1 to 365 (ignoring leap years) by sampling under Case I. All these outcomes are equally likely and the total number is $(365)^n$. Now we must count those cases in which some of the balls drawn bear the same number. This sounds complicated but it is easy to figure out the "opposite event," namely when all n balls are different. This falls under Case II and the number is $(365)_n$. Hence the desired probability is

$$p_n = 1 - \frac{(365)_n}{(365)^n}.$$

What comes as a surprise is the numerical fact that this probability exceeds 1/2 as soon as $n \geq 23$; see table below.* What would you have guessed?

n	p_n
5	.03
10	.12
15	.25
20	.41
21	.44
22	.48
23	.51
24	.54
25	.57
30	.71
35	.81
40	.89
45	.94
50	.97
55	.99

One can do this problem by a naive argument which turns out to be correct. To get the probability that n people have all different birthdays, we order them in some way and consider each one's "choice," as in the case of the six dice which show different faces (Example 2, §3.1). The first person can have any day of the year for his birthday, hence probability 1; the second can have any but one, hence probability 364/365; the third any but two, hence probability 363/365; and so on. Thus the final probability is

$$\frac{365}{365}\frac{364}{365}\frac{363}{365}\cdots \quad (n \text{ factors}),$$

which is just another way of writing $(365)_n/(365)^n$. The intuitive idea of sequential conditional probabilities used here is equivalent to a splitting diagram described in Section 3.1, beginning with 365 cases, each of which splits

* Computation from $n = 2$ to $n = 55$ was done on a small calculator to five decimal places in a matter of minutes.

into 364, then again into 363, etc. If one divides out by 365 at each stage one gets the product above.

Problem 6. (Matching). Four cards numbered 1 to 4 are laid face down on a table and a person claiming clairvoyance will name them by his extrasensory power. If he is a faker and just guesses at random, what is the probability that he gets at least one right?

There is a neat solution to this famous problem by a formula to be established later in §6.2. But for a small number like 4, brute force will do and in the process we shall learn something new. Now the faker simply picks any one of the 4! permutations and these are considered equally likely. Using tip (b), we will count the number of cases in which there is *exactly* 1 match. This means the other three cards are mismatched, and so we must count the "no match" cases for three cards. This can be done by enumerating all the $3! = 6$ possible random guesses as tabulated below:

real	(abc)	(abc)	(abc)	(abc)	(abc)	(abc)
guess	(abc)	(acb)	(bac)	(bca)	(cab)	(cba)

There are two cases of no-match: the 4th and 5th above. We obtain all cases in which there is exactly one match in 4 cards by fixing that one match and mismatch the three other cards. There are 4 choices for the card to be matched, and after this is chosen, there are 2 ways to mismatch the other three by the tabulation above. Hence by the modified Fundamental Rule there are $4 \cdot 2 = 8$ cases of exactly one match in 4 cards. Next, fix two matches and mismatch the other two. There is only one way to do the latter, hence the number of cases of exactly two matches in 4 cards is equal to that of choosing two cards (to be matched) out of 4, which is $\binom{4}{2} = 6$.

Finally, it is clear that if three cards match the remaining one must also, and there is just one way of matching them all. The results are tabulated as follows:

Exact number of matches	Number of cases	Probability
4	1	1/24
3	0	0
2	6	1/4
1	8	1/3
0	9	3/8

The last row above, for the number of cases of no-match, is obtained by subtracting the sum of the other cases from the total number:

$$24 - (1 + 6 + 8) = 9.$$

The probability of at least one match is $15/24 = 5/8$; of at least two is $7/24$.

You might propose to do the counting without any reasoning by listing all 24 cases for 4 cards, as we did for 3 cards. That is a fine thing to do, not only for your satisfaction but also to check the various cases against our reasoning above. But our step leading from 3 cards to 4 cards is meant to be an illustration of the empirical inductive method, and can lead also from 4 to 5, etc. In fact, that is the way the computing machines do things. They are really not very smart, and always do things step by step, but they are organized and tremendously fast. In our case a little neat algebra does it better, and we can establish the following general formula for the number of cases of at least one match for n cards:

$$n!\left(1 - \frac{1}{2!} + \frac{1}{3!} - + \cdots + (-1)^{n-1}\frac{1}{n!}\right).$$

Problem 7. In how many ways can n balls be put into n numbered boxes so that exactly one box is empty? This problem is instructive as it illustrates several points made above. First of all, it is ambiguous whether the balls are distinguishable or not. Using my tip (d), we will treat both hypotheses.

Hypothesis 1. The balls are indistinguishable. Then it is clearly just a matter of picking the empty box and the one which must have two balls. This is a sampling problem under Case II and the answer is $(n)_2 = n(n - 1)$.

This easy solution would probably be acceptable granted the ambiguous wording, but we learn more if we try the harder way too.

Hypothesis 2. The balls are distinguishable. Then after the choice of the two boxes as under Hypothesis 1 (call it step 1), we still have the problem as to which ball goes into which box. This is a problem of partition under Case IIIb with $m = n$, $m_1 = 2$, $m_2 = \cdots = m_{n-1} = 1$, the empty box being left out of consideration. Hence the answer is

$$\frac{n!}{2!1! \cdots 1!} = \frac{n!}{2!}. \tag{3.4.2}$$

You don't have to know about that formula, since you can argue directly as follows. The question is how to put n numbered balls into $n - 1$ numbered boxes with 2 balls going into a certain box (already chosen by step 1) and 1 ball each into all the rest. There are $\binom{n}{2}$ ways of choosing the two balls to go into that particular box, after that the remaining $n - 2$ balls can go into the other $n - 2$ boxes in $(n - 2)!$ ways. The product of these two numbers is the same as (3.4.2). Finally the total number of ways under Hypothesis 2 is given by

$$n(n-1)\cdot\frac{n!}{2}. \tag{3.4.3}$$

We have argued in two steps above. One may be tempted to argue in three steps as follows. First choose the empty box, then choose $n-1$ balls and put them one each into the other $n-1$ boxes, finally throw the last ball into any one of the latter. The number of possible choices for each step is equal respectively to n, $(n)_{n-1}$ (by sampling under II) and $n-1$. If they are multiplied together the result is

$$n\cdot n!(n-1) \tag{3.4.4}$$

which is twice as large as (3.4.3). Which is correct?

This is the kind of situation my tip (a) is meant to help. Take $n = 3$ and suppose the empty box has been chosen, so the problem is to put balls 1, 2, 3 into box A and B. For the purpose of the illustration let A be square and B round. Choose two balls to put into these two boxes; there are six cases as shown:

Now throw the last ball into one of the two boxes, so that each case above splits into two according as which box gets it:

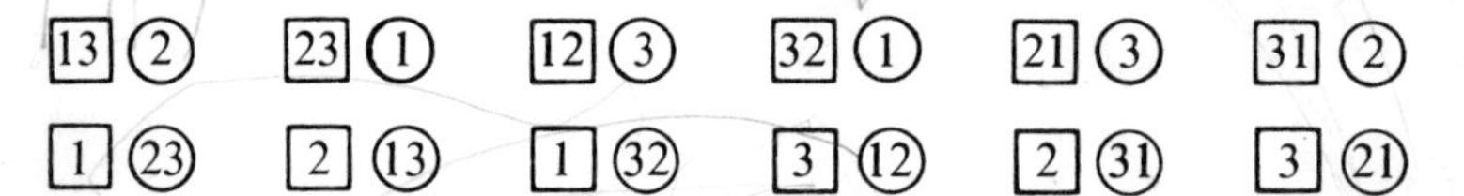

You see what the trouble is. Each final case is counted twice because the box which gets two balls can get them in two orders! The trouble is the same in the general case and so we must divide the number in formula (3.4.4) by 2 to eliminate double-counting, which makes it come out as in formula (3.4.3). All is harmony.

Exercises

(When probabilities are involved in the problems below, the equally likely cases should be "obvious" from the context. In case you demur, follow my tip (d).)

1. A girl decides to choose either a shirt or a tie for a birthday present. There are 3 shirts and 2 ties to choose from. How many choices does she have if she will get only one of them? if she may get both a shirt and a tie?
2. There are 3 kinds of shirts on sale. (a) If two men buy one shirt each

how many possibilities are there? (b) If two shirts are sold, how many possibilities are there?

3. As in No. 2 make up a good question with 3 shirts and 2 men, to which the answer is 2^3, or $\binom{3+2-1}{3}$.
4. If on the menu shown in §3.1 there are 3 kinds of ice cream and 2 kinds of pie to choose from, how many different dinners are there? If we take into account that the customer may skip the vegetable or the dessert or both, how many different dinners are there?
5. How many different initials can be formed with 2 or 3 letters of the alphabet? How large must the alphabet be in order that one million people can be identified by 3-letter initials?
6. How many integers are there between one million and ten million, in whose decimal form no two consecutive digits are the same?
7. In a "true or false" test there are 12 questions. If a student decides to check six of each at random, in how many ways can he do it?
8. In how many ways can 4 boys and 4 girls pair off? In how many ways can they stand in a row in alternating sex?
9. In how many ways can a committee of three be chosen from 20 people? In how many ways can a president, a secretary and a treasurer be chosen?
10. If you have 2 dollars, 2 quarters and 3 nickels, how many different sums can you pay without making change? Change the quarters into dimes and answer again.
11. Two screws are missing from a machine which has screws of three different sizes. If three screws of different sizes are sent over what is the probability that they are what's needed?
12. There are two locks on the door and the keys are among the six different ones you carry in your pocket. In a hurry you dropped one somewhere. What is the probability that you can still open the door? What is the probability that the first two keys you try will open the door?
13. A die is rolled three times. What is the probability that you get a larger number each time? (I gave this simple problem in a test but used inadvertently the words ". . . that the numbers you obtain increase steadily." Think of a possible *mis*interpretation of the words!)
14.* Three dice are rolled twice. What is the probability that they show the same numbers (a) if the dice are distinguishable, (b) if they are not. [Hint: divide into cases according to the pattern of the first throw: a pair, a triple or all different; then match the second throw accordingly.]
15. You walk into a party without knowing anyone there. There are 6 women and 4 men and you know there are 4 married couples. In how many ways can you guess who the couples are? What if you know there are exactly 3 couples?
16. Four shoes are taken at random from five different pairs. What is the probability that there is at least one pair among them?

17. A California driver decides that he must switch lanes every minute to get ahead. If he is on a 4-lane divided highway and does this at random, what is the probability that he is back on his original lane after 4 minutes (assuming no collision)? [Hint: the answer depends on whether he starts on an outside or inside lane.]
18. In sampling under Case I or II of §3.2, what is the probability that in n drawings a particular ball is never drawn? Assume $n < m$.
19. You are told that of the four cards face down on the table, two are red and two are black. If you guess all four at random, what is the probability that you get 0, 2, 4 right?
20. An airport shuttle bus makes 4 scheduled stops for 15 passengers. What is the probability that all of them get off at the same stop? What is the probability that someone (at least one person) gets off at each stop?
21. Ten books are made into 2 piles. In how many ways can this be done if books as well as piles may or may not be distinguishable? Treat all 4 hypotheses and require that neither pile be empty.
22. Ten different books are to be given to Daniel, Phillip, Paul and John who will get in the order given 3, 3, 2, 2 books respectively. In how many ways can this be done? Since Paul and John screamed "no fair" it is decided that they draw lots to determine which two get 3 and which two get 2. How many ways are there now for a distribution? Finally, Marilda and Corinna also want a chance and so it is decided that the six kids should draw lots to determine which two get 3, which two get 2 and which two get none. Now how many ways are there? (There is real semantical difficulty in formulating these distinct problems in general. It is better to be verbose than concise in such a situation. Try putting tokens into boxes.)
23. In a draft lottery containing the 366 days of the year (including February 29), what is the probability that the first 180 days drawn (without replacement of course) are evenly distributed among the 12 months? What is the probability that the first 30 days drawn contain none from August or September? [Hint: first choose 15 days from each month.]
24. At a certain resort the travel bureau finds that tourists occupy the 20 hotels there as if they were so many look-alike tokens (fares) placed in numbered boxes. If this theory is correct, what is the probability that when the first batch of 30 tourists arrive, no hotel is left vacant? [This model is called the *Bose-Einstein statistic* in physics. If the tourists are treated as distinct persons it is the older *Boltzmann-Maxwell statistic*; see [Feller1; §II.5].]
25. 100 trout are caught in a little lake and returned after they are tagged. Later another 100 are caught and found to contain 7 tagged ones. What is the probability of this if the lake contains n trout? (What is your best guess as to the true value of n? The latter is the kind of question asked in *statistics*.)
26. Program a one-to-one correspondence between the various possible

cases under the two counting methods in IIIa and IIIb by taking $m = 4$, $m_1 = 2$, $m_2 = m_3 = 1$.

27.* (For poker players only.) In a poker hand assume all "hands" are equally likely as under Sampling Case III. Compute the probability of (a) flush (b) straight (c) straight flush (d) four of a kind (e) full house.

28.* Show that

$$\binom{2n}{n} = \sum_{k=0}^{\infty} \binom{n}{k}^2.$$

[Hint: apply (3.3.9).]

29.* The number of different ways in which a positive integer n can be written as the sum of positive integers not exceeding n is called (in number theory) the "partition number" of n. For example,

$6 = 6$	sextuple
$= 5 + 1$	quintuple
$= 4 + 2$	quadruple and pair
$= 4 + 1 + 1$	quadruple
$= 3 + 3$	two triples
$= 3 + 2 + 1$	triple and pair
$= 3 + 1 + 1 + 1$	triple
$= 2 + 2 + 2$	three pairs
$= 2 + 2 + 1 + 1$	two pairs
$= 2 + 1 + 1 + 1 + 1$	one pair
$= 1 + 1 + 1 + 1 + 1 + 1$	all different ("no same")

Thus the partition number of 6 is equal to 11; compare this with the numbers 46656 and 462 given in Examples 2 and 3. This may be called the total number of distinguishable "coincidence patterns" when six dice are thrown. We have indicated simpler (but vaguer) names for these patterns in the listing above. Compute their respective probabilities. (It came to me as a surprise that "two pairs" is more probable than "one pair" and has a probability exceeding 1/3. My suspicion of an error in the computation was allayed only after I had rolled six dice one hundred times. It was an old custom in China to play this game over the New Year holidays, and so far as I can remember, "two pairs" were given a higher rank (prize) than "one pair." This is unfair according to their probabilities. Subsequently to my own experiment I found that Feller* had listed analogous probabilities for 7 dice. His choice of this "random

* William Feller (1906–1970), leading exponent of probability.

number" 7 in disregard or ignorance of a time-honored game had probably resulted in my overlooking his tabulation.)

30.* (Banach's match problem.) The Polish mathematician Banach kept two match boxes, one in each pocket. Each box contains n matches. Whenever he wanted a match he reached out at random into one of his pockets. When he found that the box he picked was empty, what is the distribution of the number of matches left in the other box? [Hint: divide into two cases according as the left or right box is empty, but be careful about the case when both are empty.]

Chapter 4

Random Variables

4.1 What is a random variable?

We have seen that the points of a sample space may be very concrete objects such as apples, molecules and people. As such they possess various qualities some of which may be measurable. An apple has its weight and volume; its juicy content can be scientifically measured, even its taste may be graded by expert tasters. A molecule has mass and velocity, from which we can compute its momentum and kinetic energy by formulas from physics. For a human being there are physiological characteristics such as age, height and weight. But there are many other numerical data attached to him (or her) like I.Q., number of years of schooling, number of brothers and sisters, annual income earned and taxes paid, and so on. We will examine some of these illustrations and then set up a mathematical description in general terms.

Example 1. Let Ω be a human population containing n individuals. These may be labeled as

$$\Omega = \{\omega_1, \omega_2, \ldots, \omega_n\}. \tag{4.1.1}$$

If we are interested in their age distribution, let $A(\omega)$ denote the age of ω. Thus to each ω is associated a number $A(\omega)$ in some unit, such as "year." So the mapping

$$\omega \to A(\omega)$$

is a *function* with Ω as its domain of definition. The range is a set of integers but can be made more precise by fractions or decimals or spelled out as e.g., "18 years, 5 months and 1 day." There is no harm if we take all positive integers or all positive real numbers as the range, although only a very small portion of it will be needed. Accordingly, we say A is an integer-valued or real-valued function. Similarly, we may denote the height, weight and income by the functions:

$$\omega \to H(\omega),$$

$$\omega \to W(\omega),$$

$$\omega \to I(\omega).$$

In the last case I may take negative values! Now for some medical purposes, a linear combination of height and weight may be an appropriate measure:

$$\omega \to \lambda H(\omega) + \mu W(\omega)$$

where λ and μ are two numbers. This then is also a function of ω. Similarly, if ω is a "head of family" alias breadearner, the census bureau may want to compute the function:

$$\omega \longrightarrow \frac{I(\omega)}{N(\omega)}$$

where $N(\omega)$ is the number of persons in his family, namely the number of mouths to be fed. The ratio above represents then the "income per capita" for the family.

Let us introduce some convenient symbolism to denote various sets of sample points derived from random variables. For example, the set of ω in Ω for which the age is between 20 and 40 will be denoted by

$$\{\omega \mid 20 \leq A(\omega) \leq 40\}$$

or more briefly when there is no danger of misunderstanding by

$$\{20 \leq A \leq 40\}.$$

The set of ω for which the height is between 65 and 75 (in inches) and the weight is between 120 and 180 (in pounds) can be denoted in several ways as follows:

$$\begin{aligned}\{\omega \mid 65 \leq H(\omega) \leq 75\} &\cap \{\omega \mid 120 \leq W(\omega) \leq 180\} \\ &= \{\omega \mid 65 \leq H(\omega) \leq 75;\ 120 \leq W(\omega) \leq 180\} \\ &= \{65 \leq H \leq 75;\ 120 \leq W \leq 180\}.\end{aligned}$$

Example 2. Let Ω be gaseous molecules in a given container. We can still represent Ω as in (4.1.1) even though n is now a very large number such as 10^{25}. Let m = mass, v = velocity, M = momentum, E = kinetic energy. Then we have the corresponding functions:

$$\begin{aligned}&\omega \longrightarrow m(\omega), \\ &\omega \longrightarrow v(\omega), \\ &\omega \longrightarrow M(\omega) = m(\omega)v(\omega), \\ &\omega \longrightarrow E(\omega) = \frac{1}{2}m(\omega)v(\omega)^2.\end{aligned}$$

In experiments with gases actual measurements may be made of m and v, but the quantities of interest may be M or E which can be derived from the formulas. Similarly, if θ is the angle of the velocity relative to the x-axis, $\omega \longrightarrow \theta(\omega)$ is a function of ω, and

$$\omega \longrightarrow \cos\theta(\omega)$$

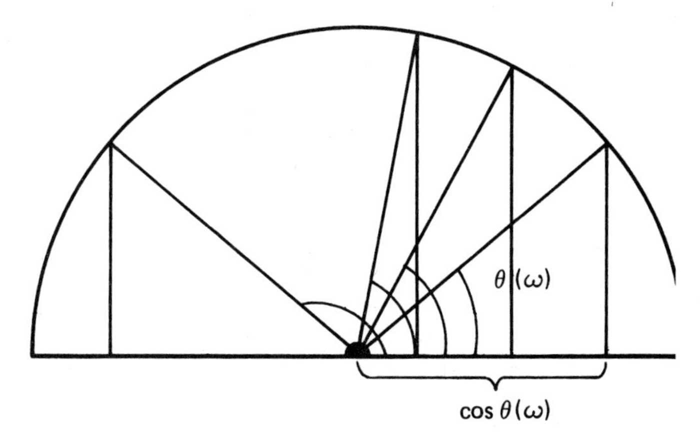

Figure 17

may be regarded as the composition of the function "cos" with the function "θ." The set of all molecules which are moving toward the right is represented by

$$\{\omega \mid \cos \theta(\omega) > 0\}.$$

Example 3. Let Ω be the outcome space of throwing a die twice. Then it consists of $6^2 = 36$ points listed below:

(1, 1)	(1, 2)	(1, 3)	(1, 4)	(1, 5)	(1, 6)
(2, 1)	(2, 2)	(2, 3)	(2, 4)	(2, 5)	(2, 6)
(3, 1)	(3, 2)	(3, 3)	(3, 4)	(3, 5)	(3, 6)
(4, 1)	(4, 2)	(4, 3)	(4, 4)	(4, 5)	(4, 6)
(5, 1)	(5, 2)	(5, 3)	(5, 4)	(5, 5)	(5, 6)
(6, 1)	(6, 2)	(6, 3)	(6, 4)	(6, 5)	(6, 6)

Thus each ω is represented by an ordered pair or a two-dimensional *vector:*

$$\omega_k = (x_k, y_k), \quad k = 1, 2, \ldots, 36$$

where x_k and y_k take values from 1 to 6. The first *coordinate* x represents the outcome of the first throw, the second coordinate y the outcome of the second throw. These two coordinates are determined by the point ω, hence they are *functions* of ω:

(4.1.2) $$\omega \to x(\omega), \quad \omega \to y(\omega).$$

On the other hand, each ω is completely determined by its two coordinates, so much so that we may say that ω *is* the pair of them:

$$\omega \equiv (x(\omega), y(\omega)).$$

This turnabout is an important concept to grasp. For example, let the die be thrown n times and the results of the successive throws be denoted by

$$x_1(\omega), x_2(\omega), \ldots, x_n(\omega),$$

then not only each $x_k(\omega)$, $k = 1, 2, \ldots, n$, is a function of ω which may be called its *kth coordinate*, but the totality of these n functions in turn determines ω, and therefore ω is nothing more or less than the n-dimensional vector

$$\omega \equiv (x_1(\omega), x_2(\omega), \ldots, x_n(\omega)). \tag{4.1.3}$$

In general, each $x_k(\omega)$ represents a certain numerical characteristic of the sample ω, and although ω may possess many, many characteristics, in most questions only a certain set of them is taken into account. Then a representation like (4.1.3) is appropriate. For example, in a traditional beauty contest, only three bodily measurements given in inches, are considered such as (36, 29, 38). In such a contest (no "song and dance") each contestant is reduced to such an ordered triple:

$$\text{contestant} = (x, y, z).$$

Another case of this kind is when a student takes a number of tests, say 4, which are graded on the usual percentage basis. Let the student be ω, his score on the 4 tests be $x_1(\omega)$, $x_2(\omega)$, $x_3(\omega)$, $x_4(\omega)$. For the grader (or the computing machine if all the tests can be machine processed), each ω is just the 4 numbers: $(x_1(\omega), \ldots, x_4(\omega))$. Two students who have the same scores are not distinguished. Suppose the criterion for success is that the total should exceed 200; then the set of successful candidates is represented by

$$\{\omega \mid x_1(\omega) + x_2(\omega) + x_3(\omega) + x_4(\omega) > 200\}.$$

A variation is obtained if different weights λ_1, λ_2, λ_3, λ_4 are assigned to the 4 tests, then the criterion will depend on the linear combination $\lambda_1 x_1(\omega) + \cdots + \lambda_4 x_4(\omega)$. Another possible criterion for passing the tests is given by

$$\{\omega \mid \min (x_1(\omega), x_2(\omega), x_3(\omega), x_4(\omega)) > 35\}.$$

What does this mean in plain English?

4.2. How do random variables come about?

We can now give a general formulation for numerical characteristics of sample points. *Assume first that* Ω *is a countable space.* This assumption makes an essential simplification which will become apparent; other spaces will be discussed later.

Definition of Random Variable. A numerically valued function X of ω with domain Ω:

$$\omega \in \Omega: \quad \omega \rightarrow X(\omega) \tag{4.2.1}$$

is called a random variable [on Ω].

The term "random variable" is well established and so we will use it in this book but "chance variable" or "stochastic variable" would have been good too. The adjective "random" is just to remind us that we are dealing with a sample space and trying to describe certain things which are commonly called "random events" or "chance phenomena." What might be said to have an element of randomness in $X(\omega)$ is the sample point ω which is picked "at random," such as in a throw of dice or the polling of an individual from a population. Once ω is picked $X(\omega)$ is thereby determined and there is nothing vague, indeterminate or chancy about it anymore. For instance after an apple ω is picked from a bushel its weight $W(\omega)$ can be measured and may be considered as known. In this connection the term "variable" should also be understood in the broad sense as a "dependent variable," namely a function of ω, as discussed in §4.1. We can say that the sample point ω serves here as an "independent variable" in the same way the variable x in $\sin x$ does, but it is better not to use this language since "independent" has a very different and more important meaning in probability theory (see §5.5).

Finally it is a custom (not always observed) to use a capital letter to denote a random variable, such as X, Y, N or S, but there is no reason why we cannot use small letters x or y as we did in the Examples of §4.1.

Observe that random variables can be defined on a sample space before any probability is mentioned. Later we shall see that they acquire their probability distributions through a probability measure imposed on the space.

Starting with some random variables, we can at once make new ones by operating on them in various ways. Specific examples have already been given in the examples in §4.1. The general proposition may be stated as follows:

Proposition 1. *If X and Y are random variables, then so are*

$$X + Y, \quad X - Y, \quad XY, \quad X/Y\,(Y \neq 0), \tag{4.2.2}$$

and $aX + bY$ where a and b are two numbers.

This is immediate from the general definition, since, e.g.

$$\omega \to X(\omega) + Y(\omega)$$

is a function on Ω as well as X and Y. The situation is exactly the same as in calculus: if f and g are functions then so are

$$f + g, \quad f - g, \quad fg, \quad f/g \; (g \neq 0), \quad af + bg.$$

The only difference is that in calculus these are functions of x, a real number; while here the functions in (4.2.2) are those of ω, a sample point. Also, as in calculus where a constant is regarded as a very special kind of function, so is a constant a very special kind of random variable. For example it is quite possible that in a class in elementary school, all the pupils are of the same age. Then the random variable $A(\omega)$ discussed in Example 1 of §4.1 is equal to a constant, say $= 9$ (years) in a fourth grade class.

In calculus a *function of a function* is still a function such as $x \to \log(\sin x)$ or $x \to f(\varphi(x)) = (f \circ \varphi)(x)$. A function of a random variable is still a random variable such as the $\cos\theta$ in Example 2 of §4.1. More generally we can have a function of several random variables.

Proposition 2. *If φ is a function of two (ordinary) variables and X and Y are random variables, then*

$$\omega \to \varphi(X(\omega), Y(\omega)) \tag{4.2.3}$$

is also a random variable, which is denoted more concisely as $\varphi(X, Y)$.

A good example is the function $\varphi(x, y) = \sqrt{x^2 + y^2}$. Suppose $X(\omega)$ and $Y(\omega)$ denote respectively the horizontal and vertical velocities of a gas molecule, then

$$\varphi(X, Y) = \sqrt{X^2 + Y^2}$$

will denote its absolute *speed.*

Let us note in passing that Proposition 2 contains Proposition 1 as a particular case. For instance if we take $\varphi(x, y) = x + y$, then $\varphi(X, Y) = X + Y$. It also contains functions of a single random variable as a particular case such as $f(X)$. Do you see why? Finally, extension of Proposition 2 to more than two variables is obvious. A particularly important case is the sum of n random variables:

$$S_n(\omega) = X_1(\omega) + \cdots + X_n(\omega) = \sum_{k=1}^{n} X_k(\omega). \tag{4.2.4}$$

For example if $X_1, \ldots, X_n$ denote the successive outcomes of a throw of a die, then S_n is the total obtained in n throws. We shall have much to do with these *partial sums* S_n.

We will now illustrate the uses of random variables in some everyday situations. Quite often the intuitive notion of some random quantity precedes that of a sample space. Indeed one can often talk about random variables X, Y, etc. without bothering to specify Ω. The rather formal (and formidable?) mathematical set-up serves as a necessary logical backdrop, but it need not be dragged into the open on every occasion when the language of probability can be readily employed.

Example 4. The cost of manufacturing a certain book is \$3 per book up to 1000 copies, \$2 per copy between 1000 and 5000 copies, and \$1 per copy afterwards. In reality of course books are printed in round lots and not on demand "as you go." What we assume here is tantamount to selling all over-stock at cost, with no loss of business due to understock. Suppose we print 1000 copies initially and price the book at \$5. What is "random" here is the number of copies that will be sold; call it X. It should be evident that once X is known, we can compute the profit or loss from the sales; call this Y. Thus Y is a function of X and is random only because X is so. The formula connecting Y with X is given below (See Fig. 18 on page 78):

$$Y = \begin{cases} 5X - 3000 & \text{if } X \le 1000, \\ 2000 + 3(X - 1000) & \text{if } 1000 < X \le 5000, \\ 14000 + 4(X - 5000) & \text{if } X > 5000. \end{cases}$$

What is the probability that the book is a financial loss? It is that of the event represented by the set

$$\{5X - 3000 < 0\} = \{X < 600\}.$$

What is the probability that the profit will be at least \$10000? It is that of the set

$$\begin{aligned} \{2000 + 3(X - 1000) \ge 10000\} &\cup \{X > 5000\} \\ &= \left\{X \ge \frac{8000}{3} + 1000\right\} \cup \{X > 5000\} \\ &= \{X \ge 3667\}. \end{aligned}$$

But what are these probabilities? They will depend on a knowledge of X. One can only guess at it in advance; so it is a random phenomena. But after the sales are out, we shall know the exact value of X; just as after a die is cast we shall know the outcome. The various probabilities are called the distribution of X and will be discussed in §4.3 below.

What is the sample space here? Since the object of primary interest is X, we may very well take it as our sample point and call it ω instead to conform with our general notation. Then each ω is some positive integer and $\omega \to Y(\omega)$ is a random variable with Ω the space of positive integers. To pick an ω means in this case to hazard a guess (or make a hypothesis) on the number of sales, from which we can compute the profit by the preceding formula. There is

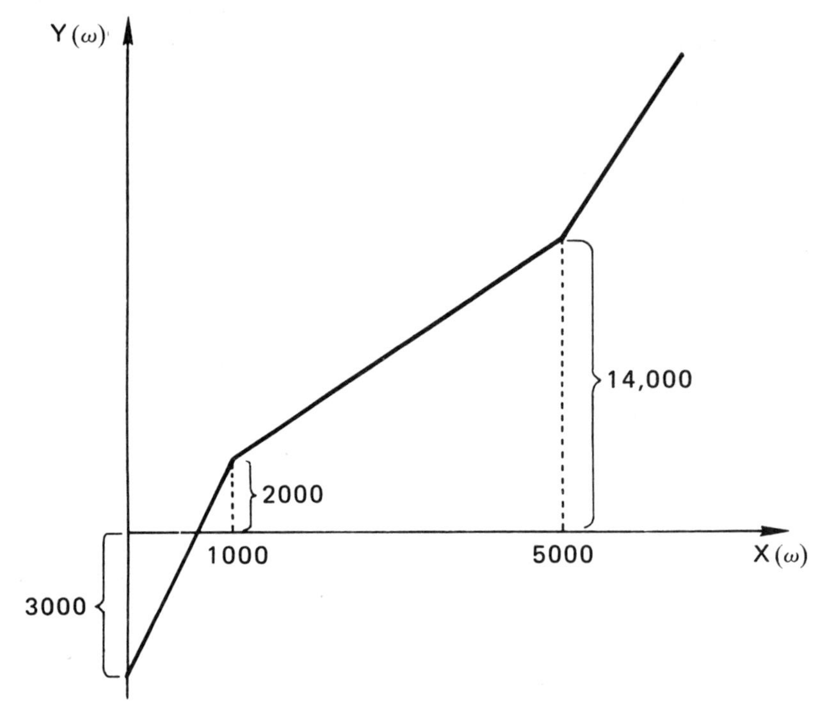

Figure 18

nothing wrong about this model of a sample space, though it seems a bit superfluous.

A more instructive way of thinking is to consider each ω as representing "a possible sales-record for the book." A publisher is sometimes interested in other information than the total number of sales. An important factor which has been left out of consideration above is the time element involved in the sale. Surely it makes a difference whether 5000 copies are sold in one or ten years. If the book is a college text like this one, it may be important to know how it does in different types of schools and in different regions of the country. If it is fiction or drama it may mean a great deal (even only from the profit motive) to know what the critics say about it, though this would be in a promotions rather than sales record. All these things may be contained in a capsule which is the sample point ω. You can imagine it to be a *complete record* of every bit of information pertaining to the book, of which $X(\omega)$ and $Y(\omega)$ represent only two facets. Then what is Ω? It is the totality of all such conceivable records. This concept of a sample space may sound weird and is unwieldy (can we say that Ω is countable?), but it gives the appropriate picture when one speaks of e.g. the path of a particle in Brownian motion or the evolution of a stochastic process (see Chapter 8). On the other hand, it also shows the expediency of working with some specific random variables rather than worrying about the whole universe.

Example 5. An insurance company receives claims for indemnification from time to time. Both the times of arrival of such claims and their amounts are unknown in advance and determined by chance; ergo, random. The total amount of claims in one year, say, is of course also random, but clearly it will be determined as soon as we know the "when" and "how much" of the claims. Let the claims be numbered as they arrive and let S_n denote the date of the nth claim. Thus $S_3 = 33$ means the third claim arrives on February the second. So we have

$$1 \leq S_1 \leq S_2 \leq \cdots \leq S_n \leq \cdots$$

and there is equality whenever several claims arrive on the same day. Let the amount of the nth claim be C_n (in dollars). What is the total number of claims received in the year? It is given by N where

$$N = \max \{n \mid S_n \leq 365\}.$$

Obviously N is also random but it is determined by the sequence of S_n's; in theory we need to know the entire sequence because N may be arbitrarily large. Knowing N, and the sequence of C_n's, we can determine the total amount of claims in that year:

$$C_1 + \cdots + C_N \tag{4.2.5}$$

in the notation of (4.2.4). Observe that in (4.2.5) not only each term on the right side is a random variable but also the number of terms. Of course the sum is also a random variable. It depends on the S_n's as well as the C_n's.

In this case we can easily imagine that the claims arrive at the office one after another and a complete record of them is kept in a ledger printed like a diary. Under some dates there may be no entry, under others there may be many in various different amounts. Such a ledger is kept over the years and will look quite different from one period of time to another. Another insurance company will have another ledger which may be similar in some respects and different in others. Each conceivable account kept in such a ledger may be considered as a sample point, and a reasonably large collection of them may serve as the sample space. For instance, an account in which one million claims arrive on the same day may be left out of the question, or a claim in the amount of 95 cents. In this way we can keep the image of a sample space within proper bounds of realism.

If we take such a view, other random variables come easily to mind. For example, we may denote by Y_k the total amount of claims on the kth day. This will be the number which is the sum of all the entries under the date, possibly zero. The total claims from the first day of the account to the nth day can then be represented by the sum

$$Z_n = \sum_{k=1}^{n} Y_k = Y_1 + Y_2 + \cdots + Y_n. \tag{4.2.6}$$

The total claims over any period of time $[s, t]$ can then be represented as

(4.2.7) $$Z_t - Z_{s-1} = \sum_{k=s}^{t} Y_k = Y_s + Y_{s+1} + \cdots + Y_t.$$

We can plot the *accumulative* amount of claims Z_t against the time t by a graph of the following kind.

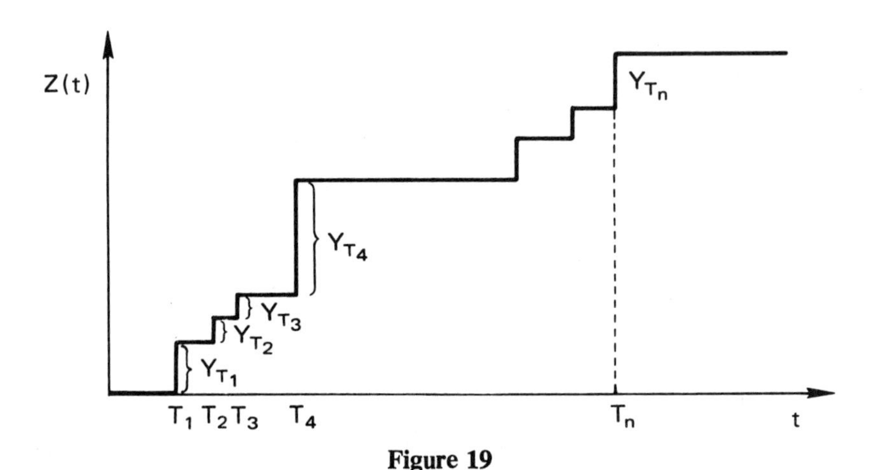

Figure 19

There is a jump at t when the entry for the tth day is not empty, and the size of the rise is the total amount of claims on that day. Thus the successive rises correspond to the Y_k's which are greater than 0. Clearly you can read off from such a graph the total claim over any given period of time, and also e.g. the lengths of the "free periods" between the claims, but you cannot tell what the individual claims are when several arrive on the same day. If all the information you need can be got from such a graph then you may regard each conceivable graph as a sample point. This will yield a somewhat narrower sample space than the one described above, but it will serve our purpose. From the mathematical point of view, the identification of a sample point with a graph (also called a *sample curve*, *path*, or *trajectory*) is very convenient, since a curve is a more precise (and familiar!) object than a ledger or some kind of sales-and-promotions-record.

4.3. Distribution and expectation

In Chapter 2 we discussed the probabilities of sets of sample points. These sets are usually determined by the values of random variables. A typical example is

(4.3.1) $$\{a \leq X \leq b\} = \{\omega \mid a \leq X(\omega) \leq b\}$$

where X is a random variable, a and b are two constants. Particular cases of this have been indicated among the examples in §4.1. Since every subset of Ω has a probability assigned to it when Ω is countable, the set above has a probability which will be denoted by

$$P(a \leq X \leq b). \tag{4.3.2}$$

More generally let A be a set of real numbers, alias a set of points on the real line $R^1 = (-\infty, +\infty)$; then we can write

$$P(X \in A) = P(\{\omega \mid X(\omega) \in A\}). \tag{4.3.3}$$

For instance, if A is the closed interval $[a, b]$ then this is just the set in (4.3.2); but A may be the open interval (a, b), half-open interval $(a, b]$ or $[a, b)$, infinite intervals $(-\infty, b)$ or $(a, +\infty)$; the union of several intervals, or a set of integers say $\{m, m + 1, \ldots, m + n\}$. An important case occurs when A reduces to a single point x; it is then called the *singleton* $\{x\}$. The distinction between the point x and the set $\{x\}$ may seem academic. Anyway, the probability

$$P(X = x) = P(X \in \{x\}) \tag{4.3.4}$$

is "the probability that X takes (or assumes) the value x." If X is the age of a human population, $\{X = 18\}$ is the subpopulation of 18-year-olds—a very important set!

Now the hypothesis that Ω is countable will play an essential simplifying role. Since X has Ω as domain of definition, it is clear that the range of X must be finite when Ω is finite, and at most countably infinite when Ω is so. Indeed the *exact range* of X is just the set of real numbers below:

$$V_X = \bigcup_{\omega \in \Omega} \{X(\omega)\}, \tag{4.3.5}$$

and many of these numbers may be the same. For the mapping $\omega \to X(\omega)$ is in general many-to-one, not necessarily one-to-one. In the extreme case when X is a constant random variable, the set V_X reduces to a single number. Let the distinct values in V_X be listed in any order as

$$\{v_1, v_2, \ldots, v_n, \ldots\}.$$

The sequence may be finite or infinite. Clearly if $x \notin V_X$, namely if x is not one of the values v_n, then $P(X = x) = 0$. On the other hand, we do not forbid that some v_n may have zero probability. This means that *some sample points may have probability zero*. You may object: why don't we throw such nuisance points out of the sample space? Because it is often hard to know in advance which ones to throw out. It is easier to leave them in since they do no harm.

[In an uncountable Ω, every single point ω may have probability zero! But we are not talking about this at present; see §4.5 below.]

Let us introduce the notation

$$p_n = P(X = v_n), \quad v_n \in V_X. \tag{4.3.6}$$

It should be obvious that if we know all the p_n's, then we can calculate all probabilities concerning the random variable X, *alone*. Thus, the probabilities in (4.3.2) and (4.3.3) can be expressed in terms of the p_n's as follows:

$$P(a \leq X \leq b) = \sum_{a \leq v_n \leq b} p_n; \quad P(X \in A) = \sum_{v_n \in A} p_n. \tag{4.3.7}$$

The first is a particular case of the second, and the last-written sum reads this way: "the sum of the p_n's for which the corresponding v_n's belong to A."

When A is the infinite interval $(-\infty, x]$ for any real number x, we can introduce a function of x as follows:

$$F_X(x) = P(X \leq x) = \sum_{v_n \leq x} p_n. \tag{4.3.8}$$

This function $x \to F_X(x)$ defined on R^1 is called the *distribution function of* X. Its value at x "picks up" all the probabilities of values of X up to x (inclusive); for this reason the adjective "cumulative" is sometimes added to its name. For example if X is the annual income (in \$'s) of a breadwinner, then $F_X(10000)$ is the probability of the income group earning anywhere up to ten thousand, and can theoretically include all those whose incomes are negative.

The distribution function F_X is determined by the v_n's and p_n's as shown in (4.3.8). Conversely if we know F_X, namely we know $F_X(x)$ for all x, we can "recover" the v_n's and p_n's. We will not prove this fairly obvious assertion here. For the sake of convenience, we shall say that the two sets of numbers $\{v_n\}$ and $\{p_n\}$ determine the *probability distribution* of X, where any v_n for which $p_n = 0$ may be omitted. It is easy to see that if $a < b$, then

$$P(a < X \leq b) = P(X \leq b) - P(X \leq a) = F_X(b) - F_X(a); \tag{4.3.9}$$

but how do we get $P(a \leq X \leq b)$, or $P(X = x)$ from F_X? (see Exercise 7 below).

Now to return to the p_n's, which are sometimes called the "elementary probabilities" for the random variable X. In general they have the following two properties:

$$\begin{aligned} &\text{(i)} \qquad \forall n: \quad p_n \geq 0, \\ &\text{(ii)} \qquad \sum_n p_n = 1. \end{aligned} \tag{4.3.10}$$

Compare this with (2.4.1). The sum in (ii) may be over a finite or infinite sequence according as V_X is a finite or infinite set. The property (i) is obvious, apart from the observation already made that some p_n may $= 0$. The property (ii) says that the values $\{v_n\}$ in V_X exhaust all possibilities for X, hence their probabilities must add up to that of the "whole universe." This is a fine way to say things, but let us learn to be more formal by converting the verbal argument into a symbolic proof. We begin with

$$\bigcup_n \{X = v_n\} = \Omega.$$

Since the v_n's are distinct, the sets $\{X = v_n\}$ must be disjoint. Hence by countable additivity (see §2.3) we have

$$\sum_n P(X = v_n) = P(\Omega) = 1.$$

This is property (ii).

Before making further specialization on the random variables, let us formulate a fundamental new definition in its full generality. It is motivated by the intuitive notion of the average of a random quantity.

Definition of Mathematical Expectation. For a random variable X defined on a countable sample space Ω, its *mathematical expectation* is the number $E(X)$ given by the formula

$$E(X) = \sum_{\omega \in \Omega} X(\omega)P(\{\omega\}), \tag{4.3.11}$$

provided that the series converges absolutely, namely

$$\sum_{\omega \in \Omega} |X(\omega)|P(\{\omega\}) < \infty. \tag{4.3.12}$$

In this case we say that the mathematical expectation of X *exists*. The process of "taking expectations" may be described in words as follows: take the value of X at each ω, multiply it by the probability of that point, and sum over all ω in Ω. If we think of $P(\{\omega\})$ as the weight attached to ω then $E(X)$ is the weighted average of the function X. Note that if we label the ω's as $\{\omega_1, \omega_2, \ldots, \omega_n, \ldots\}$, then we have

$$E(X) = \sum_n X(\omega_n)P(\{\omega_n\}).$$

But we may as well use ω itself as label, and save a subscript, which explains the cryptic notation in (4.3.11).

Example 6. Let $\Omega = \{\omega_1, \ldots, \omega_7\}$ be a parcel of land sub-divided into seven "lots for sale." These lots have percentage areas and prices as follows:

5%, 10%, 10%, 10%, 15%, 20%, 30%;
$800, $900, $1000, $1200, $800, $900, $800.

Define $X(\omega)$ to be the price per acre of the lot ω. Then $E(X)$ is the average price per acre of the whole parcel and is given by

$$(800)\frac{5}{100} + (900)\frac{10}{100} + (1000)\frac{10}{100} + (1200)\frac{10}{100} + (800)\frac{15}{100} + (900)\frac{20}{100} + (800)\frac{30}{100} = 890;$$

namely \$890 per acre. This can also be computed by first lumping together all acreage at the same price, and then summing over the various prices;

$$(800)\left(\frac{5}{100} + \frac{15}{100} + \frac{30}{100}\right) + (900)\left(\frac{10}{100} + \frac{20}{100}\right) + (1000)\frac{10}{100} + (1200)\frac{10}{100}$$
$$= (800)\frac{50}{100} + (900)\frac{30}{100} + (1000)\frac{10}{100} + (1200)\frac{10}{100} = 890.$$

The adjective in "mathematical expectation" is frequently omitted, and it is also variously known as "expected value," "mean (value)" or "first moment" (see §6.3 for the last term). In any case, do not *expect* the value $E(X)$ when X is observed. For example, if you toss a fair coin to win \$1 or nothing according as it falls heads or tails, you will never get the expected value \$.50! However, if you do this n times and n is large, then you can expect to get about $n/2$ dollars with a good probability. This is the implication of the Law of Large Numbers, to be made precise in §7.6.

We shall now amplify on the condition given in (4.3.12). Of course it is automatically satisfied when Ω is a finite space, but it is essential when Ω is countably infinite. For it allows us to calculate the expectation in any old way by rearranging and regrouping the terms in the series in (4.3.11), without fear of getting contradictory results. In other words, if the series is absolutely convergent, then it has a uniquely defined "sum" which in no way depends on how the terms are picked out and added together. The fact that contradictions can indeed arise if this condition is dropped may be a surprise to you. If so, you will do well to review your knowledge of the convergence and absolute convergence of a numerical series. This is a part of the calculus course which is often poorly learned (and taught), but will be essential for probability theory, not only in this connection but generally speaking. Can you, for instance, think of an example where the series in (4.3.11) converges but the one in (4.3.12) does not? [Remember that the p_n's must satisfy the conditions in (4.3.10), though the v_n's are quite arbitrary. So the question is a little harder than just to find an arbitrary non-absolutely convergent series; but see Exercise 21.] In such a case the expectation is *not* defined at all. The reason why we are being so strict is: absolutely convergent series can be manipulated in ways that non-absolutely [conditionally] convergent series cannot be. Surely the definition of $E(X)$ would not make sense if its value could be altered simply by shuffling around the various terms in the series in

(4.3.11), which merely means that we enumerate the sample points in a different way. Yet this can happen without the condition (4.3.12)!

Let us state explicitly a general method of calculating $E(X)$ which is often expedient. Suppose the sample space Ω can be decomposed into disjoint sets A_n:

$$\Omega = \bigcup_n A_n \tag{4.3.13}$$

in such a way that X takes the same value on each A_n. Thus we may write

$$X(\omega) = a_n \quad \text{for} \quad \omega \in A_n, \tag{4.3.14}$$

where the a_n's need not be all different. We have then

$$E(X) = \sum_n P(A_n)a_n = \sum_n P(X = a_n)a_n. \tag{4.3.15}$$

This is obtained by regrouping the ω's in (4.3.11) first into the subsets A_n, and then summing over all n. In particular if $(v_1, v_2, \ldots, v_n, \ldots)$ is the range of X, and we group the sample points ω according to the values of $X(\omega)$, i.e., putting

$$A_n = \{\omega \mid X(\omega) = v_n\}, \quad P(A_n) = p_n,$$

then we get

$$E(X) = \sum_n p_n v_n, \tag{4.3.16}$$

where the series will automatically converge absolutely because of (4.3.12). In this form it is clear that the expectation of X is determined by its probability distribution.

Finally, it is worthwhile to point out that the formula (4.3.11) contains an expression for the expectation of any function of X:

$$E(\varphi(X)) = \sum_{\omega \in \Omega} \varphi(X(\omega))P(\{\omega\})$$

with a proviso like (4.3.12). For by Proposition 2 or rather a simpler analogue, $\varphi(X)$ is also a random variable. It follows that we have

$$E(\varphi(X)) = \sum_n p_n \varphi(v_n), \tag{4.3.17}$$

where the v_n's are as in (4.3.16), but note that the $\varphi(v_n)$'s need not be distinct. Thus the expectation of $\varphi(X)$ is already determined by the probability distribution of X (and of course also by the function φ), without the intervention of the probability distribution of $\varphi(X)$ itself. This is most convenient in calculations. In particular, for $\varphi(x) = x^r$ we get the rth *moment* of X:

$$E(X^r) = \sum_n p_n v_n^r, \tag{4.3.18}$$

see §6.3.

4.4. Integer-valued random variables

In this section we consider random variables which take only nonnegative integer values. In this case it is convenient to consider the range to be the entire set of such numbers:

$$N^0 = \{0, 1, 2, \ldots, n, \ldots\}$$

since we can assign probability zero to those which are not needed. Thus we have, as specialization of (4.3.6), (4.3.8) and (4.3.11):

$$\begin{aligned} p_n &= P(X = n), \; n \in N^0; \\ F_X(x) &= \sum_{0 \leq n \leq x} p_n; \\ E(X) &= \sum_{n=0}^{\infty} n p_n. \end{aligned} \tag{4.4.1}$$

Since all terms in the last-written series are nonnegative, there is no difference between convergence and absolute convergence. Furthermore, since such a series either converges to a finite sum or diverges to $+\infty$, we may even allow $E(X) = +\infty$ in the latter case. This is in contrast to our general definition in the last section, but is a convenient extension.

In many problems there is practical justification to consider the random variables to take only integer values, provided a suitably small unit of measurement is chosen. For example, monetary values can be expressed in cents rather than dollars, or one tenth of a cent if need be; if "inch" is not a small enough unit for lengths we can use one hundredth or one thousandth of an inch. There is a unit called *angstrom* (Å) which is equal to 10^{-7} of a millimeter, used to measure electromagnetic wavelengths. For practical purposes, of course, incommensurable magnitudes (irrational ratios) do not exist; at one time π was legally defined to be 3.14 in some state of the United States! But one can go too far in this kind of justification!

We proceed to give some examples of (4.4.1).

Example 7. Suppose L is a positive integer, and

$$p_n = \frac{1}{L}, \quad 1 \leq n \leq L. \tag{4.4.2}$$

Then automatically all other p_n's must be zero because $\sum_{n=1}^{L} p_n = L \cdot \frac{1}{L} = 1$ and the conditions in (4.3.10) must be satisfied. Next, we have

$$E(X) = \frac{1}{L}\sum_{n=1}^{L} n = \frac{1}{L}\cdot\frac{L(L+1)}{2} = \frac{L+1}{2}.$$

The sum above is done by a formula for *arithmetical progression* which you have probably learned in school.

We say in this case that X has a *uniform distribution* over the set $\{1, 2, \ldots, L\}$. In the language of Chapter 3, the L possible cases $\{X = 1\}$, $\{X = 2\}, \ldots, \{X = L\}$ are all *equally likely*. The expected value of X is equal to the arithmetical mean [average] of the L possible values. Here is an illustration of its meaning. Suppose you draw at random a token X from a box containing 100 tokens valued at 1¢ to 100¢. Then your expected prize is given by $E(X) = 50.5$¢. Does this sound reasonable to you?

Example 8. Suppose you toss a perfect coin repeatedly until a head turns up. Let X denote the number of tosses it takes until this happens, so that $\{X = n\}$ means $n - 1$ tails before the first head. It follows from the discussion in Example 8 of §2.4 that

(4.4.3) $$p_n = P(T = n) = \frac{1}{2^n}.$$

because the favorable outcome is just the specific sequence $\underbrace{TT\cdots T}_{n-1 \text{ times}}H$. What is the expectation of X? According to (4.4.1), it is given by the formula

(4.4.4) $$\sum_{n=1}^{\infty}\frac{n}{2^n} = ?$$

Let us learn how to sum this series, though properly speaking this does not belong to this course. We begin with the *fountainhead* of many of such series:

(4.4.5) $$\frac{1}{1-x} = 1 + x + x^2 + \cdots + x^n + \cdots = \sum_{n=0}^{\infty} x^n \quad \text{for} \quad |x| < 1.$$

This is a geometric series of the simplest kind which you surely have seen. Now differentiate it term by term:

(4.4.6) $$\frac{1}{(1-x)^2} = 1 + 2x + 3x^2 + \cdots + nx^{n-1} + \cdots = \sum_{n=0}^{\infty}(n+1)x^n \quad \text{for } |x| < 1.$$

This is valid because the radius of convergence of the power series in (4.4.5) is equal to 1, so such manipulations are legitimate for $|x| < 1$. [Absolute and uniform convergence of the power series is involved here.] If we substitute $x = 1/2$ in (4.4.6) we obtain

(4.4.7) $$4 = \sum_{n=0}^{\infty} (n+1)\left(\frac{1}{2}\right)^n.$$

There is still some difference between (4.4.4) and the series above, so a little algebraic manipulation is needed. One way is to split up the terms above:

$$4 = \sum_{n=0}^{\infty} \frac{n}{2^n} + \sum_{n=0}^{\infty} \frac{1}{2^n} = \sum_{n=1}^{\infty} \frac{n}{2^n} + 2,$$

where we have summed the second series by substituting $x = 1/2$ into (4.4.5). Thus the answer to (4.4.4) is equal to 2. Another way to manipulate the formula is to change the index of summation: $n + 1 = \nu$. Then we have

$$4 = \sum_{n=0}^{\infty} \frac{n+1}{2^n} = \sum_{\nu=1}^{\infty} \frac{\nu}{2^{\nu-1}} = 2 \sum_{\nu=1}^{\infty} \frac{\nu}{2^\nu},$$

which of course yields the same answer. Both techniques are very useful!

The expectation $E(X) = 2$ seems eminently fair on intuitive grounds. For if the probability of your obtaining a head is $1/2$ on one toss, then two tosses should get you $2 \cdot 1/2 = 1$ head, *on the average*. This plausible argument [which was actually given in a test paper by a smart student] can be made rigorous, but the necessary reasoning involved is far more sophisticated than you might think. It is a case of *Wald's equation*† or *martingale theorem* [for advanced reader].

Let us at once generalize this problem to the case of a biased coin, with probability p for head and $q = 1 - p$ tail. Then (4.4.3) becomes

(4.4.8) $$p_n = \underbrace{(q \cdots q)}_{n-1 \text{ times}} p = q^{n-1}p,$$

and (4.4.4) becomes

(4.4.9) $$\sum_{n=1}^{\infty} nq^{n-1}p = p \sum_{n=0}^{\infty} (n+1)q^n = \frac{p}{(1-q)^2} = \frac{p}{p^2} = \frac{1}{p}.$$

The random variable X is called the *waiting time*, for head to fall, or more generally for a "success." The distribution $\{q^{n-1}p; n = 1, 2, \ldots\}$ will be called the *geometrical distribution with success probability p*.

Example 9. A perfect coin is tossed n times. Let S_n denote the number of heads obtained. In the notation of §2.4, we have $S_n = X_1 + \cdots + X_n$. We know from §3.2 that

(4.4.10) $$p_k = P(S_n = k) = \frac{1}{2^n}\binom{n}{k}, \quad 0 \le k \le n.$$

† Named after Abraham Wald (1902-1950), leading U.S. statistician.

If we believe in probability, then we know $\sum_{k=0}^{\infty} p_k = 1$ from (4.3.10). Hence

$$\text{(4.4.11)} \qquad \sum_{k=0}^{n} \frac{1}{2^n} \binom{n}{k} = 1 \quad \text{or} \quad \sum_{k=0}^{n} \binom{n}{k} = 2^n.$$

This has been shown in (3.3.7) and can also be obtained from (4.4.13) below by putting $x = 1$ there, but we have done it by an argument based on probability. Next we have

$$\text{(4.4.12)} \qquad E(S_n) = \sum_{k=0}^{n} \frac{k}{2^n} \binom{n}{k}.$$

Here again we must sum a series, a finite one. We will do it in two different ways, both useful for other calculations. First by direct manipulation, the series may be rewritten as

$$\sum_{k=0}^{n} \frac{k}{2^n} \frac{n!}{k!(n-k)!} = \frac{n}{2^n} \sum_{k=1}^{n} \frac{(n-1)!}{(k-1)!(n-k)!} = \frac{n}{2^n} \sum_{k=1}^{n} \binom{n-1}{k-1}.$$

What we have done above is to cancel k from $k!$, split off n from $n!$ and omit a zero term for $k = 0$. Now change the index of summation by putting $k - 1 = j$ (we have done this kind of thing in Example 8):

$$\frac{n}{2^n} \sum_{k=1}^{n} \binom{n-1}{k-1} = \frac{n}{2^n} \sum_{j=1}^{n-1} \binom{n-1}{j} = \frac{n}{2^n} \cdot 2^{n-1} = \frac{n}{2},$$

where the step before the last is obtained by using (4.4.11) with n replaced by $n - 1$. Hence the answer is $n/2$.

This method is highly recommended if you enjoy playing with combinatorial formulas such as the binomial coefficients. But most of you will probably find the next method easier because it is more like a cook-book recipe. Start with Newton's binomial theorem in the form:

$$\text{(4.4.13)} \qquad (1 + x)^n = \sum_{k=0}^{n} \binom{n}{k} x^k.$$

Observe that this is just an expression of a polynomial in x and is a special case of Taylor's series, just as the series in (4.4.5) and (4.4.6) are. Now differentiate to get

$$\text{(4.4.14)} \qquad n(1 + x)^{n-1} = \sum_{k=0}^{n} \binom{n}{k} kx^{k-1}.$$

Substitute $x = 1$:

$$n2^{n-1} = \sum_{k=0}^{n} \binom{n}{k} k;$$

divide through by 2^n and get the answer $n/2$ again for (4.4.12). So the expected number of heads in n tosses is $n/2$. Once more, what could be more reasonable since heads are expected half of the time!

We can generalize this problem to a biased coin too. Then (4.4.10) becomes

$$P(S_n = k) = \binom{n}{k} p^k q^{n-k}, \quad 0 \leq k \leq n. \tag{4.4.15}$$

There is a preview of the above formula in §2.4. We now see that it gives the probability distribution of the random variable $S_n = \sum_{i=1}^{n} X_i$. It is called the *binomial distribution* $B(n; p)$. The random variable X_i here as well as its distribution is often referred to as *Bernoullian;* and when $p = 1/2$ the adjective *symmetric* is added. Next, (4.4.12) becomes

$$\sum_{k=0}^{n} \binom{n}{k} k p^k q^{n-k} = np. \tag{4.4.16}$$

Both methods used above still work. The second is quicker: setting $x = \frac{p}{q}$ in (4.4.14), we obtain since $p + q = 1$,

$$n\left(1 + \frac{p}{q}\right)^{n-1} = \frac{n}{q^{n-1}} = \sum_{k=0}^{n} \binom{n}{k} k \left(\frac{p}{q}\right)^{k-1};$$

multiplying through by pq^{n-1} we establish (4.4.16).

Example 10. In Problem 1 of §3.4, if we denote by X the number of defective items, then $P(X = j)$ is given by the formula in (3.4.1). This is called the *hypergeometric distribution*.

4.5. Random variables with densities

In the preceding sections we have given a quite rigorous discussion of random variables which take only a countable set of values. But even at an elementary level there are many important questions in which we must consider random variables not subject to such a restriction. This means that we need a sample space which is not countable. Technical questions of "measurability" then arise which cannot be treated satisfactorily without more advanced mathematics. As we have mentioned in Chapter 2, this kind of difficulty stems from the impossibility of assigning a probability to every subset of the sample space when it is uncountable. The matter is resolved by confining ourselves to sample sets belonging to an adequate class called a Borel field; see Appendix 1. Without going into this here we will take up a particular but very important situation which covers most applications and requires little mathematical abstraction. This is the case of random variable with a "density."

Consider a function f defined on $R^1 = (-\infty, +\infty)$:

$$u \to f(u)$$

and satisfying two conditions:

$$\text{(4.5.1)} \qquad \begin{aligned} &\text{(i)} \qquad \forall u: \quad f(u) \geq 0; \\ &\text{(ii)} \qquad \int_{-\infty}^{\infty} f(u)\, du = 1. \end{aligned}$$

Such a function is called a *density function* on R^1. The integral in (ii) is the Riemann integral taught in calculus. You may recall that if f is continuous or just *piecewise continuous*, then the definite integral

$$\int_a^b f(u)\, du$$

exists for any interval $[a, b]$. But in order that the "improper integral" over the infinite range $(-\infty, +\infty)$ should exist, further conditions are needed to make sure that $f(u)$ is pretty small for large $|u|$. In general, such a function is said to be "integrable over R^1." The requirement that the total integral be equal to one is less serious than it might appear, because if

$$\int_{-\infty}^{\infty} f(u)\, du = M < \infty,$$

we can just divide through by M and use f/M instead of f. Here are some possible pictures of density functions, some smooth, some not so.

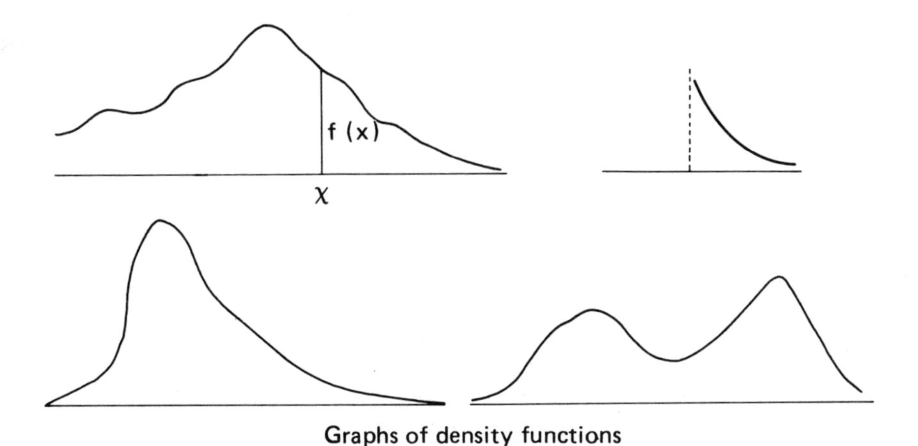

Graphs of density functions

Figure 20

You see what a variety they can be. The only constraints are that the curve should not lie below the x-axis anywhere, that the area under the curve should have a meaning, and the total area should be equal to one. You may agree that this is not asking for too much.

We can now define a class of random variables on a general sample space as follows. As in §4.2, X is a function on Ω: $\omega \to X(\omega)$, but its probabilities are prescribed by means of a density function f so that for any interval $[a, b]$ we have

$$P(a \leq X \leq b) = \int_a^b f(u)\, du. \tag{4.5.2}$$

More generally, if A is the union of intervals not necessarily disjoint and some of which may be infinite, we have

$$P(X \in A) = \int_A f(u)\, du. \tag{4.5.3}$$

Such a random variable is said to *have a density*, and its density function is f. [In some books this is called a "continuous" random variable, whereas the kind discussed in §2 is called "discrete." Both adjectives are slightly misleading so we will not use them here.]

If A is a finite union of intervals, then it can be split up into *disjoint* ones, some of which may abut on each other, such as

$$A = \bigcup_{j=1}^{k} [a_j, b_j],$$

and then the right hand side of (4.5.3) may be written as

$$\int_A f(u)\, du = \sum_{j=1}^{k} \int_{a_j}^{b_j} f(u)\, du.$$

This is a property of integrals which is geometrically obvious when you consider them as areas. Next if $A = (-\infty, x]$, then we can write

$$F(x) = P(X \leq x) = \int_{-\infty}^{x} f(u)\, du; \tag{4.5.4}$$

compare with (4.3.8). This formula defines the *distribution function* F of X as a *primitive* [*indefinite integral*] of f. It follows from the fundamental theorem of calculus that *if f is continuous*, then f is the *derivative* of F:

$$F'(x) = f(x). \tag{4.5.5}$$

Thus in this case the two functions f and F mutually determine each other. If f is not continuous everywhere, (4.5.5) is still true for every x at which f is continuous. These things are proved in calculus.

Let us observe that in the definition above of a random variable with a density, it is *implied* that the sets $\{a \leq X \leq b\}$ and $\{X \in A\}$ have probabilities assigned to them, in fact they are specified in (4.5.2) and (4.5.3) by means of the density function. This is a subtle point in the wording that should be brought out, but will not be elaborated on. [Otherwise we shall be getting into the difficulties that we are trying to circumvent here. But see Appendix 1.] Rather, let us remark on the close resemblance between the formulas above and corresponding ones in §4.3. This will be amplified by a definition of mathematical expectation in the present case and listed below for comparison.

	Countable case	Density Case
Range	$v_n, n = 1, 2, \ldots$	$-\infty < u < +\infty$
element of probability	p_n	$f(u)\,du = dF(u)$
$P(a \leq X \leq b)$	$\sum_{a \leq v_n \leq b} p_n$	$\int_a^b f(u)\,du$
$P(X \leq x) = F(x)$	$\sum_{v_n \leq x} p_n$	$\int_{-\infty}^x f(u)\,du$
$E(X)$	$\sum_n p_n v_n$	$\int_{-\infty}^{\infty} uf(u)\,du$
proviso	$\sum_n p_n \lvert v_n \rvert < \infty$	$\int_{-\infty}^{\infty} \lvert u \rvert f(u)\,du < \infty$

More generally, the analogue of (4.3.17) is

$$E(\varphi(X)) = \int_{-\infty}^{\infty} \varphi(u)f(u)\,du. \tag{4.5.6}$$

You may ignore the second item in the density case above involving a *differential* if you don't know what it means.

Further insight into the analogy is gained by looking at the following picture:

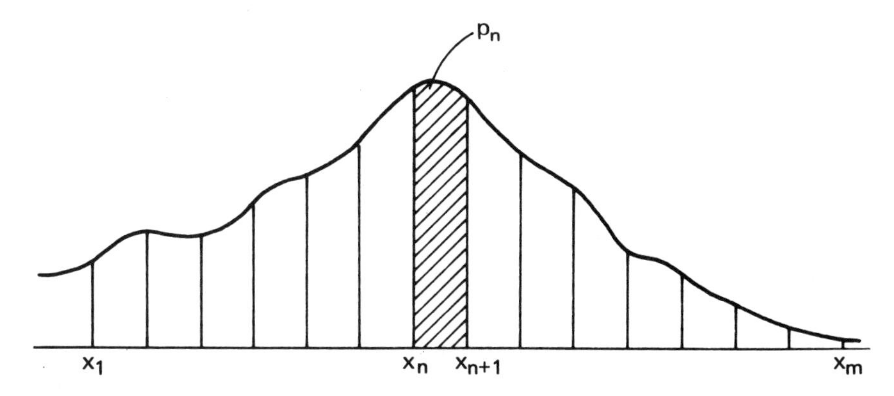

Figure 21

The curve is the graph of a density function f. We have divided up the x-axis into $m + 1$ pieces, not necessarily equal and not necessarily small, and denote the area under the curve between x_n and x_{n+1} by p_n, thus:

$$p_n = \int_{x_n}^{x_{n+1}} f(u)\,du, \quad 0 \leq n \leq m.$$

where $x_0 = -\infty$, $x_{m+1} = +\infty$. It is clear that we have

$$\forall n\colon \ p_n \geq 0; \ \sum_n p_n = 1.$$

Hence the numbers p_n satisfy the conditions in (4.3.10). Instead of a finite partition we may have a countable one by suitable labeling such as . . . , $p_{-2}, p_{-1}, p_0, p_1, \ldots$. Thus we can derive a set of "elementary probabilities" from a density function, in infinitely many ways. This process may be called *discretization*. If X has the density f, we may consider a random variable Y such that

$$P(Y = x_n) = p_n,$$

where we may replace x_n by any other number in the subinterval $[x_n, x_{n+1}]$. Now if f is continuous and the partition is sufficiently fine, namely if the pieces are sufficiently small, then it is geometrically evident that Y is in some sense a discrete approximation of X. For instance

$$E(Y) = \sum_n p_n x_n$$

will be an approximation of $E(X) = \int_{-\infty}^{\infty} uf(u)\,du$. Remember the *Riemann sums* defined in calculus to lead to a Riemann integral? There the strips with curved tops in Figure 21 are replaced by flat-tops (rectangles), but the ideas involved are quite similar. From a practical point of view, it is the discrete approximations that can be really measured, whereas the continuous density is only a mathematical idealization. We shall return to this in a moment.

Having dwelled on the similarity of the two cases of random variable, we will pause to stress a fundamental difference between them. If X has a density, then by (4.5.2) with $a = b = x$, we have

(4.5.7) $$P(X = x) = \int_x^x f(u)\,du = 0.$$

Geometrically speaking, this merely states the trivial fact that a line segment has zero area. Since x is arbitrary in (4.5.7), it follows that X takes any preassigned value with probability zero. This is in direct contrast to a random variable taking a countable set of values, for then it must take some of these values with positive probability. It seems paradoxical that on the one hand, $X(\omega)$ *must be some number* for every ω, and on the other hand *any given number has probability zero.* The following simple concrete example should clarify this point.

Example 10. Spin a needle on a circular dial. When it stops it points at a random angle θ (measured from the horizontal, say). Under normal conditions it is reasonable to suppose that θ is *uniformly distributed* between 0° and 360° (cf. Example 7 of §4.4). This means it has the following density function:

$$f(u) = \begin{cases} \dfrac{1}{360} & \text{for } 0 \le u \le 360, \\ 0 & \text{otherwise.} \end{cases}$$

Thus for any $\theta_1 < \theta_2$ we have

$$P(\theta_1 \le \theta \le \theta_2) = \int_{\theta_1}^{\theta_2} \frac{1}{360}\, du = \frac{\theta_2 - \theta_1}{360}. \tag{4.5.8}$$

This formula says that the probability of the needle pointing between any two directions is proportional to the angle between them. If the angle $\theta_2 - \theta_1$ shrinks to zero, then so does the probability. Hence in the limit the probability of the needle pointing exactly at θ is equal to zero. From an empirical point of view, this event does not really make sense because the needle itself must have a width. So in the end it is the mathematical fiction or idealization of a "line without width" that is the root of the paradox.

There is a deeper way of looking at this situation which is very rich. It should be clear that instead of spinning a needle we may just as well "pick a number at random" from the interval [0, 1]. This can be done by bending the circle into a line segment and changing the unit. Now every point in [0, 1] can be represented by a decimal such as

$$.141592653589793 \cdots. \tag{4.5.9}$$

There is no real difference if the decimal terminates because then we just have all digits equal to 0 from a certain place on, and 0 is no different from any other digit. Thus, to pick a number in [0, 1] amounts to picking all its decimal digits one after another. That is the kind of thing a computing machine churns out. Now the chance of picking any prescribed digit, say the first digit "1" above, is equal to 1/10 and the successive pickings form totally independent trials (see §2.4). Hence the chance of picking the 15 digits shown in (4.5.9) is equal to

$$\underbrace{\frac{1}{10} \cdot \frac{1}{10} \cdots \frac{1}{10}}_{15 \text{ times}} = \left(\frac{1}{10}\right)^{15}.$$

If we remember that 10^9 is a billion this probability is already so small that according to Emile Borel [1871–1956; great French mathematician and one of the founders of modern probability theory], it is *terrestrially negligible* and should be equated to zero! But we have only gone 15 digits in the decimals of the number $\pi - 3$, so there can be no question whatsoever of picking this number itself and yet if you can imagine going on forever, you will end up with some number which is just as *impossible à priori* as this $\pi - 3$. So here again we are up against a mathematical fiction—the real number system.

We may generalize this example as follows. Let $[a, b]$ be any finite, nondegenerate interval in R^1 and put

$$f(u) = \begin{cases} \dfrac{1}{b - a} & \text{for } a \leq u \leq b, \\ 0 & \text{otherwise.} \end{cases}$$

This is a density function and the corresponding distribution is called *the uniform distribution on* $[a, b]$. We can write the latter explicitly:

$$F(x) = \frac{[(a \vee x) \wedge b] - a}{b - a}$$

if you have a taste for such tricky formulas.

Example 11. A chord is drawn at random in a circle. What is the probability that its length exceeds that of a side of an inscribed equilateral triangle?

Let us draw such a triangle in a circle with center 0 and radius R, and make the following observations. The side is at distance $R/2$ from 0; its midpoint is on a concentric circle of radius $R/2$; it subtends an angle of 120 degrees at 0. You ought to know how to compute the length of the side, but this will not be needed. Let us denote by A the desired event that a random chord be longer than that side. Now the length of any chord is determined

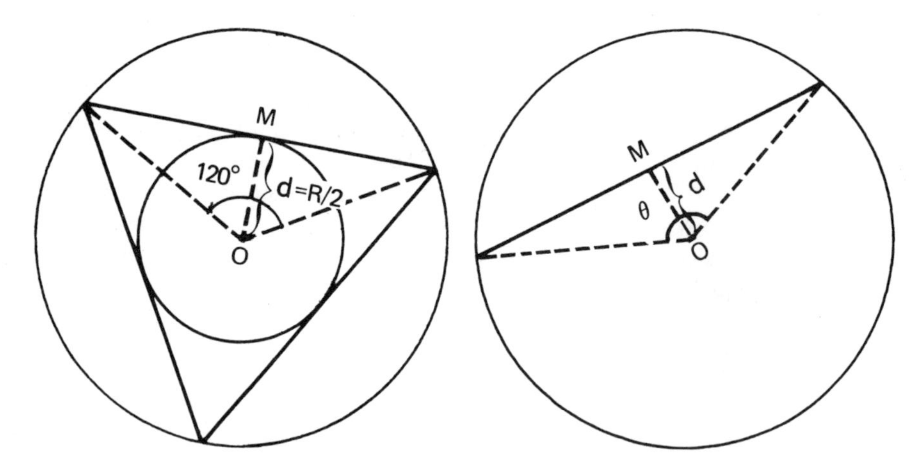

Figure 22

by any one of the three quantities: its distance d from 0; the location of its midpoint M; the angle θ it subtends at 0. We are going to assume in turn that each of these has a uniform distribution over its range and compute the probability of A under each assumption.

(1) Suppose that d is uniformly distributed in $[0, R]$. This is a plausible assumption if we move a ruler parallel to itself with constant speed from a tangential position towards the center, stopping somewhere to intersect the circle in a chord. It is geometrically obvious that the event A will occur if and only if $d < R/2$. Hence $P(A) = 1/2$.

(2) Suppose that M is uniformly distributed over the disk D formed by the given circle. This is a plausible assumption if a tiny dart is thrown at D and a chord is then drawn perpendicular to the line joining the hitting point to 0. Let D' denote the concentric disk of radius $R/2$. Then the event A will occur if and only if M falls within D'. Hence $P(A) = P(M \in D') =$ (area of D')/(area of D) $= 1/4$.

(3) Suppose that θ is uniformly distributed between zero and 360 degrees. This is plausible if one endpoint of the chord is arbitrarily fixed and the other is obtained by rotating a radius at constant speed to stop somewhere on the circle. Then it is clear from the picture that A will occur if and only if θ is between 120 and 240 degrees. Hence $P(A) = (240 - 120)/360 = 1/3$.

Thus the answer to the problem is 1/2, 1/4 or 1/3 according to the different hypotheses made. It follows that these hypotheses are not compatible with one another. Other hypotheses are possible and may lead to still other answers. Can you think of a good one? This problem was known as *Bertrand's paradox* in the earlier days of discussions of probability theory. But of course the paradox is due only to the fact that the problem is not well-posed without specifying the underlying nature of the randomness. It is not surprising that the different ways of randomization should yield different probabilities, which can be verified experimentally by the mechanical procedures described. Here is a facile analogy. Suppose that you are asked how long it takes to go from your dormitory to the classroom without specifying whether we are talking about "walking," "biking," or "driving" time. Would you call it paradoxical that there are different answers to the question?

We end this section with some other simple examples of random variables with densities. Another important case, the normal distribution, will be discussed in Chapter 6.

Example 12. Suppose you station yourself at a spot on a relatively serene country road and watch the cars that pass by that spot. With your stopwatch you can clock the time before the first car passes. This is a random variable T called the *waiting time*. Under certain circumstances it is a reasonable hypothesis that T has the density function below with a certain $\lambda > 0$:

$$f(u) = \lambda e^{-\lambda u}, \quad u \geq 0. \tag{4.5.10}$$

It goes without saying that $f(u) = 0$ for $u < 0$. The corresponding distribution function is called the *exponential distribution with parameter* λ, obtained by integrating f as in (4.5.4):

$$F(x) = \int_{-\infty}^{x} f(u)\,du = \int_{0}^{x} \lambda e^{-\lambda u}\,du = 1 - e^{-\lambda x}.$$

In particular if we put $x = +\infty$, or better, let $x \to \infty$ in the above, we see that f satisfies the conditions in (4.5.1), so it is indeed a density function. We have

$$P(T \leq x) = F(x) = 1 - e^{-\lambda x}; \tag{4.5.11}$$

but in this case it is often more convenient to use the *tail probability:*

$$P(T > x) = 1 - F(x) = e^{-\lambda x}. \tag{4.5.12}$$

This can be obtained directly from (4.5.3) with $A = (x, \infty)$, thus:

$$P(T \in (x, \infty)) = \int_{(x,\infty)} \lambda e^{-\lambda u}\,du = \int_{x}^{\infty} \lambda e^{-\lambda u}\,du = e^{-\lambda x}.$$

For every given x, say 5 (seconds), the probability $e^{-5\lambda}$ in (4.5.12) decreases as λ increases. This means your waiting time tends to be shorter if λ is larger. On a busy highway λ will be large indeed. The expected waiting time is given by

$$E(T) = \int_{0}^{\infty} u\lambda e^{-\lambda u}\,du = \frac{1}{\lambda}\int_{0}^{\infty} te^{-t}\,dt = \frac{1}{\lambda}. \tag{4.5.13}$$

[Can you compute the integral above using "integration by parts" without recourse to a table?] This result supports our preceding observation that T tends on the average to be smaller when λ is larger.

The exponential distribution is a very useful model for various types of waiting time problems such as telephone calls, service times, splitting of radioactive particles, etc.; see §7.2.

Example 13. Suppose in a problem involving the random variable T above, what we really want to measure is its logarithm (to the base e):

$$S = \log T. \tag{4.5.14}$$

This is also a random variable (cf. Proposition 2 in §4.2); it is negative if $T < 1$, zero if $T = 1$ and positive if $T > 1$. What are its probabilities? We may be interested in $P(a \leq S \leq b)$ but it is clear that we need only find $P(S \leq x)$, namely the distribution function F_S of S. Now the function

$$x \to \log x$$

is monotone and its inverse is

$$x \longrightarrow e^x.$$

So that

$$S \leq x \Leftrightarrow \log T \leq x \Leftrightarrow T \leq e^x.$$

Hence by (4.5.11)

$$F_S(x) = P\{S \leq x\} = P\{T \leq e^x\} = 1 - e^{-\lambda e^x}$$

The density function f_S is obtained by differentiating:

$$f_S(x) = F_S'(x) = \lambda e^x e^{-\lambda e^x} = \lambda e^{x - \lambda e^x}$$

This looks formidable but you see it is easily derived.

Example 14. A certain river floods every year. Suppose the low water mark is set at 1, and the high water mark Y has the distribution function

$$F(y) = P(Y \leq y) = 1 - \frac{1}{y^2}, \quad 1 \leq y < \infty. \tag{4.5.15}$$

Observe that $F(1) = 0$, that $F(y)$ increases with y and $F(y) \to 1$ as $y \to \infty$. This is as it should be from the meaning of $P(Y \leq y)$. To get the density function we differentiate:

$$f(y) = F'(y) = \frac{2}{y^3}, \quad 1 \leq y < \infty. \tag{4.5.16}$$

It is not necessary to check that $\int_{-\infty}^{\infty} f(y)\,dy = 1$, because this is equivalent to $\lim_{y \to \infty} F(y) = 1$. The expected value of Y is given by

$$E(Y) = \int_1^{\infty} u \cdot \frac{2}{u^3}\,du = \int_1^{\infty} \frac{2}{u^2}\,du = 2.$$

Thus the maximum of Y is twice that of the minimum, on the average.

What happens if we set the low water mark at 0 instead of 1, and use a unit of measuring the height which is $1/10$ of that used above? This means we set

$$Z = 10(Y - 1). \tag{4.5.17}$$

As in Example 13 we have

$$Z \leq z \Leftrightarrow 10(Y - 1) \leq z \Leftrightarrow Y \leq 1 + \frac{z}{10}, \quad 0 \leq z < \infty.$$

From this we can compute:

$$F_Z(z) = 1 - \frac{100}{(10 + z)^2},$$

$$f_Z(z) = \frac{200}{(10 + z)^3}.$$

The calculation of $E(Z)$ from f_Z is tedious but easy. The answer is $E(Z) = 10$ and comparing with $E(Y) = 2$ we see that

$$E(Z) = 10(E(Y) - 1). \tag{4.5.18}$$

Thus the *means* of Y and Z are connected by the same linear relation as the random variables themselves. Does this seem obvious to you? The general proposition will be discussed in §6.1.

4.6. General case

The most general random variable is a function X defined on the sample space Ω such that *for any real x, the probability $P(X \leq x)$ is defined.*

To be frank, this statement has put the cart before the horse. What comes first is a probability measure P defined on a class of subsets of Ω. This class is called the *sample Borel field* or *probability field* and is denoted by $\mathfrak{F}$. Now if a function X has the property that for every x, the set $\{\omega \mid X(\omega) \leq x\}$ belongs to the class $\mathfrak{F}$, then it is called a random variable. [We must refer to Appendix 1 for a full description of this concept; but the rest of this section should be intelligible without the formalities.] In other words, an arbitrary function must pass a test to become a member of the club. The new idea here is that P is defined only for subsets in $\mathfrak{F}$, not necessarily for all subsets of Ω. If it happens to be defined for all subsets, then of course the test described above becomes a nominal one and every function is automatically a random variable. This is the situation for a countable space Ω discussed in §4.1. In general, as we have hinted several times before, it is impossible to define a probability measure on all subsets of Ω, and so we must settle for a certain class $\mathfrak{F}$. Since only sets in $\mathfrak{F}$ have probabilities assigned to them, and since we wish to discuss sample sets of the sort "$X \leq x$," we are obliged to require that these belong to $\mathfrak{F}$. Thus the necessity of such a test is easy to understand. What may be a little surprising is that this test is all we need. Namely, once we have made this requirement, we can then go on to discuss the probabilities of a whole variety of sample sets such as $\{a \leq X \leq b\}$, $\{X = x\}$, $\{X$ takes a rational value$\}$, or some crazy thing like $\{e^X > X^2 + 1\}$.

Next, we define for every real x:

$$F(x) = P(X \leq x) \tag{4.6.1}$$

or equivalently for $a < b$:

$$F(b) - F(a) = P(a < X \leq b);$$

and call the function F the *distribution function of* X. This has been done in previous cases but we no longer have the special representative in (4.3.8) or (4.5.4):

$$F(x) = \sum_{v_n \leq x} p_n, \quad F(x) = \int_{-\infty}^{x} f(u)\,du$$

in terms of elementary probability or a density function. As a matter of fact, the general F turns out to be a mixture of these two kinds together with a more weird kind (the *singular* type). But we can operate quite well with the F as defined by (4.6.1) without further specification. The mathematical equipment required to handle the general case, however, is somewhat more advanced (at the level of a course like "Fundamental concepts of analysis"). So we cannot go into this but will just mention two easy facts about F.

(i) F is monotone nondecreasing: namely $x \leq x' \Rightarrow F(x) \leq F(x')$;
(ii) F has limits 0 and 1 at $-\infty$ and $+\infty$ respectively:

$$F(-\infty) = \lim_{x \to -\infty} F(x) = 0, \quad F(+\infty) = \lim_{x \to +\infty} F(x) = 1.$$

Property (i) holds because if $x \leq x'$, then $\{X \leq x\} \subset \{X \leq x'\}$. Property (ii) is intuitively obvious because the event $\{X \leq x\}$ becomes impossible as $x \to -\infty$, and certain as $x \to +\infty$. This argument may satisfy you but the rigorous proofs are a bit more sophisticated and depend on the countable additivity of P (see §2.3). Let us note that the existence of the limits in (ii) follows from the monotonicity in (i) and a fundamental theorem in calculus: a bounded monotone sequence of real numbers has a limit.

The rest of the section is devoted to a brief discussion of some basic notions concerning random vectors. This material may be postponed until it is needed in Chapter 6.

For simplicity of notation we will consider only two random variables X and Y, but the extension to any finite number is straightforward. We first consider the case where X and Y are countably-valued. Let X take the values $\{x_i\}$, Y take the values $\{y_j\}$, and put

$$P(X = x_i, Y = y_j) = p(x_i, y_j). \tag{4.6.2}$$

When x_i and y_j range over all possible values, the set of "elementary probabilities" above gives the *joint probability distribution* of the *random vector* (X, Y). To get the probability distribution of X alone, we let y_j range over all possible values in (4.6.2), thus:

$$P(X = x_i) = \sum_{y_j} p(x_i, y_j) = p(x_i, *) \tag{4.6.3}$$

where the last quantity is defined by the middle sum. When x_i ranges over all possible values, the set of $p(x_i, *)$ gives the *marginal distribution* of X. The marginal distribution of Y is similarly defined. Let us observe that these marginal distributions do not in general determine the joint distribution.

Just as we can express the expectation of any function of X by means of its probability distribution (see (4.3.17)), we can do the same for any function of (X, Y) as follows:

$$E(\varphi(X, Y)) = \sum_{x_i} \sum_{y_j} \varphi(x_i, y_j)p(x_i, y_j). \tag{4.6.4}$$

It is instructive to see that this results from a rearrangement of terms in the definition of the expectation of $\varphi(X, Y)$ as *one* random variable as in (4.3.11):

$$E(\varphi(X, Y)) = \sum_{\omega} \varphi(X(\omega), Y(\omega))\, P(\omega).$$

Next, we consider the density case extending the situation in §4.5. The random vector (X, Y) is said to have a joint density function f in case

$$P(X \leq x, Y \leq y) = \int_{-\infty}^{x} \int_{-\infty}^{y} f(u, v)\, du\, dv \tag{4.6.5}$$

for all (x, y). It then follows that for any "reasonable" subset S of the Cartesian plane (called a *Borel set*) we have

$$P((X, Y) \in S) = \iint_S f(u, v)\, du\, dv. \tag{4.6.6}$$

For example S may be polygons, disks, ellipses and unions of such shapes. Note that (4.6.6) contains (4.6.5) as a very particular case and we can, at a pinch, accept the more comprehensive condition (4.6.6) as the *definition* of f as density for (X, Y). However, here is a heuristic argument from (4.6.5) to (4.6.6). Let us denote by $R(x, y)$ the infinite rectangle in the plane with sides parallel to the coordinate axes and lying to the southwest of the point (x, y). The picture below shows that for any $\delta > 0$ and $\delta' > 0$:

$$R(x + \delta, y + \delta') - R(x + \delta, y) - R(x, y + \delta') + R(x, y)$$

is the shaded rectangle

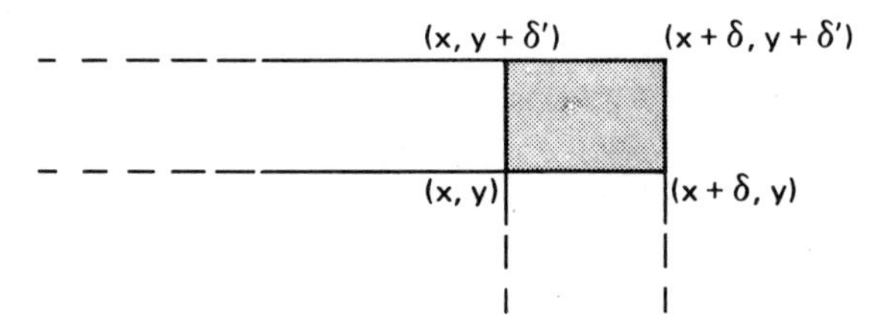

It follows that if we manipulate the relation (4.6.5) in the same way, we get

$$P(x \leq X \leq x + \delta, y \leq Y \leq y + \delta') = \int_{x}^{x+\delta} \int_{y}^{y+\delta'} f(u, v)\, du\, dv.$$

This means (4.6.6) is true for the shaded rectangle. By varying x, y as well as δ, δ' we see that the formula is true for any rectangle of this shape. Now any reasonable figure can be approximated from inside and outside by a number of such small rectangles (even just squares)—a fact known already to the ancient Greeks. Hence in the limit we can get (4.6.6) as asserted.

The curious reader may wonder why a similar argument was not given earlier for the case of one random variable in (4.5.3)? The answer is: heuristically speaking, there are hardly any sets in R^1 other than intervals, points and their unions! Things are pretty tight in one dimension and our geometric intuition does not work well. This is one reason why classical measure theory is a sophisticated business.

The joint density function f satisfies the following conditions:

(i) $f(u, v) \geq 0$ for all (u, v);

(ii) $\int_{-\infty}^{\infty} \int_{-\infty}^{\infty} f(u, v)\, du\, dv = 1.$

Of course (ii) implies that f is integrable over the whole plane. Frequently we assume also that f is continuous. Now the formulas analogous to (4.6.3) are

$$(4.6.7)\qquad \begin{aligned} P(X \leq x) &= \int_{-\infty}^{x} f(u, *)\, du, \quad \text{where } f(u, *) = \int_{-\infty}^{\infty} f(u, v)\, dv \\ P(Y \leq y) &= \int_{-\infty}^{y} f(*, v)\, dv, \quad \text{where } f(*, v) = \int_{-\infty}^{\infty} f(u, v)\, du. \end{aligned}$$

The functions $u \to f(u, *)$ and $v \to f(*, v)$ are called respectively the *marginal density functions* of X and Y. They are derived from the joint density function after "integrating out" the variable which is not in question.

The formula corresponding to (4.6.4) becomes in the density case: for any "reasonable" [Borel] function φ:

$$(4.6.8)\qquad E(\varphi(X, Y)) = \int_{-\infty}^{\infty} \int_{-\infty}^{\infty} \varphi(u, v) f(u, v)\, du\, dv.$$

The class of reasonable functions includes all bounded continuous functions in (u, v), indicators of reasonable sets, and functions which are continuous except across some smooth boundaries, for which the integral above exists, etc.

In the most general case the *joint distribution function* F of (X, Y) is defined by

$$(4.6.9)\qquad F(x, y) = P(X \leq x, Y \leq y) \quad \text{for all } (x, y).$$

If we denote $\lim_{y \to \infty} F(x, y)$ by $F(x, \infty)$, we have

$$F(x, \infty) = P(X \leq x, Y < \infty) = P(X \leq x)$$

since "$Y < \infty$" puts no restriction on Y. Thus $x \to F(x, \infty)$ is the *marginal distribution function* of X. The marginal distribution function of Y is similarly defined.

Although these general concepts form the background whenever several random variables are discussed, explicit use of them will be rare in this book.

Exercises

1. If X is a random variable [on a countable sample space], is it true that

$$X + X = 2X, \; X - X = 0?$$

Explain in detail.

2. Let $\Omega = \{\omega_1, \omega_2, \omega_3\}$, $P(\omega_1) = P(\omega_2) = P(\omega_3) = \frac{1}{3}$, and define X, Y and Z as follows:

$$X(\omega_1) = 1, \; X(\omega_2) = 2, \; X(\omega_3) = 3;$$
$$Y(\omega_1) = 2, \; Y(\omega_2) = 3, \; Y(\omega_3) = 1;$$
$$Z(\omega_1) = 3, \; Z(\omega_2) = 1, \; Z(\omega_3) = 2.$$

Show that these three random variables have the same probability distribution. Find the probability distributions of $X + Y$, $Y + Z$ and $Z + X$.

3. In No. 2 find the probability distribution of

$$X + Y - Z, \quad \sqrt{(X^2 + Y^2)Z}, \quad \frac{Z}{|X - Y|}.$$

4. Take Ω to be a set of 5 real numbers. Define a probability measure and a random variable X on it which takes the values 1, 2, 3, 4, 5 with probabilities $\frac{1}{10}, \frac{1}{10}, \frac{1}{5}, \frac{1}{5}, \frac{2}{5}$ respectively; another random variable Y which takes the value $\sqrt{2}$, $\sqrt{3}$, π with probabilities $\frac{1}{5}, \frac{3}{10}, \frac{1}{2}$. Find the probability distribution of XY. [Hint: the answer depends on your choice and is not unique.]
5. Generalize No. 4 by constructing Ω, P, X so that X takes the values $v_1, v_2, \ldots, v_n$ with probabilities $p_1, p_2, \ldots, p_n$ where the p_n's satisfy (4.3.10).
6. In Example 3 of §1, what do the following sets mean?

$$\{X + Y = 7\}, \; \{X + Y \leq 7\}, \; \{X \vee Y > 4\}, \; \{X \neq Y\}.$$

List all the ω's in each set.

7.* Let X be integer valued and let F be its distribution function. Show that for every x and $a < b$:

$$P(X = x) = \lim_{\epsilon \downarrow 0} [F(x + \epsilon) - F(x - \epsilon)],$$

$$P(a < X < b) = \lim_{\epsilon \downarrow 0} [F(b - \epsilon) - F(a + \epsilon)].$$

[The results are true for any random variable, but require more advanced proofs even when Ω is countable.]

8. In Example 4 of §4.2, suppose that

$$X = 5000 + X'$$

where X' is uniformly distributed over the set of integers from 1 to 5000. What does this hypothesis mean? Find the probability distribution and mean of Y under this hypothesis.

9. As in No. 8 but now suppose that

$$X = 4000 + X'$$

where X' is uniformly distributed from 1 to 10000.

10.* As in No. 8 but now suppose that

$$X = 3000 + X'$$

and X' is the exponential distribution with mean 7000. Find $E(Y)$.

11. Let $\lambda > 0$ and define f as follows:

$$f(u) = \begin{cases} \frac{1}{2} \lambda e^{-\lambda u}, & \text{if } u \geq 0, \\ \frac{1}{2} \lambda e^{+\lambda u}, & \text{if } u < 0. \end{cases}$$

This f is called *bilateral exponential*. If X has density f, find the density of $|X|$. [Hint: begin with the distribution function.]

12. If X is a positive random variable with density f, find the density of $+\sqrt{X}$. Apply this to the distribution of the side-length of a square when its area is uniformly distributed in $[a, b]$.

13. If X has density f, find the density of (i) $aX + b$ where a and b are constants; (ii) X^2.

14. Prove (4.4.5) in two ways: (a) by multiplying out $(1 - x)(1 + x + \cdots + x^n)$, (b) by using Taylor's series.

15. Suppose that

$$p_n = cq^{n-1}p, \; 1 \leq n \leq m;$$

where c is a constant and m is a positive integer, cf. (4.4.8). Determine c so that $\sum_{n=1}^{m} p_n = 1$. (This scheme corresponds to the waiting time for a success when it is supposed to occur within m trials.)

16. A perfect coin is tossed n times. Let Y_n denote the number of heads obtained minus the number of tails. Find the probability distribution of Y_n, and its mean. [Hint: there is a simple relation between Y_n and the S_n in Example 9 of §4.4.]
17. Refer to Problem 1 in §3.4. Suppose there are 11 rotten apples in a bushel of 550, and 25 apples are picked at random. Find the probability distribution of the number X of rotten apples among those picked.
18.* Generalize No. 17 to arbitrary numbers and find the mean of X. [Hint: this requires some expertise in combinatorics but becomes trivial after §6.1.]
19. Let

$$P(X = n) = p_n = \frac{1}{n(n+1)}, \quad n \geq 1.$$

Is this a probability distribution for X? Find $P(X \geq m)$ for any m and $E(X)$.

20. If all the books in a library have been upset and a monkey is hired to put them all back on the shelves, it can be shown that a good approximation for the probability of having exactly n books put back in their original places is

$$\frac{e^{-1}}{n!}, \quad n \geq 0.$$

Find the expected number of books returned to their original places. [This oft-quoted illustration is a variant on the matching problem discussed in Problem 6 of §3.4.]

21. Find an example in which the series $\sum_n p_n v_n$ in (4.3.11) converges but not absolutely. [Hint: there is really nothing hard about this: choose $p_n = 1/2^n$ say, and now choose v_n so that $p_n v_n$ is the general term of any nonabsolutely convergent series you know.]
22. If f and g are two density functions, show that $\lambda f + \mu g$ is also a density function, where $\lambda + \mu = 1$, $\lambda \geq 0$, $\mu \geq 0$.
23. Find the probability that a random chord drawn in a circle is longer than the radius. As in Example 11 of §4.5 work this out under the three different hypotheses discussed there.
24. Let

$$f(u) = ue^{-u}, \quad u \geq 0.$$

Show that f is a density function. Find $\int_0^\infty uf(u)\,du$.

25. In the figure below an equilateral triangle, a trapezoid and a semi-disk are shown:

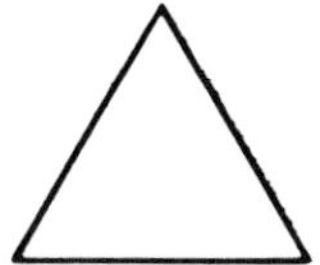  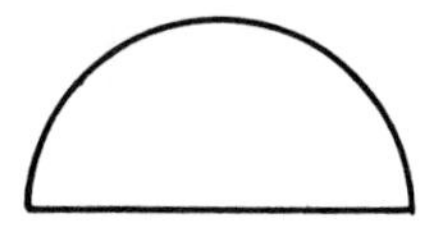

Determine numerical constants for the sides and radius to make these the graphs of density functions.

26. Suppose a target is a disk of radius 10 feet and suppose that the probability of hitting within any concentric disk is proportional to the area of the disk. Let R denote the distance of the bullet from the center. Find the distribution function, density function and mean of R.
27. Agent 009 was trapped between two narrow abysmal walls. He swung his gun around in a vertical circle touching the walls as shown in Fig. 23, and fired a wild [random] shot. Assume that the angle which his pistol makes with the horizontal is uniformly distributed between 0° and 90°. Find the distribution of the height where the bullet landed and its mean.
28. [St. Petersburg Paradox] You play a game with your pal by tossing a perfect coin repeatedly and betting on the waiting time X until a head is tossed up. You agree to pay him 2^X¢ when the value of X is known, namely 2^n¢ if $X = n$. If you figure that a fair price for him to pay you in advance in order to win this random prize should be equal to the maticians as Daniel Bernoulli, D'Alembert, Poisson, Borel, to name only *honestly* would you accept to play this game? [If you do not see any paradox in this, then you do not agree with such illustrious mathematicians as Daniel Bernoulli, D'Alembert, Poisson, Borel, to name only a few. For a brief account see [Keynes]. Feller believed that the paradox would go away if more advanced mathematics were used to reformulate the problem. You will have to decide for yourself whether it is not more interesting as a philosophical and psychological challenge.]
29. One objection to the scheme in No. 28 is that "time must have a stop." So suppose that only m tosses at most are allowed and your pal gets nothing if head does not show up in m tosses. Try $m = 10$ and $m = 100$. What is now a fair price for him to pay? and do you feel more comfortable after this change of rule? In this case Feller's explanation melts away but the psychological element remains.
30.* A number μ is called the *median* of the random variable X iff $P(X \geq \mu) \geq 1/2$ and $P(X \leq \mu) \geq 1/2$. Show that such a number always exists but need not be unique. Here is a practical example. After n examination papers have been graded, they are arranged in descending order. There is one in the middle if n is odd, two if n is even, corresponding to the median(s). Explain the probability model used.
31. An urn contains n tickets numbered from 1 to n. Two tickets are drawn (without replacement). Let X denote the smaller, Y the larger of the

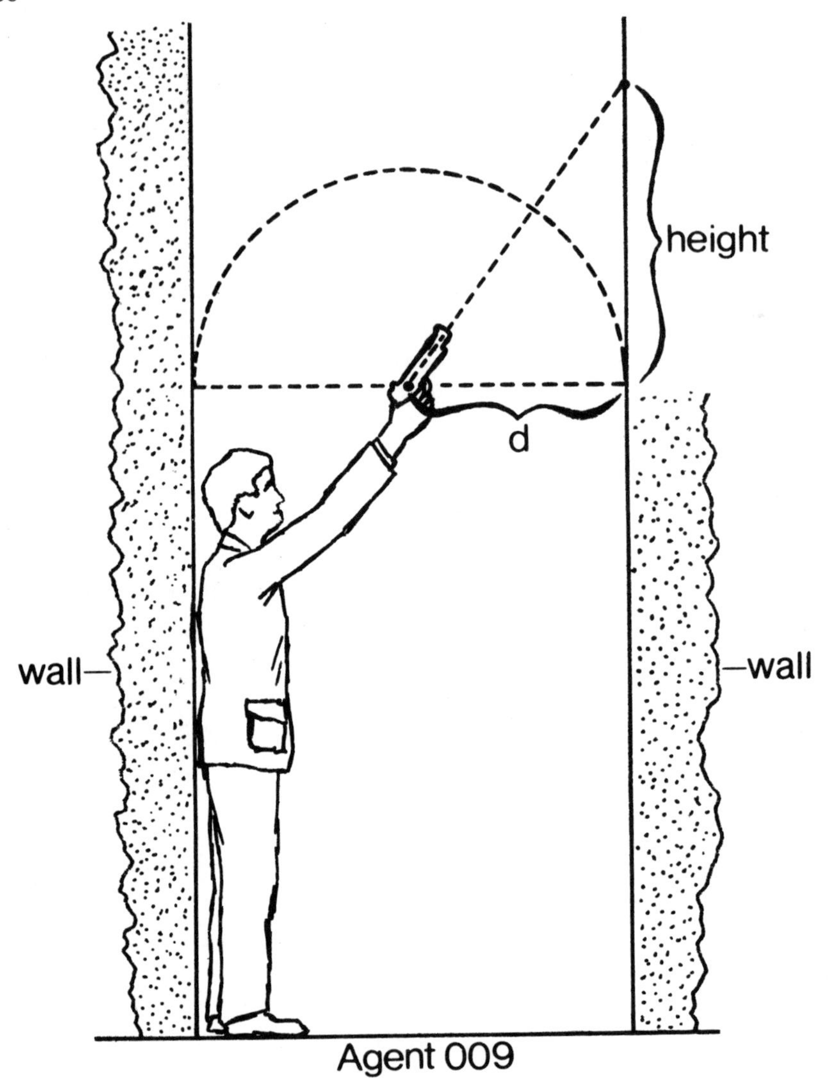

Figure 23

two numbers so obtained. Describe the joint distribution of (X, Y), and the marginal ones. Find the distribution of $Y - X$ from the joint distribution.

32. Pick two numbers at random from $[0, 1]$. Define X and Y as in No. 31 and answer the same questions. [Hint: draw the picture and compute areas.]

Appendix 1

Borel Fields and General Random Variables

When the sample space Ω is uncountable it may not be possible to define a probability measure for all its subsets, as we did for a countable Ω in §2.4. We must restrict the measure to sets of a certain family which must however be comprehensive enough to allow the usual operations with sets. Specifically, we require the family $\mathfrak{F}$ to have two properties:

(a) if a set A belongs to $\mathfrak{F}$, then its complement $A^c = \Omega - A$ also belongs to $\mathfrak{F}$;
(b) if a countable number of sets $A_1, A_2, \ldots$ all belong to $\mathfrak{F}$, then their union $\bigcup_n A_n$ also belongs to $\mathfrak{F}$.

It follows from De Morgan's laws that the union in (b) may be replaced by the intersection $\bigcap_n A_n$ as well. Thus if we operate on the members of the family with the three basic operations mentioned above, for a countable number of times, in any manner or order (see e.g. p. 9), the result is still a member of the family. In this sense the family is said to be *closed under* these operations, and so also under other derived operations such as differences. Such a family of subsets of Ω is called a *Borel field* on Ω. In general there are many such fields, for example the family of all subsets which is certainly a Borel field but may be too large to have a probability defined on it; or the family of 2 sets $\{\emptyset, \Omega\}$, or four sets $\{\emptyset, A, A^c, \Omega\}$ with a fixed set A, which are too small for most purposes. Now suppose that a reasonable Borel field $\mathfrak{F}$ has been chosen and a probability measure P has been defined on it, then we have a *probability triple* $(\Omega, \mathfrak{F}, P)$ with which we can begin our work. The sets in $\mathfrak{F}$ are said to be *measurable* and they alone have probabilities.

Let X be a real-valued function defined on Ω. Then X is called a *random variable* iff for any real number x, we have

$$\{\omega \mid X(\omega) \leq x\} \in \mathfrak{F}. \tag{A.1.1}$$

Hence $P\{X \leq x\}$ is defined, and as a function of x it is the distribution function F given in (4.6.1). Furthermore if $a < b$, then the set

$$\{a < X \leq b\} = \{X \leq b\} - \{X \leq a\} \tag{A.1.2}$$

belongs to $\mathfrak{F}$ since $\mathfrak{F}$ is closed under difference. Thus its probability is defined and is in fact given by $F(b) - F(a)$.

When Ω is countable and we take $\mathfrak{F}$ to be the Borel field of all the subsets of Ω, then of course the condition (A.1.1) is satisfied for any function X. Thus in this case an arbitrary function on Ω is a random variable, as defined in §4.2.

In general, the condition in (A.1.1) is imposed mainly because we wish to define the *mathematical expectation* by a procedure which requires such a condition. Specifically, if X is a bounded random variable, then it has an expectation given by the formula below:

$$E(X) = \lim_{\delta \downarrow 0} \sum_{n=-\infty}^{\infty} n\delta P\{n\delta < X \leq (n+1)\delta\}, \tag{A.1.3}$$

where the probabilities in the sum are well defined by the remark about (A.1.2). The existence of the limit in (A.1.3), and the consequent properties of the expectation which extend those discussed in Chapters 5 and 6, are part of a general theory known as that of *Lebesgue integration* [Henri Lebesgue (1875–1941), co-founder with Borel of the modern school of measure and integration.] We must refer the reader to standard treatments of the subject except to exhibit $E(X)$ as an integral as follows:

$$E(X) = \int_\Omega X(\omega)P(d\omega);$$

cf. the discrete analogue (4.3.11) in a countable Ω.

Chapter 5

Conditioning and Independence

5.1. Examples of conditioning

We have seen that the probability of a set A is its weighted proportion relative to the sample space Ω. When Ω is finite and all sample points have the same weight (therefore equally likely), then

$$P(A) = \frac{|A|}{|\Omega|}$$

as in Example 4 of §2.2. When Ω is countable and each point ω has the weight $P(\omega) = P(\{\omega\})$ attached to it, then

$$P(A) = \frac{\sum_{\omega \in A} P(\omega)}{\sum_{\omega \in \Omega} P(\omega)} \tag{5.1.1}$$

from (2.4.3), since the denominator above is equal to 1. In many questions we are interested in the proportional weight of one set A relative to another set S. More accurately stated, this means the proportional weight of the part of A in S, namely the intersection $A \cap S$, or AS, relative to S. The formula analogous to (5.1.1) is then

$$\frac{\sum_{\omega \in AS} P(\omega)}{\sum_{\omega \in S} P(\omega)}. \tag{5.1.2}$$

Thus we are switching our attention from Ω to S as a new universe, and considering a new proportion or probability with respect to it. We introduce the notation

$$P(A \mid S) = \frac{P(AS)}{P(S)} \tag{5.1.3}$$

and call it the *conditional probability of A relative to S*. Other phrases such as "*given S*," "*knowing S*," or "*under the hypothesis* [*of*] *S*" may also be used to describe this relativity. Of course if $P(S) = 0$ then the ratio in (5.1.3) becomes the "indeterminate" $0/0$ which has neither meaning nor utility; so whenever we write a conditional probability such as $P(A \mid S)$ we shall impose the proviso that $P(S) > 0$ even if this is not explicitly mentioned. Observe that the ratio in (5.1.3) reduces to that in (5.1.2) when Ω is countable, but is meaningful in the general context where the probabilities of A and S are defined. The following preliminary examples will illustrate the various possible motivations and interpretations of the new concept.

Example 1. All the students on a certain college campus are polled as to their reaction to a certain presidential candidate. Let D denote those who favor him. Now the student population Ω may be cross-classified in various ways, for instance according to sex, age, race, etc. Let

$$A = \text{female},\ B = \text{black},\ C = \text{of voting age}.$$

Then Ω is partitioned as in (1.3.5) into 8 subdivisions $ABC, ABC^c, \ldots, A^cB^cC^c$. Their respective numbers will be known if a complete poll is made, and the set D will in general cut across the various divisions. For instance

$$P(D \mid A^cBC) = \frac{P(A^cBCD)}{P(A^cBC)}$$

denotes the proportion of male black students of voting age who favor the candidate;

$$P(D^c \mid A^cC) = \frac{P(A^cCD^c)}{P(A^cC)}$$

denotes the proportion of male students of voting age who do not favor the candidate, etc.

Example 2. A perfect die is thrown twice. Given [knowing] that the total obtained is 7, what is the probability that the first point obtained is k, $1 \leq k \leq 6$?

Look at the list in Example 3 of §4.1. The outcomes with total equal to 7 are those on the "second diagonal" and their number is six. Among these there is one case in which the first throw is k. Hence the conditional probability is equal to $1/6$. In symbols, let X_1 and X_2 denote respectively the point obtained in the first and second throw. Then we have as a case of (5.1.3), with $A = \{X_1 = k\}$, $S = \{X_1 + X_2 = 7\}$:

$$P\{X_1 = k \mid X_1 + X_2 = 7\} = \frac{P\{X_1 = k;\ X_1 + X_2 = 7\}}{P\{X_1 + X_2 = 7\}} = \frac{1}{6}.$$

The fact that this turns out to be the same as the *unconditional probability* $P\{X_1 = k\}$ is an accident due to the lucky choice of the number 7. It is the only value of the total which allows all six possibilities for each throw. As other examples, we have

$$P\{X_1 = k \mid X_1 + X_2 = 6\} = \frac{1}{5}, \quad 1 \le k \le 5,$$

$$P\{X_1 = k \mid X_1 + X_2 = 9\} = \frac{1}{4}, \quad 3 \le k \le 6.$$

Here it should be obvious that the conditional probabilities will be the same if X_1 and X_2 are interchanged. Why?

Next, we ask the apparently simpler question: given $X_1 = 4$, what is the probability that $X_2 = k$? You may jump to the answer that this must be $1/6$ since the second throw is not affected by the first, so the conditional probability $P\{X_2 = k \mid X_1 = 4\}$ must be the same as the unconditional one $P\{X_2 = k\}$. This is certainly correct provided we use the *independence* between the two trials (see §2.4). For the present we can use (5.1.3) to get

$$(5.1.4) \qquad P\{X_2 = k \mid X_1 = 4\} = \frac{P\{X_1 = 4; X_2 = k\}}{P\{X_1 = 4\}} = \frac{\frac{1}{36}}{\frac{1}{6}} = \frac{1}{6}.$$

Finally, we have

$$(5.1.5) \qquad P\{X_1 + X_2 = 7 \mid X_1 = 4\} = \frac{P\{X_1 = 4; X_1 + X_2 = 7\}}{P\{X_1 = 4\}}.$$

Without looking at the list of outcomes, we observe that the event $\{X_1 = 4; X_1 + X_2 = 7\}$ is exactly the same as $\{X_1 = 4; X_2 = 7 - 4 = 3\}$; so in effect (5.1.5) is a case of (5.1.4). This argument may seem awfully devious at this juncture, but is an essential feature of a *random walk* (see Chapter 8).

Example 3. Consider the waiting time X in Example 8 of §4.4, for a biased coin. Knowing that it has fallen tails three times, what is the probability that it will fall heads within the next two trials?

This is the conditional probability

$$(5.1.6) \qquad P(X \le 5 \mid X \ge 4) = \frac{P(4 \le X \le 5)}{P(X \ge 4)}.$$

We know that

$$(5.1.7) \qquad P(X = n) = q^{n-1}p, \quad n = 1, 2, \ldots;$$

from which we can calculate

$$(5.1.8) \qquad P(X \ge 4) = \sum_{n=4}^{\infty} q^{n-1}p = \frac{q^3p}{1-q} = q^3;$$

(how do we sum the series?) Again from (5.1.7),

$$P(4 \le X \le 5) = q^3p + q^4p.$$

Thus the answer to (5.1.6) is $p + qp$. Now we have also from (5.1.7) the probability that the coin falls heads (at least once in two trials):

$$P(1 \le X \le 2) = p + qp.$$

Comparing these two results, we conclude that the three previous failures do not affect the future waiting time. This may seem obvious to you *a priori*, but it is a consequence of independence of the successive trials. By the way, many veteran gamblers at the roulette game believe that "if reds have appeared so many times in a row, then it is smart to bet on the black on the next spin because in the long run red and black should balance out." On the other hand, you might argue (with Lord Keynes† on your side) that if red has appeared say ten times in a row, in the absence of other evidence, it would be a natural presumption that the roulette wheel or the croupier is biased toward the red, namely $p > 1/2$ in the above, and therefore the smart money should be on *it*. See Example 8 in §5.2 below for a similar discussion.

Example 4. We shall bring out an analogy between the geometrical distribution given in (5.1.7) [see also (4.4.8)] and the exponential distribution in (4.5.11). If X has the former distribution, then for any non-negative integer n we have

$$P(X > n) = q^n. \tag{5.1.9}$$

This can be shown by summing a geometrical series as in (5.1.8), but is obvious if we remember that "$X > n$" means that the first n tosses all show tails. It now follows from (5.1.9) that for any non-negative integers m and n, we have

$$\begin{aligned} P(X > n + m \mid X > m) &= \frac{P(X > n + m)}{P(X > m)} = \frac{q^{m+n}}{q^m} \\ &= q^n = P(X > n). \end{aligned} \tag{5.1.10}$$

Now let T denote the waiting time in Example 12 of §4.5; then we have analogously for any non-negative real values of s and t:

$$\begin{aligned} P(T > s + t \mid T > s) &= \frac{P(T > s + t)}{P(T > s)} = \frac{e^{-\lambda(s+t)}}{e^{-\lambda s}} \\ &= e^{-\lambda t} = P(T > t). \end{aligned} \tag{5.1.11}$$

This may be announced as follows: if we have already spent some time in waiting, the distribution of further waiting time is the same as that of the initial waiting time as if we have waited in vain! A suggestive way of saying this is that the random variable *T has no memory*. This turns out to be a fundamental property of the exponential distribution which is not shared by any other, and is basic for the theory of Markov processes. Note that although the

† John Maynard Keynes [1883-1946], English economist and writer.

geometrical distribution is a discrete analogue as shown in (5.1.10), strictly speaking it does not have the "memoryless" property because (5.1.10) may become false when n and m are not integers: take e.g. $n = m = 1/2$.

Example 5. Consider all families with two children and assume that boys and girls are equally likely. Thus the sample space may be denoted schematically by 4 points:

$$\Omega = \{(bb), (bg), (gb), (gg)\}$$

where b = boy, g = girl; the order in each pair is the order of birth; and the 4 points have probability 1/4 each. We may of course use instead a space of $4N$ points, where N is a large number, in which the four possibilities have equal numbers. This will be a more realistic population model but the arithmetic below will be the same.

If a family is chosen at random from Ω, and found to have a boy in it, what is the probability that it has another boy, namely that it is of the type (b, b)? A quickie answer might be 1/2 if you jumped to the conclusion from the equal likelihood of the sexes. This is a mistake induced by a misplaced "relative clause" for the conditional probability in question. Here is the detailed explanation.

Let us put

$$A = \{\omega \mid \text{there is a boy in } \omega\}$$
$$B = \{\omega \mid \text{there are two boys in } \omega\}.$$

Then $B \subset A$ and so $AB = B$, thus

$$P(B \mid A) = \frac{P(B)}{P(A)} = \frac{\frac{1}{4}}{\frac{3}{4}} = \frac{1}{3}.$$

This is the correct answer to the question. But now let us ask a similar sounding but really different question. If a child is chosen at random from these families and is found to be a boy, what is the probability that the other child in his family is also a boy? This time the appropriate representation of the sample space should be

$$\tilde{\Omega} = \{g_g, g_b, b_g, b_b\},$$

where the sample points are not families but the children of these families, and g_g = a girl who has a sister, g_b = a girl who has a brother, etc. Now we have

$$\tilde{A} = \{\tilde{\omega} \mid \tilde{\omega} \text{ is a boy}\}$$
$$\tilde{B} = \{\tilde{\omega} \mid \tilde{\omega} \text{ has a brother}\}$$

so that

$$\tilde{A}\tilde{B} = \{\tilde{\omega} \mid \tilde{\omega} = b_b\}.$$

Therefore

$$P(\tilde{B} \mid \tilde{A}) = \frac{P(\tilde{A}\tilde{B})}{P(\tilde{A})} = \frac{\frac{1}{4}}{\frac{1}{2}} = \frac{1}{2}.$$

This is a wonderful and by no means artificial illustration of the importance of understanding "*what* we are sampling" in statistics.

5.2. Basic formulas

Generally speaking, most problems of probability have to do with several events or random variables and it is their *mutual relation* or *joint action* that must be investigated. In a sense all probabilities are conditional because nothing happens in a vacuum. We omit the stipulation of conditions which are implicit or taken for granted, or if we feel that they are irrelevant to the situation in hand. For instance, when a coin is tossed we usually ignore the possibility that it will stand on its edge, and do not even specify whether it is Canadian or American. The probability that a certain candidate will win an election is certainly conditioned on his surviving the campaign—an assumption which has turned out to be premature in recent American history.

Let us begin by a few simple but fundamental propositions involving conditional probabilities:

Proposition 1. *For arbitrary events $A_1, A_2, \ldots, A_n$, we have*

(5.2.1)
$$P(A_1A_2 \ldots A_n) = P(A_1)P(A_2 \mid A_1)P(A_3 \mid A_1A_2) \ldots P(A_n \mid A_1A_2 \ldots A_{n-1})$$

provided $P(A_1A_2 \ldots A_{n-1}) > 0$.

Proof: Under the proviso all conditional probabilities in (5.2.1) are well defined since

$$P(A_1) \geq P(A_1A_2) \geq \cdots \geq P(A_1A_2 \ldots A_{n-1}) > 0.$$

Now the right side of (5.2.1) is explicitly:

$$\frac{P(A_1)}{P(\Omega)} \frac{P(A_1A_2)}{P(A_1)} \frac{P(A_1A_2A_3)}{P(A_1A_2)} \cdots \frac{P(A_1A_2 \cdots A_n)}{P(A_1A_2 \cdots A_{n-1})}$$

which reduces to the left side by successive cancellation. Q.E.D.

By contrast with the additivity formula (2.3.3) for a disjoint union, the formula (5.2.1) may be called the *general multiplicative formula* for the probability of an intersection. But observe how the conditioning events are also "multiplied" step by step. A much simpler formula has been given in §2.4 for independent events. As an important application of (5.2.1), suppose the random variables $X_1, X_2, \ldots, X_n, \ldots$ are all countably-valued; this is surely the case when Ω is countable. Now for arbitrary possible values $x_1, x_2, \ldots, x_n, \ldots$, we put

$$A_k = \{X_k = x_k\}, \quad k = 1, 2, \ldots,$$

and obtain

$$\begin{aligned}(5.2.2)\quad P\{X_1 = x_1; X_2 = x_2; \ldots; X_n = x_n\} &= P\{X_1 = x_1\}P\{X_2 = x_2 \mid X_1 = x_1\}P\{X_3 = x_3 \mid X_1 = x_1, X_2 = x_2\} \\ &\quad \cdots P\{X_n = x_n \mid X_1 = x_1, \ldots, X_{n-1} = x_{n-1}\}.\end{aligned}$$

The first term above is called the *joint probability* of $X_1, X_2, \ldots, X_n$; so the formula expresses this by successive conditional probabilities. Special cases of this will be discussed later.

Proposition 2. *Suppose that*

$$\Omega = \sum_n A_n$$

is a partition of the sample space into disjoint sets. Then for any set B we have

$$(5.2.3)\qquad P(B) = \sum_n P(A_n)P(B \mid A_n).$$

Proof: First we write

$$B = \Omega B = \left(\sum_n A_n\right) B = \sum_n A_n B$$

by simple set theory, in particular (1.3.6); then we deduce

$$P(B) = P\left(\sum_n A_n B\right) = \sum_n P(A_n B)$$

by countable additivity of P. Finally we substitute

$$P(A_n B) = P(A_n)P(B \mid A_n)$$

from the definition (5.1.3). This establishes (5.2.3); note that if $P(A_n) = 0$ for some n, the corresponding term in the sum there may be taken to be 0 even though $P(B \mid A_n)$ is undefined. Q.E.D.

From now on we shall adopt the convention that $x \cdot 0 = 0$ if x is undefined, in order to avoid repetition of such remarks as in the preceding sentence.

The formula (5.2.3) will be referred to as that of *total probability*. Here is a useful interpretation. Suppose that the event B may occur under a number of mutually exclusive circumstances (or "*causes*"). Then the formula shows how its "total probability" is compounded from the probabilities of the various circumstances, and the corresponding conditional probabilities figured under the respective hypotheses.

Suppose X and Y are two integer-valued random variables and k is an integer. If we apply (5.2.3) to the sets

$$A_n = \{X = n\}, \ B = \{Y = k\},$$

we obtain

$$P(Y = k) = \sum_n P(X = n)P(Y = k \mid X = n) \tag{5.2.4}$$

where the sum is over all integers n, and if $P(X = n) = 0$ the corresponding term may be taken to be 0. It is easy to generalize the formula when X takes values in any countable range, and when "$Y = k$" is replaced by e.g., "$a \leq Y \leq b$" for a more general random variable, not necessarily taking integer values.

Proposition 3. *Under the assumption and notation of Proposition 2, we have also*

$$P(A_n \mid B) = \frac{P(A_n)P(B \mid A_n)}{\sum_n P(A_n)P(B \mid A_n)} \tag{5.2.5}$$

provided $P(B) > 0$.

Proof: The denominator above is equal to $P(B)$ by Proposition 2, so the equation may be multiplied out to read

$$P(B)P(A_n \mid B) = P(A_n)P(B \mid A_n).$$

This is true since both sides are equal to $P(A_nB)$. Q.E.D.

This simple proposition with an easy proof is very famous under the name of *Bayes' Theorem*, published in 1763. It is supposed to yield an "inverse probability," or probability of the "cause" A, on the basis of the observed "effect" B. Whereas $P(A_n)$ is the *a priori*, $P(A_n \mid B)$ is the *a posteriori* probability of the cause A_n. Numerous applications were made in all areas of natural phenomena and human behavior. For instance, if B is a "body" and the A_n's are the several suspects of the murder, then the theorem will help the jury or court to decide the whodunit. [Jurisprudence was in fact a major field of early speculations on probability.] If B is an earthquake and the A_n's are the different physical theories to explain it, then the theorem will help the scientists to choose between them. Laplace [1749–1827; one of the great mathematicians of all time who wrote a monumental treatise on probability around 1815] used the theorem to estimate the probability that "the sun will also rise tomorrow" (see Example 9 below). In modern times Bayes lent his name to a school of statistics. For our discussion here let us merely comment that Bayes has certainly hit upon a remarkable turn-around for conditional probabilities, but the practical utility of his formula is limited by our usual lack of knowledge on the various *a priori* probabilities.

The following simple examples are given to illustrate the three propositions above. Others will appear in the course of our work.

Example 6. We have actually seen several examples of Proposition 1 before in Chapter 3. Let us re-examine them using the new notion.

What is the probability of throwing six perfect die and getting six different faces? [See Example 2 of §3.1.] Number the dice from 1 to 6, and put:

$$\begin{aligned}
A_1 &= \text{any face for Die 1,}\\
A_2 &= \text{Die 2 shows a different face from Die 1,}\\
A_3 &= \text{Die 3 shows a different face from Die 1 and Die 2,}
\end{aligned}$$

etc. Then we have, assuming that the dice act independently:

$$P(A_1) = 1;\, P(A_2 \mid A_1) = \frac{5}{6};\, P(A_3 \mid A_1A_2) = \frac{4}{6};\, \ldots;\, P(A_6 \mid A_1A_2 \cdots A_5) = \frac{1}{6}.$$

Hence an application of Proposition 1 gives

$$P(A_1A_2 \cdots A_6) = \frac{6}{6}\cdot\frac{5}{6}\cdot\frac{4}{6}\cdot\frac{3}{6}\cdot\frac{2}{6}\cdot\frac{1}{6} = \frac{6!}{6^6}.$$

The birthday problem [Problem 5 of §3.4] is now seen to be practically the same problem, where the number 6 above is replaced by 365. The sequential method mentioned there is just another case of Proposition 1.

Example 7. The family dog is missing after the picnic. Three hypotheses are suggested:

(A) it has gone home;
(B) it is still worrying that big bone in the picnic area;
(C) it has wandered off into the woods.

The *a priori* probabilities, which are assessed from the habits of the dog, are estimated respectively to be $\frac{1}{4}, \frac{1}{2}, \frac{1}{4}$. A child each is sent back to the picnic ground and the edge of the woods to look for the dog. If it is in the former area, it is a cinch (90%) that it will be found; if it is in the latter, the chance is only a toss-up (50%). What is the probability that the dog will be found in the park?

Let A, B, C be the hypotheses above, and let D = "dog will be found in the park." Then we have the following data:

$$P(A) = \frac{1}{4}, \qquad P(B) = \frac{1}{2}, \qquad P(C) = \frac{1}{4};$$

$$P(D \mid A) = 0, \quad P(D \mid B) = \frac{90}{100}, \quad P(D \mid C) = \frac{50}{100}.$$

Hence by (5.2.3),

$$\begin{aligned} P(D) &= P(A)P(D \mid A) + P(B)P(D \mid B) + P(C)P(D \mid C) \\ &= \frac{1}{4} \cdot 0 + \frac{1}{2} \cdot \frac{90}{100} + \frac{1}{4} \cdot \frac{50}{100} = \frac{115}{200}. \end{aligned}$$

What is the probability that the dog will be found at home? Call this D', and assume that $P(D' \mid A) = 1$, namely that if it is home it will be there to greet the family. Clearly $P(D' \mid B) = P(D' \mid C) = 0$ and so

$$\begin{aligned} P(D') &= P(A)P(D' \mid A) + P(B)P(D' \mid B) + P(C)P(D' \mid C) \\ &= \frac{1}{4} \cdot 1 + \frac{1}{2} \cdot 0 + \frac{1}{4} \cdot 0 = \frac{1}{4}. \end{aligned}$$

What is the probability that the dog is "lost"? It is

$$1 - P(D) - P(D') = \frac{35}{200}.$$

Example 8. Urn one contains 2 black and 3 red balls; urn two contains 3 black and 2 red balls. We toss an unbiased coin to decide on the urn to draw from but we do not know which is which. Suppose the first ball drawn is black and it is put back, what is the probability that the second ball drawn from the same urn is also black?

Call the two urns U_1 and U_2; the *a priori* probability that either one is chosen by the coin-tossing is $1/2$:

$$P(U_1) = \frac{1}{2}, \quad P(U_2) = \frac{1}{2}.$$

Denote the event that the first ball is black by B_1, that the second ball is black by B_2. We have by (5.2.5)

$$P(U_1 \mid B_1) = \frac{\frac{1}{2} \cdot \frac{2}{5}}{\frac{1}{2} \cdot \frac{2}{5} + \frac{1}{2} \cdot \frac{3}{5}} = \frac{2}{5}; \quad P(U_2 \mid B_1) = \frac{3}{5}.$$

Note that the two probabilities must add up to one (why?) so we need only compute one of them. Note also that the two *a posteriori* probabilities are directly proportional to the probabilities $P(B_1 \mid U_1)$ and $P(B_1 \mid U_2)$. That is, the black ball drawn is more likely to have come from the urn which is more likely to yield a black ball, and in the proper ratio. Now use (5.2.3) to compute the probability that the second ball is also black. Here A_1 = "B_1 is from U_1," A_2 = "B_1 is from U_2" are the two alternative hypotheses. Since the second drawing is conditioned on B_1, the probabilities of the hypotheses are really conditional ones:

$$P(A_1) = P(U_1 \mid B_1) = \frac{2}{5}, \quad P(A_2) = P(U_2 \mid B_1) = \frac{3}{5}.$$

On the other hand, it is obvious that

$$P(B_2 \mid A_1) = \frac{2}{5}, \quad P(B_2 \mid A_2) = \frac{3}{5}.$$

Hence we obtain the conditional probability

$$P(B_2 \mid B_1) = \frac{2}{5} \cdot \frac{2}{5} + \frac{3}{5} \cdot \frac{3}{5} = \frac{13}{25}.$$

Compare this with

$$P(B_2) = P(U_1)P(B_2 \mid U_1) + P(U_2)P(B_2 \mid U_2) = \frac{1}{2} \cdot \frac{2}{5} + \frac{1}{2} \cdot \frac{3}{5} = \frac{1}{2}.$$

Note that $P(B_2 \mid U_1) = P(B_1 \mid U_1)$, why? We see that the knowledge of the first ball drawn being black has strengthened the probability of drawing a second black ball, because it has increased the likelihood that we have picked the urn with more black balls. To proceed one more step, given that the first two balls drawn are both black and put back, what is the probability of drawing a third black ball from the same urn? We have in notation similar to the above:

$$P(U_1 \mid B_1B_2) = \frac{\frac{1}{2}\left(\frac{2}{5}\right)^2}{\frac{1}{2} \cdot \left(\frac{2}{5}\right)^2 + \frac{1}{2}\left(\frac{3}{5}\right)^2} = \frac{4}{13}, \quad P(U_2 \mid B_1B_2) = \frac{9}{13}$$

$$P(B_3 \mid B_1B_2) = \frac{4}{13} \cdot \frac{2}{5} + \frac{9}{13} \cdot \frac{3}{5} = \frac{35}{65}.$$

This is greater than $\frac{13}{25}$, so a further strengthening has occurred. Now it is easy to see that we can extend the result to any number of drawings. Thus,

$$P(U_1 \mid B_1B_2 \cdots B_n) = \frac{\frac{1}{2}\left(\frac{2}{5}\right)^n}{\frac{1}{2}\left(\frac{2}{5}\right)^n + \frac{1}{2}\left(\frac{3}{5}\right)^n} = \frac{1}{1 + \left(\frac{3}{2}\right)^n},$$

where we have divided the denominator by the numerator in the middle term. It follows that as n becomes larger and larger, the *a posteriori* probability of U_1 becomes smaller and smaller, in fact it decreases to zero and consequently the *a posteriori* probability of U_2 increases to one in the limit. Thus we have

$$\lim_{n \to \infty} P(B_{n+1} \mid B_1B_2 \cdots B_n) = \frac{3}{5} = P(B_1 \mid U_2).$$

This simple example has important implications on the empirical viewpoint of probability. Replace the two urns above by a coin which may be biased (as all real coins are). Assume that the probability p of heads is either $\frac{2}{5}$ or $\frac{3}{5}$ but we do not know which is the true value. The two possibilities are then two alternative hypotheses between which we must decide. If they both have the *a priori* probability $\frac{1}{2}$, then we are in the situation of the two urns. The outcome of each toss will affect our empirical estimate of the value of p. Suppose for some reason we believe that $p = \frac{2}{5}$. Then if the coin falls heads 10 times in a row, can we still maintain that $p = \frac{2}{5}$ and give probability $\left(\frac{2}{5}\right)^{10}$ to this rare event? Or shall we concede that really $p = \frac{3}{5}$ so that the same event will have probability $\left(\frac{3}{5}\right)^{10}$? This is very small but still $\left(\frac{3}{2}\right)^{10}$ larger than the other. In certain problems of probability theory it is customary to consider the value of p as fixed and base the rest of our calculations on it. So the query is what reason do we have to maintain such a fixed stance in the face of damaging evidence given by observed outcomes? Keynes made a point of this criticism on the foundations of probability. From the axiomatic point of view, as followed in this book, a simple answer is this: our formulas are correct for each arbitrary value of p, but axioms of course do not tell us what this value is, nor even whether it makes sense to assign any value at all. The latter may be the case when one talks about the probability of the existence of some "big living creatures somewhere in outer space." [It used to be the *moon!*] In other words, mathematics proper being a deductive science, the problem of evaluating, estimating or testing the value of p lies outside its eminent domain. Of course, it is of the utmost importance in practice, and *statistics* was invented to cope with this kind of problem. But it need not concern us too much here. [The author had the authority of Dr. Albert

Einstein on this point, while on a chance stroll on Mercer Street in Princeton, N.J., sometime in 1946 or 1947. Here is the gist of what he said: in any branch of science which has applications, there is always a gap, which needs a bridge between theory and practice. This is so for instance in geometry or mechanics; and probability is no exception.]

The preceding example has a natural extension when the unknown p may take values in a finite or infinite range. Perhaps the most celebrated illustration is *Laplace's Law of Succession* below.

Example 9. Suppose that the sun has risen n times in succession, what is the probability that it will rise once more?

It is assumed that the *a priori* probability for a sunrise on any day is a constant whose value is unknown to us. Owing to our total ignorance it will be assumed to take all possible values in $[0, 1]$ with equal likelihood. That is to say, this probability will be treated as a random variable ξ which is uniformly distributed over $[0, 1]$. Thus ξ has the density function f such that $f(p) = 1$ for $0 \le p \le 1$. This can be written heuristically as

$$P(p \le \xi \le p + dp) = dp, \quad 0 \le p \le 1. \tag{5.2.6}$$

Cf. the discussion in Example 10 of §4.5. Now if the true value of ξ is p, then under this hypothesis the probability of n successive sunrises is equal to p^n, because they are assumed to be independent events. Let S^n denote the event that "the sun rises n times in succession," then we may write heuristically:

$$P(S^n \mid \xi = p) = p^n. \tag{5.2.7}$$

The analogue to (5.2.3) should then be

$$P(S^n) = \sum_{0 \le p \le 1} P(\xi = p) P(S^n \mid \xi = p). \tag{5.2.8}$$

This is of course meaningless as it stands, but if we pass from the sum into an integral and use (5.2.6), the result is

$$P(S^n) = \int_0^1 P(S^n \mid \xi = p)\, dp = \int_0^1 p^n\, dp = \frac{1}{n+1}. \tag{5.2.9}$$

This continuous version of (5.2.3) is in fact valid, although its derivation above is not quite so. Accepting the formula and applying it for both n and $n + 1$, then taking the ratio, we obtain

$$P(S^{n+1} \mid S^n) = \frac{P(S^n S^{n+1})}{P(S^n)} = \frac{P(S^{n+1})}{P(S^n)} = \frac{\dfrac{1}{n+2}}{\dfrac{1}{n+1}} = \frac{n+1}{n+2}. \tag{5.2.10}$$

This is Laplace's answer to the sunrise problem.

In modern parlance, Laplace used an "urn model" to study successive sunrise as a random process. A sunrise is assimilated to the drawing of a black ball from an urn of unknown composition. The various possible compositions are assimilated to so many different urns containing various proportions of black balls. Finally, the choice of the true value of the proportion is assimilated to the picking of a random number in $[0, 1]$. Clearly, these are weighty assumptions calling forth serious objections at several levels. Is sunrise a random phenomenon or is it deterministic? Assuming that it can be treated as random, is the preceding simple urn model adequate to its description? Assuming that the model is appropriate in principle, why should the *a priori* distribution of the true probability be uniformly distributed, and if not how could we otherwise assess it?

Leaving these great questions aside, let us return for a moment to (5.2.7). Since $P(\xi = p) = 0$ for every p (see §4.5 for a relevant discussion), the so-called conditional probability in that formula is *not* defined by (5.1.3). Yet it makes good sense from the interpretation given before (5.2.7). In fact, it can be made completely legitimate by a more advanced theory [Radon-Nikodym derivative]. Once this is done, the final step (5.2.9) follows without the intervention of the heuristic (5.2.8). Although a full explanation of these matters lies beyond the depth of this textbook, it seems proper to mention it here as a natural extension of the notion of conditional probability. A purely discrete approach to Laplace's formula is also possible but the calculations are harder (see Exercise 35 below).

We end this section by introducing the notion of conditional expectation. In a countable sample space consider a random variable Y with range $\{y_k\}$ and an event S with $P(S) > 0$. Suppose that the expectation of Y exists, then its *conditional expectation relative to* S is defined to be

$$E(Y \mid S) = \sum_k y_k P(Y = y_k \mid S). \tag{5.2.11}$$

Thus, we simply replace in the formula $E(Y) = \sum_k y_k P(Y = y_k)$ the probabilities by conditional ones. The series in (5.2.11) converges absolutely because the last-written series does so. In particular if X is another random variable with range $\{x_j\}$, then we may take $S = \{X = x_j\}$ to obtain $E(Y \mid X = x_j)$. On the other hand, we have as in (5.2.4):

$$P(Y = y_k) = \sum_j P(X = x_j)P(Y = y_k \mid X = x_j).$$

Multiplying through by y_k, summing over k and rearranging the double series, we obtain

$$E(Y) = \sum_j P(X = x_j)E(Y \mid X = x_j). \tag{5.2.12}$$

The rearrangement is justified by absolute convergence.

The next two sections contain somewhat special material. The reader may read the beginnings of §§5.3 and 5.4 up to the statements of Theorems 1 and 3 to see what they are about, but postpone the rest and go to §5.5.

5.3.* Sequential sampling

In this section we study an urn model in some detail. It is among the simplest schemes that can be handled by elementary methods. Yet it presents rich ideas involving conditioning which are important in both theory and practice.

An urn contains b black balls and r red balls. One ball is drawn at a time without replacement. Let $X_n = 1$ or 0 according as the nth ball drawn is black or red. Each sample point ω is then just the sequence $\{X_1(\omega), X_2(\omega), \ldots, X_{b+r}(\omega)\}$, briefly $\{X_n, 1 \leq n \leq b + r\}$; see the discussion around (4.1.3). Such a sequence is called a *stochastic process*, which is a fancy name for any family of random variables. [According to the dictionary, "stochastic" comes from a Greek word meaning "to aim at."] Here the family is the finite sequence indexed by n from 1 to $b + r$. This index n may be regarded as a *time parameter* as if one drawing is made per unit time. In this way we can speak of the gradual evolution of the process as time goes on by observing the successive X_n's.

You may have noticed that our model is nothing but sampling without replacement and with ordering, discussed in §3.2. You are right but our viewpoint has changed and the elaborate description above is meant to indicate this. Not only do we want to know e.g., how many black balls are drawn after so many drawings, as we would previously, but now we want also to know how the sequential drawings affect each other, how the composition of the urn changes with time, etc. In other words, we want to investigate the mutual dependence of the X_n's, and that's where conditional probabilities come in. Let us begin with the easiest kind of question.

Problem. A ball is drawn from the urn and discarded. Without knowing its color, what is the probability that a second ball drawn is black?

For simplicity let us write the events $\{X_n = 1\}$ as B_n and $\{X_n = 0\}$ as $R_n = B_n^c$. We have then from Proposition 2 of §5.2,

$$P(B_2) = P(B_1)P(B_2 \mid B_1) + P(B_1^c)P(B_2 \mid B_1^c). \tag{5.3.1}$$

Clearly we have

$$P(B_1) = \frac{b}{b + r}, \quad P(B_1^c) = \frac{r}{b + r}; \tag{5.3.2}$$

whereas

$$P(B_2 \mid B_1) = \frac{b - 1}{b + r - 1}, \quad P(B_2 \mid B_1^c) = \frac{b}{b + r - 1}$$

since there are $b + r - 1$ balls left in the urn after the first drawing, and among these are $b - 1$ or b black balls according as the first ball drawn is or is not black. Substituting into (5.3.1) we obtain

$$P(B_2) = \frac{b}{b+r}\frac{b-1}{b+r-1} + \frac{r}{b+r}\frac{b}{b+r-1} = \frac{b(b+r-1)}{(b+r)(b+r-1)} = \frac{b}{b+r}.$$

Thus $P(B_2) = P(B_1)$; namely if we take into account both possibilities for the color of the first ball, then the probabilities for the second ball are the same as if no ball had been drawn (and left out) before. Is this surprising or not? Anyone with curiosity would want to know whether this result is an accident or has a theory behind it. An easy way to test this is to try another step or two: suppose 2 or 3 balls have been drawn but their colors not noted, what then is the probability that the next ball will be black? You should carry out the simple computations by all means. The general result can be stated succinctly as follows.

Theorem 1. *We have for each n,*

$$P(B_n) = \frac{b}{b+r}, \quad 1 \le n \le b + r. \tag{5.3.3}$$

It is essential to pause here and remark on the economy of this mathematical formulation, in contrast to the verbose verbal description above. The *condition* that "we do not know" the colors of the $n - 1$ balls previously drawn is observed as it were in silence, namely by the *absence of conditioning* for the probability $P(B_n)$. What should we have if we *know* the colors? It would be something like $P(B_2 \mid B_1)$ or $P(B_3 \mid B_1B_2^c)$ or $P(B_4 \mid B_1B_2^cB_3)$. These are trivial to compute (why?); but we can also have something like $P(B_4 \mid B_2)$ or $P(B_4 \mid B_1B_3^c)$ which is slightly less trivial. See Exercise 33.

There are many different ways to prove the beautiful theorem above; each method has some merit and is useful elsewhere. We will give two now, a third one in a tremendously more general form (Theorem 5 in §5.4) later. But there are others and perhaps you can think of one later. The first method may be the toughest for you; if so skip it and go at once to the second.†

First Method. This may be called "direct confrontation" or "brute force" and employs heavy (though standard) weaponry from combinatory arsenal. Its merit lies in that it is bound to work provided that we have guessed the answer in advance, as we can in the present case after a few trials. In other words, it is a sort of experimental verification. We introduce a new random variable Y_n = the number of black balls drawn in the first n drawings. This gives the proportion of black balls when the $n + 1$st drawing is made since the total number of balls then is equal to $b + r - n$, regardless of the outcomes of the previous n drawings. Thus we have

† A third method is to make mathematical induction on n.

$$P(B_{n+1} \mid Y_n = j) = \frac{b-j}{b+r-n}, \quad 0 \le j \le b. \tag{5.3.4}$$

On the other hand, the probability $P(Y_n = j)$ can be computed as in Problem 1 of §3.4, with $m = b + r$, $k = b$ in (3.4.1):

$$P(Y_n = j) = \frac{\binom{b}{j}\binom{r}{n-j}}{\binom{b+r}{n}}. \tag{5.3.5}$$

We now apply (5.2.4):

$$\begin{aligned} P(B_{n+1}) &= \sum_{j=0}^{b} P(Y_n = j)P(B_{n+1} \mid Y_n = j) \\ &= \sum_{j=0}^{b} \frac{\binom{b}{j}\binom{r}{n-j}}{\binom{b+r}{n}} \frac{b-j}{b+r-n}. \end{aligned} \tag{5.3.6}$$

This will surely give the answer, but how in the world are we going to compute a sum like that? Actually it is not so hard, and there are excellent mathematicians who make a career out of doing such (and much harder) things. The beauty of this kind of computation is that it's got to unravel if our guess is correct. This faith lends us strength. Just write out the several binomial coefficients above explicitly, cancelling and inserting factors with a view to regrouping them into *new* binomial coefficients:

$$\begin{aligned} &\frac{b!}{j!(b-j)!} \frac{r!}{(n-j)!(r-n+j)!} \frac{n!(b+r-n)!}{(b+r)!} \frac{b-j}{b+r-n} \\ &= \frac{b!r!}{(b+r)!} \frac{(b+r-n-1)!}{(r-n+j)!(b-j-1)!} \frac{n!}{j!(n-j)!} \\ &= \frac{1}{\binom{b+r}{b}} \binom{b+r-n-1}{b-j-1} \binom{n}{j}. \end{aligned}$$

Hence

$$P(B_{n+1}) = \frac{1}{\binom{b+r}{b}} \sum_{j=0}^{b-1} \binom{n}{j} \binom{b+r-1-n}{b-1-j}, \tag{5.3.7}$$

where the term corresponding to $j = b$ has been omitted since it yields zero in (5.3.6). The new sum in (5.3.7) is a well-known identity for binomial coefficients and is equal to $\binom{b+r-1}{b-1}$; see (3.3.9). Thus

$$P(B_{n+1}) = \binom{b+r-1}{b-1} \bigg/ \binom{b+r}{b} = \frac{b}{b+r}$$

as asserted in (5.3.3).

Second Method. This is purely combinatorial and can be worked out as an example in §3.2. Its merit is simplicity; but it cannot be easily generalized to apply to the next urn model we shall consider.

Consider the successive outcomes in $n + 1$ drawings: $X_1(\omega), X_2(\omega), \ldots, X_n(\omega), X_{n+1}(\omega)$. Each $X_j(\omega)$ is 1 or 0 depending on the particular ω; even the numbers of 1's and 0's among them depend on ω when $n + 1 < b + r$. Two different outcome-sequences such as 0011 and 0101 will not have the same probability in general. But now let us put numerals on the balls, say 1 to b for the black ones and $b + 1$ to $b + r$ for the red ones so that all balls become distinguishable. We are then in the case of sampling without replacement and with ordering discussed in §3.2. The total number of possibilities with the new labeling is given by (3.2.1) with $b + r$ for m and $n + 1$ for n: $(b + r)_{n+1}$. These are now all equally likely! We are interested in the cases where the n + 1st ball is black; how many are there for these? There are b choices for the n + 1st ball, and after this is chosen there are $(b + r - 1)_n$ ways of arranging the first n balls, by another application of (3.2.1). Hence by the Fundamental Rule in §3.1, the number of cases where the n + 1st ball is black is equal to $b(b + r - 1)_n$. Now the classical ratio formula for probability applies to yield the answer

$$P(B_{n+1}) = \frac{b(b + r - 1)_n}{(b + r)_{n+1}} = \frac{b}{b + r}.$$

Undoubtedly this argument is easier to follow after it is explained, and there is little computation. But it takes a bit of perception to hit upon the counting method. Poisson [1781–1840; French mathematician for whom a distribution, a process, a limit theorem and an integral were named, among other things] gave this solution but his explanation is more brief than ours. We state his general result as follows.

Theorem 2 [*Poisson's Theorem*]. *Suppose in an urn containing b black and r red balls, n balls have been drawn first and discarded without their colors being noted. If m balls are drawn next, the probability that there are k black balls among them is the same as if we had drawn these m balls at the outset* [*without having discarded the n balls previously drawn*].

Briefly stated: The probabilities are not affected by the preliminary drawing *so long as we are in the dark as to what those outcomes are.* Obviously if we know the colors of the balls discarded, the probabilities will be affected in general. To quote [Keynes, p. 349]: "This is an exceedingly good example . . . that a probability cannot be influenced by the *occurrence* of a material event but only by such *knowledge* as we may have, respecting the occurrence of the event."

Here is Poisson's quick argument: If $n + m$ balls are drawn out, the probability of a combination which is made up of n black and red balls in given proportions followed by m balls of which k are black and $m - k$ are red, must be the same as that of a similar combination in which the m balls precede the n balls. Hence the probability of k black balls in m drawings given that n balls have already been drawn out, must be equal to the probability of the same result when no balls have been previously drawn out.

Is this totally convincing to you? The more explicit combinatorial argument given above for the case $m = 1$ can be easily generalized to settle any doubt. The doubt is quite justified despite the authority of Poisson. As we may learn from Chapter 3, in these combinatorial arguments one must do one's own thinking.

5.4.* Pólya's urn scheme

To pursue the discussion in the preceding section a step further, we will study a famous generalization due to G. Pólya [1887–; professor emeritus at Stanford University, one of the most eminent analysts of modern times who also made major contributions to probability and combinatorial theories and their applications]. As before the urn contains b black and r red balls to begin with, but after a ball is drawn each time, it is returned to the urn and c balls of the same color are added to the urn, where c is an integer and when $c < 0$ adding c balls means subtracting $-c$ balls. This may be done whether we observe the color of the ball drawn or not; in the latter case, e.g., we may suppose that it is performed by an automaton. If $c = 0$ this is just sampling with replacement, while if $c = -1$ we are in the situation studied in §5.3. In general if c is negative the process has to stop after a number of drawings, but if c is zero or positive it can be continued forever. This scheme can be further generalized (you know generalization is a mathematician's bug!) if after each drawing we add to the urn not only c balls of the color drawn but also d balls of the other color. But we will not consider this, and furthermore we will restrict ourselves to the case $c \geq -1$, referring to the scheme as *Pólya's urn model*. This model was actually invented by him to study a problem arising in medicine; see the last paragraph of this section.

Problem. What is the probability that in Pólya's model the first three balls drawn have colors $\{b, b, r\}$ in this order? or $\{b, r, b\}$? or $\{r, b, b\}$?

An easy application of Proposition 1 in §5.2 yields, in the notation introduced in §5.3:

$$P(B_1B_2R_3) = P(B_1)P(B_2 \mid B_1)P(R_3 \mid B_1B_2) = \frac{b}{b+r}\,\frac{b+c}{b+r+c}\,\frac{r}{b+r+2c}. \tag{5.4.1}$$

Similarly

$$P(B_1R_2B_3) = \frac{b}{b+r}\,\frac{r}{b+r+c}\,\frac{b+c}{b+r+2c};$$

$$P(R_1B_1B_2) = \frac{r}{b+r}\,\frac{b}{b+r+c}\,\frac{b+c}{b+r+2c}.$$

Thus they are all the same, namely the probability of drawing 2 black and 1 red balls in three drawings does not depend on the *order* in which they are drawn. It follows that the probability of drawing 2 black and 1 red in the first three drawings is equal to three times the number on the right side of (5.4.1).

The general result is given below.

Theorem 3. *The probability of drawing (from the beginning) any specified sequence of k black balls and n − k red balls is equal to*

$$\text{(5.4.2)}\quad \frac{b(b+c)\cdots(b+(k-1)c)r(r+c)\cdots(r+(n-k-1)c)}{(b+r)(b+r+c)(b+r+2c)\cdots(b+r+(n-1)c)},$$

for all $n \geq 1$ *if* $c \geq 0$; *and for* $0 \leq n \leq b + r$ *if* $c = -1$.

Proof: This is really an easy application of Proposition 1 in §5.2, but in a *scrambled* way. We have shown it above in the case $k = 2$ and $n = 3$. If you will try a few more cases with say $n = 4$, $k = 2$ or $n = 5$, $k = 3$, you will probably see how it goes in the general case more quickly than it can be explained in words. The point is: at the mth drawing, where $1 \leq m \leq n$, the denominator of the term corresponding to $P(A_m \mid A_1A_2 \cdots A_{m-1})$ in (5.2.1) is $b + r + (m - 1)c$, because a total of $(m - 1)c$ balls have been added to the urn by this time, no matter what balls have been drawn. Now at the first time when a black ball is drawn, there are b black balls in the urn; at the second time a black ball is drawn, the number of black balls in the urn is $b + c$, because one black ball has been previously drawn so c black balls have been added to the urn. This is true no matter at what time (which drawing) the second black ball is drawn. Similarly when the third black ball is drawn there will be $b + 2c$ black balls in the urn, and so on. This explains the k factors involving b in the numerator of (5.4.2). Now consider the red balls: at the first time a red ball is drawn there are r red ones in the urn; at the second time a red ball is drawn, there are $r + c$ red ones in the urn, because c red balls have been added after the first red one is drawn, and so on. This explains the $n - k$ factors involving r(=red) in the numerator of (5.4.2). The whole thing there is therefore obtained by multiplying the successive ratios as the conditional probabilities in (5.2.1), and the exact order in which the factors in the numerator occur is determined by the specific order of blacks and reds in the given sequence. However, their product is the same so long as n and k are fixed. This establishes (5.4.2).

For instance if the specified sequence is *RBRRB* then the exact order in the numerator should be $rb(r + c)(r + 2c)(b + c)$.

Now suppose that only the number of black balls is given [specified!] but not the exact sequence, then we have the next result.

Theorem 4. *The probability of drawing (from the beginning) k black balls in n drawings is equal to the number in (5.4.2) multiplied by* $\binom{n}{k}$. *In terms of generalized binomial coefficients (see (5.4.4) below), it is equal to*

$$\frac{\binom{-\frac{b}{c}}{k}\binom{-\frac{r}{c}}{n-k}}{\binom{-\frac{b+r}{c}}{n}}. \tag{5.4.3}$$

This is an extension of the hypergeometric distribution; see p. 90.

Proof: There are $\binom{n}{k}$ ways of permuting k black and $n - k$ red balls; see §3.2. According to (5.4.2), every specified sequence of drawing k black and $n - k$ red balls have the same probability. These various permutations correspond to disjoint events. Hence the probability stated in the theorem is just the sum of $\binom{n}{k}$ probabilities each of which is equal to the number given in (5.4.2). It remains to express this probability by (5.4.3), which requires only a bit of algebra. Let us note that if a is a positive real number and j is a positive integer, then by definition

(5.4.4)

$$\binom{-a}{j} = \frac{(-a)(-a-1)\cdots(-a-j+1)}{j!} = (-1)^j\frac{a(a+1)\cdots(a+j-1)}{j!}.$$

Thus if we divide every factor in (5.4.2) by c, and write

$$\beta = \frac{b}{c}, \quad \gamma = \frac{r}{c}$$

for simplicity, then use (5.4.4), we obtain

$$\frac{\beta(\beta+1)\cdots(\beta+k-1)\gamma(\gamma+1)\cdots(\gamma+n-k-1)}{(\beta+\gamma)(\beta+\gamma+1)\cdots(\beta+\gamma+n-1)}$$

$$= \frac{(-1)^k k!\binom{-\beta}{k}(-1)^{n-k}(n-k)!\binom{-\gamma}{n-k}}{(-1)^n n!\binom{-\beta-\gamma}{n}} = \frac{\binom{-\beta}{k}\binom{-\gamma}{n-k}}{\binom{-\beta-\gamma}{n}\binom{n}{k}}.$$

After multiplying by $\binom{n}{k}$ we get (5.4.3) as asserted.

We can now give a far-reaching generalization of Theorems 1 and 2 in §5.3. Furthermore the result will fall out of the fundamental formula (5.4.2) like a ripe fruit. Only a bit of terminology and notation is in the way.

Recalling the definition of X_n in §5.3, we can record (5.4.2) as giving the *joint distribution* of the n random variables $\{X_1, X_2, \ldots, X_n\}$. Let us introduce the *hedge symbol* "①" to denote either "0" or "1" and use subscripts (indices) to allow arbitrary choice for each subscript, independently of each other. On the other hand, two such symbols with the same subscript must of course denote the same choice throughout a discussion. For instance, $\{①_1, ①_2, ①_3, ①_4\}$ may mean $\{1, 1, 0, 1\}$ or $\{0, 1, 0, 1\}$, but then $\{①_1, ①_3\}$ must mean $\{1, 0\}$ in the first case and $\{0, 0\}$ in the second. Theorem 3 can be stated as follows: if k of the ①'s below are 1's and $n - k$ of them are 0's, then

$$P(X_1 = ①_1, X_2 = ①_2, \ldots, X_n = ①_n) \tag{5.4.5}$$

is given by the expression in (5.4.2). There are altogether 2^n possible choices for the ①'s in (5.4.5) [why?], and if we visualize all the resulting values corresponding to these choices, the set of 2^n probabilities determines the joint distribution of $\{X_1, X_2, \ldots, X_n\}$. Now suppose $\{n_1, n_2, \ldots, n_s\}$ is a subset of $\{1, 2, \ldots, n\}$; the joint distribution of $\{X_{n_1}, \ldots, X_{n_s}\}$ is determined by

$$P(X_{n_1} = ①_{n_1}, \ldots, X_{n_s} = ①_{n_s}) \tag{5.4.6}$$

when the latter ①'s range over all the 2^s possible choices. This is called a *marginal distribution* with reference to that of the larger set $\{X_1, \ldots, X_n\}$.

We need more notation! Let $\{n_1', \ldots, n_t'\}$ be the complementary set of $\{n_1, \ldots, n_s\}$ with respect to $\{1, \ldots, n\}$, namely those indices left over after the latter set has been taken out. Of course $t = n - s$ and the union $\{n_1, \ldots, n_s, n_1', \ldots, n_t'\}$ is just some permutation of $\{1, \ldots, n\}$. Now we can write down the following formula expressing a marginal probability by means of joint probabilities of a larger set:

$$\begin{aligned} &P(X_{n_1} = ①_1, \ldots, X_{n_s} = ①_s) \\ &\quad = \sum_{①_1', \ldots, ①_t'} P(X_{n_1} = ①_1, \ldots, X_{n_s} = ①_s, \\ &\qquad\qquad X_{n_1'} = ①_1', \ldots, X_{n_t'} = ①_t'), \end{aligned} \tag{5.4.7}$$

where $\{①_1', \ldots, ①_t'\}$ is another set of hedge symbols and the sum is over all the 2^t possible choices for them. This formula follows from the obvious set relation

$$\begin{aligned} &\{X_{n_1} = ①_1, \ldots, X_{n_s} = ①_s\} \\ &\quad = \sum_{①_1', \ldots, ①_t'} \{X_{n_1} = ①_1, \ldots, X_{n_s} = ①_s, X_{n_1'} = ①_1', \ldots, X_{n_t'} = ①_t'\} \end{aligned}$$

and the additivity of P. [Clearly a similar relation holds when the X's take other values than 0 or 1, in which case the ①'s must be replaced by all possible values.]

We now come to the *pièce de résistance* of this discussion. It will sorely test your readiness to digest a general and abstract argument. If you can't swallow it now, you need not be upset but do come back and try it again later.

Theorem 5. *The joint distribution of any s of the random variables $\{X_1, X_2, \ldots, X_n, \ldots\}$ is the same, no matter which s of them is in question.*

As noted above, the sequence of X_n's is infinite if $c \geq 0$, whereas $n \leq b + r$ if $c = -1$.

Proof: What does the theorem say? Fix s and let $X_{n_1}, \ldots, X_{n_s}$ be any set of s random variables chosen from the entire sequence. To discuss its joint distribution, we must consider all possible choices of values for these s random variables. So we need a notation for an arbitrary choice of that kind, call it $①_1, \ldots, ①_s$. Now let us write down

$$P(X_{n_1} = ①_1, X_{n_2} = ①_2, \ldots, X_{n_s} = ①_s).$$

We must show that this has the same value no matter what $\{n_1, \ldots, n_s\}$ is, namely that it has the same value as

$$P(X_{m_1} = ①_1, X_{m_2} = ①_2, \ldots, X_{m_s} = ①_s)$$

where $\{m_1, \ldots, m_s\}$ is any other subset of size s. The two sets $\{n_1, \ldots, n_s\}$ and $\{m_1, \ldots, m_s\}$ may very well be overlapping, such as $\{1, 3, 4\}$ and $\{3, 2, 1\}$. Note also that we have never said that the indices must be in increasing order!

Let the maximum of the indices used above be n. As before let $t = n - s$, and

$$\{n'_1, \ldots, n'_t\} = \{1, \ldots, n\} - \{n_1, \ldots, n_s\},$$
$$\{m'_1, \ldots, m'_t\} = \{1, \ldots, n\} - \{m_1, \ldots, m_s\}.$$

Next, let $①'_1, \ldots, ①'_t$ be an arbitrary choice of t hedge symbols. We claim then

$$\begin{aligned} (5.4.8) \quad P(X_{n_1} &= ①_1, \ldots, X_{n_s} = ①_s, X_{n_1'} = ①'_1, \ldots, X_{n_t'} = ①'_t) \\ &= P(X_{m_1} = ①_1, \ldots, X_{m_s} = ①_s, X_{m_1'} = ①'_1, \ldots, X_{m_t'} = ①'_t). \end{aligned}$$

If you can read this symbolism you will see that it is just a consequence of (5.4.2)! For both $(n_1, \ldots, n_s, n'_1 \ldots, n'_t)$ and $(m_1, \ldots, m_s, m'_1, \ldots, m'_t)$ are permutations of the whole set $(1, \ldots, n)$, whereas the set of hedge symbols $(①_1, \ldots, ①_s, ①'_1, \ldots, ①'_t)$ are the same on both sides of (5.4.8). So the equation merely repeats the assertion of Theorem 3 that any two specified sequences having the same number of black balls must have the same probability, irrespective of the permutations.

Finally keeping $①_1, \ldots, ①_s$ fixed but letting $①'_1, \ldots, ①'_t$ vary over all 2^t possible choices, we get 2^t equations of the form (5.4.8). Take their sum and use (5.4.7) once as written and another time when the n's are replaced by the m's. We get

$$P(X_{n_1} = ①_1, \ldots, X_{n_s} = ①_s) = P(X_{m_1} = ①_1, \ldots, X_{m_s} = ①_s),$$

as we set out to show. Q.E.D.

There is really nothing hard or tricky about this proof. "It's just the *notation*!", as some would say.

A sequence of random variables $\{X_n; n = 1, 2, \ldots\}$ having the property given in Theorem 5 is said to be "permutable" or "exchangeable." It follows in particular that any *block* of given length s, such as $X_{s_0+1}, X_{s_0+2}, \ldots, X_{s_0+s}$, where s_0 is any nonnegative integer (and $s_0 + s \leq b + r$ if $c = -1$), have the same distribution. Since the index is usually interpreted as the *time parameter*, the distribution of such a block may be said to be "invariant under a time-shift." A sequence of random variables having this property is said to be "[strictly] stationary." This kind of process is widely used as a model in electrical oscillations, economic time series, queuing problems etc.

Pólya's scheme may be considered as a model for a fortuitous happening [a "random event" in the everyday usage] whose likelihood tends to increase with each occurrence and decrease with each non-occurrence. The drawing of a black ball from his urn is such an event. Pólya himself cited as example the spread of an epidemic in which each victim produces many more new germs and so increases the chances of further contamination. To quote him directly (my translation from the French original), "In reducing this fact to its simplest terms and adding to it a certain symmetry, propitious for mathematical treatment, we are led to the urn scheme." The added symmetry refers to the adding of red balls when a red ball is drawn, which would mean that each non-victim also increases the chances of other non-victims. This half of the hypothesis for the urn model does not seem to be warranted, and is slipped in without comment by several authors who discussed it. Professor Pólya's candor, in admitting it as a mathematical expediency, should be reassuring to scientists who invented elaborate mathematical theories to deal with crude realities such as hens pecking (mathematical psychology) and beetles crawling (mathematical biology).

5.5. Independence and relevance

An extreme and extremely important case of conditioning occurs when the condition has no effect on the probability. This intuitive notion is common experience in tossing a coin or throwing a die several times, or drawing a ball several times from an urn with replacement. The knowledge of the outcome of the previous trials should not change the "virgin" probabilities of the next trial and in this sense the trials are intuitively independent of each other. We have already defined independent events in §2.4; observe that the defining relations in (2.4.5) are just special cases of (5.2.1) when all conditional probabilities are replaced by unconditional ones. The same replacement in (5.2.2) will now lead to the fundamental definition below.

Definition of Independent Random Variables. The countably-valued random variables $X_1, \ldots, X_n$ are said to be independent iff for any real numbers $x_1, \ldots, x_n$, we have

$$(5.5.1)\qquad P(X_1 = x_1, \ldots, X_n = x_n) = P(X_1 = x_1) \ldots P(X_n = x_n).$$

This equation is trivial if one of the factors on the right is equal to zero, hence we may restrict the x's above to the countable set of all possible values of all the X's.

The deceptively simple condition (5.5.1) actually contains much more than meets the eye. To see this let us deduce at once a major extension of (5.5.1) in which single values x_i are replaced by arbitrary sets S_i. Let $X_1, \ldots, X_n$ be independent random variables in Propositions 4 to 6 below.

Proposition 4. *We have for arbitrary countable sets* $S_1, \ldots, S_n$:

$$(5.5.2)\qquad P(X_1 \in S_1, \ldots, X_n \in S_n) = P(X_1 \in S_1) \ldots P(X_n \in S_n).$$

Proof: The left member of (5.5.2) is equal to

$$\begin{aligned} &\sum_{x_1 \in S_1} \cdots \sum_{x_n \in S_n} P(X_1 = x_1, \ldots, X_n = x_n) \\ &= \sum_{x_1 \in S_1} \cdots \sum_{x_n \in S_n} P(X_1 = x_1) \ldots P(X_n = x_n) \\ &= \left\{\sum_{x_1 \in S_1} P(X_1 = x_1)\right\} \ldots \left\{\sum_{x_n \in S_n} P(X_n = x_n)\right\}, \end{aligned}$$

which is equal to the right member of (5.5.2) by simple algebra (which you should spell out if you have any doubt).

Note that independence of a set of random variables as defined above is a property of the set as a whole. Such a property is not necessarily inherited by a subset; can you think of an easy counter-example? However, as a consequence of Proposition 4, any subset of $(X_1, \ldots, X_n)$ is indeed also a set of independent random variables. To see e.g. (X_1, X_2, X_3) is such a set when $n > 3$ above, we take $S_i = R^1$ for $i > 3$ and replace the other S_i's by x_i in (5.5.2).

Next, the condition (5.5.2) will be further strengthened into its most useful form.

Proposition 5. *The events*

$$\{X_1 \in S_1\}, \ldots, \{X_n \in S_n\} \tag{5.5.3}$$

are independent.

Proof: It is important to recall that the definition of independent events requires not only the relation (5.5.2), but also similar relations for all subsets of $(X_1, \ldots, X_n)$. However, these also hold because the subsets are also sets of independent random variables, as just shown.

Before going further let us check that the notion of independent events defined in §2.4 is a special case of independent random variables defined in this section. With the arbitrary events $\{A_j, 1 \leq j \leq n\}$ we associate their indicators I_{A_j} (see §1.4), where

$$I_{A_j}(\omega) = \begin{cases} 1 & \text{if } \omega \in A_j, \\ 0 & \text{if } \omega \in A_j^c; \end{cases} \qquad 1 \leq j \leq n.$$

These are random variables [at least in a countable sample space]. Each takes only the two values 0 or 1, and we have

$$\{I_{A_j} = 1\} = A_j, \quad \{I_{A_j} = 0\} = A_j^c.$$

Now if we apply the condition (5.5.1) of independence to the random variables $I_{A_1}, \ldots, I_{A_n}$, they reduce exactly to the conditions

$$P(\tilde{A}_1 \cdots \tilde{A}_n) = P(\tilde{A}_1) \cdots P(\tilde{A}_n), \tag{5.5.4}$$

where each $\tilde{A}_j$ may be A_j or A_j^c, but of course must be the same on both sides. Now it can be shown (Exercise 36 below) that the condition (5.5.4) for all possible choices of $\tilde{A}_j$, is exactly equivalent to the condition (2.4.5). Hence the independence of the events $A_1, \ldots, A_n$ is equivalent to the independence of their indicators.

The study of independent random variables will be a central theme in any introduction to probability theory. Historically and empirically, they are known as independent trials. We have given an informal discussion of this concept in §2.4. Now it can be formulated in terms of random variables as follows: a sequence of independent trials is just a sequence of independent random variables $(X_1, \ldots, X_n)$ where X_i represents the outcome of the ith trial. Simple illustrations are given in Examples 7 and 8 of §2.4, where in Example 7 the missing random variables are easily supplied. Incidentally, these examples establish the *existence* of independent random variables so that we are assured that our theorems such as the propositons in this section are not vacuities. Actually we can even construct independent random variables with arbitrarily given distributions (see [Chung 1; Chapter 3]). [It may amuse you to know that mathematicians have been known to define and study objects which later turn out to be non-existent!] This remark will be relevant in later chapters; for the moment we shall add one more general proposition to broaden the horizon.

Proposition 6. *Let $\varphi_1, \ldots, \varphi_n$ be arbitrary real-valued functions on $(-\infty, \infty)$; then the random variables*

$$\varphi_1(X_1), \ldots, \varphi_n(X_n) \tag{5.5.5}$$

are independent.

Proof: Let us omit the subscripts on X and φ and ask the question: for a given real number y, what are the values of x such that

$$\varphi(x) = y \quad \text{and} \quad X = x?$$

The set of such values must be countable since X is countably-valued; call it S, of course it depends on y, φ and X. Then $\{\varphi(X) = y\}$ means exactly the same thing as $\{X \in S\}$. Hence for arbitrary $y_1, \ldots, y_n$, the events

$$\{\varphi_1(X_1) = y_1\}, \ldots, \{\varphi_n(X_n) = y_n\}$$

are just those in (5.5.3) for certain sets $S_1, \ldots, S_n$ specified above. So Proposition 6 follows from Proposition 5.

This proposition will be put to good use in Chapter 6. Actually there is a more general result as follows. If we separate the random variables $X_1, \ldots, X_n$ into any number of blocks, and take a function of those in each block, then the resulting random variables are independent. The proof is not so different from the special case given above, and will be omitted.

As for general random variables, they are defined to be independent iff for any real numbers $x_1, \ldots, x_n$, the events

$$\{X_1 \le x_1\}, \ldots, \{X_n \le x_n\} \tag{5.5.6}$$

are independent. In particular,

$$P(X_1 \le x_1, \ldots, X_n \le x_n) = P(X_1 \le x_1) \ldots P(X_n \le x_n). \tag{5.5.7}$$

In terms of the joint distribution function F for the random vector $(X_1, \ldots, X_n)$ discussed in §4.6, the preceding equation may be written as

$$F(x_1, \ldots, x_n) = F_1(x_1) \ldots F_n(x_n) \tag{5.5.8}$$

where F_j is the marginal distribution of X_j, $1 \le j \le n$. Thus in case of independence the marginal distributions determine the joint distribution.

It can be shown that as a consequence of the definition, events such as those in (5.5.3) are also independent, provided that the sets $S_1, \ldots, S_n$ are reasonable [Borel]. In particular if there is a joint density function f, then we have

$$P(X_1 \in S_1, \ldots, X_n \in S_n) = \left\{\int_{S_1} f_1(u)\,du\right\} \cdots \left\{\int_{S_n} f_n(u)\,du\right\}$$
$$= \int_{S_1} \cdots \int_{S_n} f_1(u_1) \cdots f_n(u_n)\,du_1 \cdots du_n$$

where $f_1, \ldots, f_n$ are the marginal densities. But the probability in the first member above is also equal to

$$\int_{S_1} \cdots \int_{S_n} f(u_1, \ldots, u_n)\,du_1 \ldots du_n$$

as in (4.6.6). Comparison of these two expressions yields the equation

(5.5.9) $$f(u_1, \ldots, u_n) = f_1(u_1) \ldots f_n(u_n).$$

This is the form that (5.5.8) takes in the density case.

Thus we see that stochastic independence makes it possible to factorize a joint probability, distribution or density. In the next chapter we shall see that it enables us to factorize mathematical expectation, generating function and other transforms.

Numerous results and applications of independent random variables will be given in Chapters 6 and 7. In fact, the main body of classical probability theory is concerned with them. So much so that in his epoch-making monograph *Foundations of the Theory of Probability*, Kolmogorov [1903–; leading Russian mathematician and one of the founders of modern probability theory] said: "Thus one comes to perceive, in the concept of independence, at least the first germ of the true nature of problems in probability theory." Here we will content ourselves with two simple examples.

Example 10. A letter from Pascal to Fermat (dated Wednesday, 29th July, 1654), contains, among many other mathematical problems, the following passage:

"M. de Méré told me that he had found a fallacy in the theory of numbers, for this reason: If one undertakes to get a six with one die, the advantage in getting it in 4 throws is as 671 is to 625. If one undertakes to throw 2 sixes with two dice, there is a disadvantage in undertaking it in 24 throws. And nevertheless 24 is to 36 (which is the number of pairings of the faces of two dice) as 4 is to 6 (which is the number of faces of one die). This is what made him so indignant and made him say to one and all that the propositions were not consistent and Arithmetic was self-contradictory: but you will very easily see that what I say is correct, understanding the principles as you do."

This famous problem, one of the first recorded in the history of probability and which challenged the intellectual giants of the time, can now be solved by a beginner.

To throw a six with one die in 4 throws means to obtain the point "six" at least once in 4 trials. Define X_n, $1 \leq n \leq 4$, as follows:

$$P(X_n = k) = \frac{1}{6}, \quad k = 1, 2, \ldots, 6,$$

and assume that X_1, X_2, X_3, X_4 are independent. Put $A_n = \{X_n = 6\}$; then the event in question is $A_1 \cup A_2 \cup A_3 \cup A_4$. It is easier to calculate the probability of its complement which is identical to $A_1^cA_2^cA_3^cA_4^c$. The trials are assumed to be independent and the dice unbiased. We have as a case of (5.5.4),

$$P(A_1^cA_2^cA_3^cA_4^c) = P(A_1^c)P(A_2^c)P(A_3^c)P(A_4^c) = \left(\frac{5}{6}\right)^4,$$

hence

$$P(A_1 \cup A_2 \cup A_3 \cup A_4) = 1 - \left(\frac{5}{6}\right)^4 = 1 - \frac{625}{1296} = \frac{671}{1296}.$$

This last number is approximately equal to 0.5177. Since $1296 - 671 = 625$, the "odds" are as 671 to 625 as stated by Pascal.

Next consider two dice; let (X_n', X_n'') denote the outcome obtained in the nth throw of the pair, and let

$$B_n = \{X_n' = 6; X_n'' = 6\}.$$

Then $P(B_n^c) = \frac{35}{36}$, and

$$P(B_1^cB_2^c \cdots B_{24}^c) = \left(\frac{35}{36}\right)^{24},$$

$$P(B_1 \cup B_2 \cup \cdots \cup B_{24}) = 1 - \left(\frac{35}{36}\right)^{24}.$$

This last number is approximately equal to 0.4914, which confirms the disadvantage.

One must give great credit to de Méré for his sharp observation and long experience at gaming tables to discern the narrow inequality

$$P(A_1 \cup A_2 \cup A_3 \cup A_4) > \frac{1}{2} > P(B_1 \cup B_2 \cup \cdots \cup B_{24}).$$

His arithmetic went wrong because of a fallacious "linear hypothesis." [According to some historians the problem was not originated with de Méré.]

Example 11. If two points are picked at random from the interval [0, 1], what is the probability that the distance between them is less than 1/2?

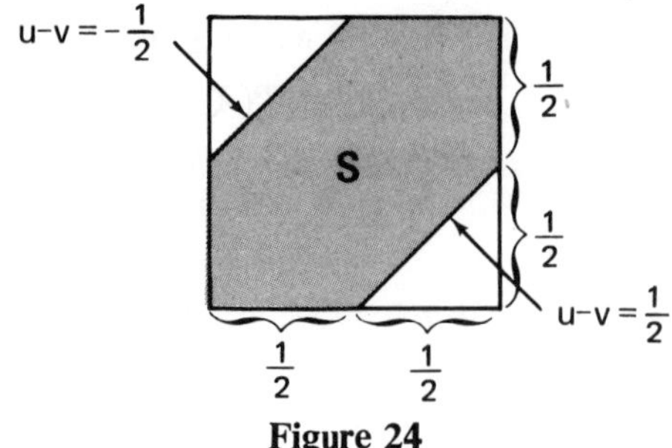

Figure 24

By now you should be able to interpret this kind of cryptogram. It means: if X and Y are two independent random variables each of which is uniformly distributed in $[0, 1]$, find the probability $P\left(|X - Y| < \frac{1}{2}\right)$. Under the hypotheses the random vector (X, Y) is uniformly distributed over the unit square U (see Figure); namely for any reasonable subset S of U, we have

$$P\{(X, Y) \in S\} = \iint_S du\, dv.$$

This is seen from the discussion after (4.6.6); in fact the $f(u, v)$ there is equal to $f_1(u)f_2(v)$ by (5.5.9) and both f_1 and f_2 are equal to one in $[0, 1]$ and zero outside. For the present problem S is the set of points (u, v) in U satisfying the inequality $|u - v| < \frac{1}{2}$. You can evaluate the double integral above over this set if you are good at calculus, but it is a lot easier to do this geometrically as follows. Draw the two lines $u - v = \frac{1}{2}$ and $u - v = -\frac{1}{2}$; then S is the area bounded by these lines and the sides of the square. The complementary area $U - S$ is the union of two triangles each of area $\frac{1}{2}\left(\frac{1}{2}\right)^2 = \frac{1}{8}$. Hence we have

$$\text{area of } S = 1 - 2 \cdot \frac{1}{8} = \frac{3}{4}$$

and this is the required probability.

Example 12. Suppose $X_1, X_2, \ldots, X_n$ are independent random variables with distributions $F_1, F_2, \ldots, F_n$ as in (4.5.4). Let

$$M = \max(X_1, X_2, \ldots, X_n)$$
$$m = \min(X_1, X_2, \ldots, X_n).$$

Find the distribution functions of M and m.

Using (5.5.7) we have for each x:

$$\begin{aligned} F_{\max}(x) = P(M \le x) &= P(X_1 \le x; X_2 \le x; \ldots ; X_n \le x) \\ &= P(X_1 \le x)P(X_2 \le x) \ldots P(X_n \le x) \\ &= F_1(x)F_2(x) \cdots F_n(x). \end{aligned}$$

In particular if all the F's are the same,

$$F_{\max}(x) = F(x)^n.$$

As for the minimum, it is convenient to introduce the "tail distribution" G_j corresponding to each F_j as follows:

$$G_j(x) = P\{X_j > x\} = 1 - F_j(x).$$

Then we have, using the analogue of (5.5.2) this time with $S_j = (x_j, \infty)$:

$$\begin{aligned} G_{\min}(x) &= P(m > x) = P(X_1 > x; X_2 > x; \ldots ; X_n > x) \\ &= P(X_1 > x)P(X_2 > x) \cdots P(X_n > x) \\ &= G_1(x)G_2(x) \cdots G_n(x). \end{aligned}$$

Hence

$$F_{\min}(x) = 1 - G_1(x)G_2(x) \cdots G_n(x).$$

If all the F's are the same, this becomes

$$G_{\min}(x) = G(x)^n; \quad F_{\min}(x) = 1 - G(x)^n.$$

Here is a concrete illustration. Suppose a town depends on 3 reservoirs for its water supply, and suppose that its daily draws from them are independent and have exponential densities $e^{-\lambda_1 x}$, $e^{-\lambda_2 x}$, $e^{-\lambda_3 x}$ respectively. Suppose each reservoir can supply a maximum of N gallons per day to that town. What is the probability that on a specified day the town will run out of water?

Call the draws X_1, X_2, X_3 on that day, the probability in question is by (4.5.12)

$$P(X_1 > N; X_2 > N; X_3 > N) = e^{-\lambda_1 N}e^{-\lambda_2 N}e^{-\lambda_3 N} = e^{-(\lambda_1+\lambda_2+\lambda_3)N}.$$

* The rest of the section is devoted to a brief study of a logical notion which is broader than pairwise independence. This notation is inherent in statistical comparison of empirical data, operational evaluation of alternative policies, etc. Some writers even base the philosophical foundation of statistics on such a qualitative notion.

An event A is said to be *favorable* to another event B iff

$$P(AB) \ge P(A)P(B). \tag{5.5.10}$$

This will be denoted symbolically by $A \parallel B$. It is thus a binary relation between two events which includes pairwise independence as a special case.

An excellent example is furnished by the divisibility by any two positive integers; see §2.4 and Exercise 17 in Chapter 2.

It is clear from (5.5.10) that the relation $\|$ is symmetric; it is also reflexive since $P(A) \geq P(A)^2$ for any A. But it is not transitive, namely $A \| B$ and $B \| C$ do not imply $A \| C$. In fact, we will show by an example that even the stronger relation of pairwise independence is not transitive.

Example 13. Consider families with 2 children as in Example 5 of §5.1: $\Omega = \{(bb), (bg), (gb), (gg)\}$. Let such a family be chosen at random and consider the three events below:

$$\begin{aligned} A &= \text{first child is a boy,} \\ B &= \text{the two children are of different sex,} \\ C &= \text{first child is a girl.} \end{aligned}$$

Then

$$AB = \{(bg)\}, \; BC = \{(gb)\}, \; AC = \emptyset.$$

A trivial computation then shows that $P(AB) = P(A)P(B)$, $P(BC) = P(B)P(C)$ but $P(AC) = 0 \neq P(A)P(C)$. Thus the pairs $\{A, B\}$ and $\{B, C\}$ are independent but the pair $\{A, C\}$ is not.

A slight modification will show that pairwise independence does not imply total independence for three events. Let

$$D = \text{second child is a boy.}$$

Then

$$AD = \{(bb)\}, \; BD = \{(gb)\}, \; ABD = \emptyset;$$

and so $P(ABD) = 0 \neq P(A)P(B)P(D) = 1/8$.

Not so long ago one could still find textbooks on probability and statistics in which total independence was confused with pairwise independence. It is easy on hindsight to think of everyday analogues of the counter-examples above. For instance, if A is friendly to B, and B is friendly to C, why should it follow that A is friendly to C? Again, if every two of three people A, B, C get along well, it is not necessarily the case that all three of them do.

These commonplace illustrations should tell us something about the use and misuse of "intuition." Pushing a bit further, let us record a few more *non-sequitors* below ("$\not\Rightarrow$" reads "does not imply"):

$$\begin{aligned} &A \| C \text{ and } B \| C \not\Rightarrow (A \cap B) \| C; \\ &A \| B \text{ and } A \| C \not\Rightarrow A \| (B \cap C); \\ &A \| C \text{ and } B \| C \not\Rightarrow (A \cup B) \| C; \\ &A \| B \text{ and } A \| C \not\Rightarrow A \| (B \cup C). \end{aligned} \tag{5.5.11}$$

You may try some verbal explanations for these; rigorous but artificial examples are also very easy to construct; see Exercise 15.

The great caution needed in making conditional evaluation is no academic matter, for much statistical analysis of experimental data depends on a critical understanding of the basic principles involved. The following illustration is taken from Colin R. Blyth, "On Simpson's paradox and the sure-thing principle," *Journal of American Statistical Association*, Vol. 67 (1972) pp. 364–366.

Example 14. A doctor has the following data on the effect of a new treatment. Because it involved extensive follow-up treatment after discharge, he could handle only a few out-of-town patients and had to work mostly with patients residing in the city.

	City-residents		Non City-residents	
	Treated	Untreated	Treated	Untreated
Alive	1000	50	95	5000
Dead	9000	950	5	5000

Let

$$A = \text{alive}$$
$$B = \text{treated}$$
$$C = \text{city-residents}$$

The sample space may be partitioned first according to A and B; then according to A, B and C. The results are shown in the diagrams:

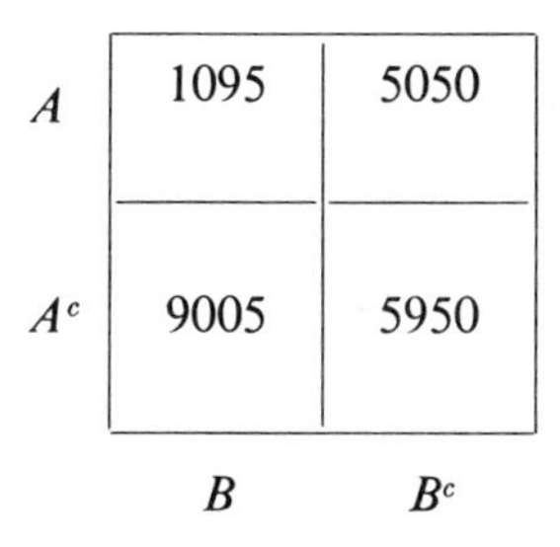

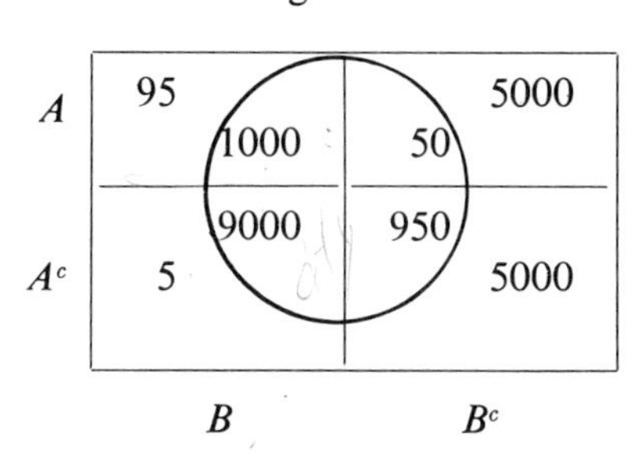

The various conditional probabilities, namely the classified proportions are as follows:

$$P(A \mid B) = \frac{1095}{10100} = \text{about } 10\% \qquad P(A \mid BC) = \frac{1000}{10000}$$

$$P(A \mid B^c) = \frac{5050}{11000} = \text{about } 50\% \qquad P(A \mid B^cC) = \frac{50}{1000}$$

$$P(A \mid BC^c) = \frac{95}{100}$$

$$P(A \mid B^cC^c) = \frac{5000}{10000}.$$

Thus if the results (a matter of life or death) are judged from the conditional probabilities in the left column, the treatment seems to be a disaster since it had decreased the chance of survival five times! But now look at the right column, for city-residents and non city-residents separately:

$$P(A \mid BC) = 10\%, \; P(A \mid B^cC) = 5\%;$$
$$P(A \mid BC^c) = 95\%, \; P(A \mid B^cC^c) = 50\%.$$

In both cases the chance of survival is doubled by the treatment.

The explanation is this: for some reason (such as air pollution), the C patients are much less likely to recover than the C^c patients, and most of those treated were C patients. Naturally, a treatment is going to show a poor recovery rate when used on the most seriously ill of the patients.

The arithmetical puzzle is easily solved by the following explicit formulas

$$\begin{aligned}
P(A \mid B) &= \frac{P(AB)}{P(B)} = \frac{P(ABC) + P(ABC^c)}{P(B)} \\
&= \frac{P(ABC)}{P(BC)}\frac{P(BC)}{P(B)} + \frac{P(ABC^c)}{P(BC^c)}\frac{P(BC^c)}{P(B)} \\
&= P(A \mid BC)P(C \mid B) + P(A \mid BC^c)P(C^c \mid B) \\
&= \frac{1000}{10000}\frac{10000}{10100} + \frac{95}{100}\frac{100}{10100}; \\
P(A \mid B^c) &= P(A \mid B^cC)P(C \mid B^c) + P(A \mid B^cC^c)P(C^c \mid B^c) \\
&= \frac{50}{1000}\frac{1000}{11000} + \frac{5000}{10000}\frac{10000}{11000}.
\end{aligned}$$

It is those "hidden coefficients" $P(C \mid B)$, $P(C^c \mid B)$, $P(C \mid B^c)$, $P(C^c \mid B^c)$ that have caused a reverse. A little parable will clarify the arithmetic involved. Suppose in two families both husbands and wives work. Husband of family 1 earns more than husband of family 2, wife of family 1 earns more than wife of family 2. For a certain good cause [or fun] both husband and wife of family 2 contribute half their monthly income; but in family 1 the husband contributes only 5% of his income, letting the wife contribute 95% of hers. Can you see why the poorer couple give more to the cause [or spend more on the vacation]?

This example should be compared with a simpler analogue in Exercise 11, where there is no paradox and intuition is a sure thing.

5.6.* Genetical models

This section treats an application to genetics. The probabilistic model discussed here is among the simplest and most successful in empirical sciences.

Hereditary characters in *diploid* organisms such as human beings are carried by genes which appear in pairs. In the simplest case each gene of a pair can assume two forms called *alleles:* A and a. For instance A may be "blue-eyed," and a "brown-eyed" in a human being; or A may be "red blossom" and a "white blossom" in garden peas, which were the original subject of experiment by Mendel [1822–1884]. We have then three *genotypes:*

$$AA, Aa, aa,$$

there being no difference between Aa and aA [nature does not order the pair]. In some characters, A may be *dominant* whereas *a recessive* so that Aa cannot be distinguished from AA in appearance so far as the character in question is concerned; in others Aa may be intermediate such as shades of green for eye color or pink for pea blossom. The reproductive cells, called *gametes,* are formed by splitting the gene pairs and have only one gene of each pair. At mating each parent therefore transmits one of the genes of the pair to the offspring through the gamete. The pure type AA or aa can of course transmit only A or a, whereas the mixed type Aa can transmit either A or a but not both. Now let us fix a gene pair and suppose that the parental genotypes AA, Aa, aa are in the proportions

$$u:2v:w \quad \text{where } u > 0,\ v > 0,\ w > 0,\ u + 2v + w = 1.$$

[The factor 2 in $2v$ is introduced to simplify the algebra below.] The total pool of these three genotypes is very large and the mating couples are formed "at random" from this pool. At each mating, each parent transmits one of the pair of genes to the offspring with probability 1/2, independently of each other, and independently of all other mating couples. Under these circumstances *random mating* is said to take place. For example, if peas are well mixed in a garden these conditions hold approximately; on the other hand if the pea patches are segregated according to blossom colors then the mating will not be quite random.

The stochastic model can be described as follows. Two urns contain a very large number of coins of three types: with A on both sides, with A on one side and a on the other, and with a on both sides. Their proportions are as $u:2v:w$ for each urn. One coin is chosen from each urn in such a way that all coins are equally likely. The two chosen coins are then tossed and the two uppermost faces determine the genotype of the offspring. What is the probability that it be AA, Aa or aa? In a more empirical vein and using the frequency interpretation, we may repeat the process a large number of times to get an actual sample of the distribution of the types. Strictly speaking, the coins must be replaced each time so that the probability of each type remains constant in the repeated trials.

Let us tabulate the cases in which an offspring of type AA will result from the mating. Clearly this is possible only if there are at least two A-genes available between the parents. Hence the possibilities are given in the first and second columns below.

Type of male	Type of female	Probability of mating of the couple	Probability of producing offspring AA from the couple	Probability of offspring AA
AA	AA	$u \cdot u = u^2$	1	u^2
AA	Aa	$u \cdot 2v = 2uv$	$\frac{1}{2}$	uv
Aa	AA	$2v \cdot u = 2uv$	$\frac{1}{2}$	uv
Aa	Aa	$2v \cdot 2v = 4v^2$	$\frac{1}{4}$	v^2

In the third column we give the probability of mating between the two designated genotypes in the first two entries of the same row; in the fourth column we give the conditional probability for the offspring to be of type AA given the parental types; in the fifth column the product of the probabilities in the third and fourth entries of the same row. By Proposition 2 of §5.2, the total probability for the offspring to be of type AA is given by adding the entries in the fifth column. Thus

$$P(\text{offspring is } AA) = u^2 + uv + uv + v^2 = (u + v)^2.$$

From symmetry, replacing u by w, we get

$$P(\text{offspring is } aa) = (v + w)^2.$$

Finally, we list all cases in which an offspring of type Aa can be produced, in a similar tabulation as the preceding one.

Type of male	Type of female	Probability of mating of the couple	Probability of producing offspring Aa from the couple	Probability of offspring Aa
AA	Aa	$u \cdot 2v = 2uv$	$\frac{1}{2}$	uv
Aa	AA	$2v \cdot u = 2uv$	$\frac{1}{2}$	uv
AA	aa	$u \cdot w = uw$	1	uw
aa	AA	$w \cdot u = uw$	1	uw
Aa	aa	$2v \cdot w = 2vw$	$\frac{1}{2}$	vw
aa	Aa	$w \cdot 2v = 2vw$	$\frac{1}{2}$	vw
Aa	Aa	$2v \cdot 2v = 4v^2$	$\frac{1}{2}$	$2v^2$

Hence we obtain by adding up the last column:

$$P(\text{offspring is } Aa) = 2(uv + uw + vw + v^2) = 2(u + v)(v + w).$$

Let us put

$$p = u + v, \quad q = v + w \tag{5.6.1}$$

so that $p > 0, q > 0, p + q = 1$. Let us also denote by $P_n(\cdots)$ the probability of the genotypes for offspring of the nth generation. Then the results obtained above are as follows:

$$P_1(AA) = p^2, P_1(Aa) = 2pq, P_1(aa) = q^2. \tag{5.6.2}$$

These give the proportions of the parental genotypes for the second generation. Hence in order to obtain P_2, we need only substitute p^2 for u, pq for v and q^2 for w in the two preceding formulas. Thus,

$$P_2(AA) = (p^2 + pq)^2 = p^2,$$

$$P_2(Aa) = 2(p^2 + pq)(pq + q^2) = 2pq,$$

$$P_2(aa) = (pq + q^2)^2 = q^2.$$

Lo and behold: P_2 is the same as P_1! Does this mean that P_3 is also the same as P_1, etc.? This is true, but only after the observation below. We have shown that $P_1 = P_2$ *for an arbitrary* P_0 [in fact, even the nit-picking conditions $u > 0$, $v > 0$, $w > 0$ may be omitted]. Moving over one generation, therefore, $P_2 = P_3$, even although P_1 may not be the same as P_0. The rest is smooth sailing, and the result is known as Hardy-Weinberg theorem. (G. H. Hardy [1877–1947] was a leading English mathematician whose main contributions were to number theory and classical analysis.)

Theorem. *Under random mating for one pair of genes, the distribution of the genotypes becomes stationary from the first generation on, no matter what the original distribution is.*

Let us assign the numerical values 2, 1, 0 to the three types AA, Aa, aa according to the number of A-genes in the pair; and let us denote by X_n the random variable which represents the numerical genotype of the nth generation. Then the theorem says that for $n \geq 1$:

$$P(X_n = 2) = p^2, P(X_n = 1) = 2pq, P(X_n = 0) = q^2. \tag{5.6.3}$$

The distribution of X_n is stationary in the sense that these probabilities do not depend on n. Actually it can be shown that the process $\{X_n, n \geq 1\}$ is *strictly stationary* in the sense described in §5.4, because it is also a Markov chain; see Exercise 40 below.

The result embodied in (5.6.2) may be reinterpreted by an even simpler model than the one discussed above. Instead of gene-pairs we may consider a pool of gametes, namely after the splitting of the pairs into individual genes. Then the A-genes and a-genes are originally in the proportion

$$(2u + 2v):(2v + 2w) = p:q$$

because there are two A-genes in the type AA, etc. Now we can think of these gametes as so many little tokens marked A or a in an urn, and assimilate the birth of an offspring to the random drawing (with replacement) of two of the gametes to form a pair. Then the probabilities of drawing AA, Aa, aa are respectively:

$$p \cdot p = p^2, \quad p \cdot q + q \cdot p = 2pq, \quad q \cdot q = q^2.$$

This is the same result as recorded in (5.6.2).

The new model is not the same as the old one, but it leads to the same conclusion. It is tempting to try to identify the two models on hindsight, but the only logical way of doing so is to go through both cases as we have done. *A priori* or *prima facie*, they are not equivalent. Consider, for instance, the case of fishes: the females lay billions of eggs first and then the males come along and fertilize them with sperm. The partners may never meet. In this circumstance the second model fits the picture better, especially if we use two urns for eggs and sperm separately. [There are in fact creatures in which sex is not differentiated and which suits the one-urn model.] Such a model may be called the *spawning model*, in contrast to the *mating model* described earlier. In more complicated cases where more than one pair of genes is involved, the two models need not yield the same result.

Example 15. It is known in human genetics that certain "bad" genes cause crippling defects or disease. If a is such a gene the genotype aa will not survive to adulthood. A person of genotype Aa is a *carrier* but appears normal because a is a recessive character. Suppose the probability of a carrier among the general population is p, irrespective of sex. Now if a person has an affected brother or sister who died in childhood, then he has a *history* in the family and cannot be treated genetically as a member of the general population. The probability of his being a carrier is a conditional one to be computed as follows. Both his parents must be carriers, namely of genotype Aa, for otherwise they could not have produced a child of genotype aa. Since each gene is transmitted with probability 1/2, the probabilities of their child to be AA,

Aa, aa are $\frac{1}{4}, \frac{1}{2}, \frac{1}{4}$ respectively. Since the person in question has survived he cannot be aa, and so the probability that he be AA or Aa is given by

$$P(AA \mid AA \cup Aa) = \frac{1}{3}, P(Aa \mid AA \cup Aa) = \frac{2}{3}.$$

If he marries a woman who is not known to have a history of that kind in the family, then she is of genotype AA or Aa with probability $1 - p$ or p as for the general population. The probabilities for the genotypes of their children are listed below.

Male	Female	Probability of the combination	Probability of producing AA	Probability of producing Aa	Probability of producing aa
AA	AA	$\frac{1}{3}(1-p)$	1	0	0
AA	Aa	$\frac{1}{3}p$	$\frac{1}{2}$	$\frac{1}{2}$	0
Aa	AA	$\frac{2}{3}(1-p)$	$\frac{1}{2}$	$\frac{1}{2}$	0
Aa	Aa	$\frac{2}{3}p$	$\frac{1}{4}$	$\frac{1}{2}$	$\frac{1}{4}$

A simple computation gives the following distribution of the genotypes for the offspring:

$$P_1(AA) = \frac{2}{3} - \frac{p}{3},$$

$$P_1(Aa) = \frac{1}{3} + \frac{p}{6},$$

$$P_1(aa) = \frac{p}{6}.$$

The probability of a surviving child being a carrier is therefore

$$P_1(Aa \mid AA \cup Aa) = \frac{2+p}{6-p}.$$

If p is negligible, this is about 1/3. Hence from the surviving child's point of view, his having an affected uncle or aunt is only half as bad a hereditary risk as his father's having an affected sibling. One can now go on computing the chances for *his* children, and so on—exercises galore left to the reader.

In concluding this example which concerns a serious human condition, it is proper to stress that the simple mathematical theory should be regarded only as a rough approximation since other genetical factors have been ignored in the discussion.

Exercises

1. Based on the data given in Example 14 of §5.5, what is the probability that (a) a living patient resides in the city? (b) a living treated patient lives outside the city?
2. All the screws in a machine come from the same factory but it is as likely to be from Factory A as from Factory B. The percentage of defective screws is 5% from A and 1% from B. Two screws are inspected; if the first is found to be good what is the probability that the second is also good?
3. There are two kinds of tubes in an electronic gadget. It will cease to function if and only if one of each kind is defective. The probability that there is a defective tube of the first kind is .1; the probability that there is a defective tube of the second kind is .2. It is known that two tubes are defective, what is the probability that the gadget still works?
4. Given that a throw of three unbiased dice shows different faces, what is the probability that (a) at least one is a six; (b) the total is eight?
5. Consider families with three children and assume that each child is equally likely to be a boy or a girl. If such a family is picked at random and the eldest child is found to be a boy, what is the probability that the other two are girls? The same question if a randomly chosen child from the family turns out to be a boy.
6. Instead of picking a family as in No. 5, suppose now a child is picked at random from all children of such families. If he is a boy, what is the probability that he has two sisters?
7. Pick a family as in No. 5, and then pick two children at random from this family. If they are found to be both girls, what is the probability that they have a brother?
8. Suppose that the probability that both twins are boys is α, and that both are girls is β; suppose also that when the twins are of different sexes the probability of the first born being a girl is $1/2$. If the first born of twins is a girl, what is the probability that the second is also a girl?
9. Three marksmen hit the target with probabilities $\frac{1}{2}, \frac{2}{3}, \frac{3}{4}$ respectively. They shoot simultaneously and there are two hits. Who missed? Find the probabilities.
10. On a flight from Urbana to Paris my luggage did not arrive with me. It had been transferred three times and the probabilities that the transfer

was not done in time were estimated to be $\frac{4}{10}, \frac{2}{10}, \frac{1}{10}$ respectively in the order of transfer. What is the probability that the first airline goofed?

11. Prove the "sure-thing principle": if

$$P(A \mid C) \geq P(B \mid C),$$
$$P(A \mid C^c) \geq P(B \mid C^c),$$

then $P(A) \geq P(B)$.

12. Show that if $A \parallel B$, then

$$A^c \parallel B^c, \; A \not\parallel B^c, \; A^c \not\parallel B.$$

13. Show that if $A \cap B = \emptyset$, then

    (i) $A \parallel C$ and $B \parallel C \Rightarrow (A \cup B) \parallel C$;
    (ii) $C \parallel A$ and $C \parallel B \Rightarrow C \parallel (A \cup B)$;
    (iii) A and C are independent, B and C are independent $\Rightarrow A \cup B$ and C are independent.

14. Suppose $P(H) > 0$. Show that the set function

$$S \rightarrow P(S \mid H) \quad \text{for} \quad S \subset \Omega \text{ (countable)}$$

is a probability measure.

15.* Construct examples for all the assertions in (5.5.11). [Hint: a systematic but tedious way to do this is to assign $p_1, \ldots, p_8$ to the eight atoms $ABC, \ldots, A^cB^cC^c$ (see (1.3.5)) and express the desired inequalities by means of them. The labor can be reduced by preliminary simple choices among the p's, such as making some of them zero and others equal. One can also hit upon examples by using various simple properties of a small set of integers; see an article by the author: "On mutually favorable events," *Annals of Mathematical Statistics*, Vol. 13 (1942), pp. 338–349.]

16. Suppose that A_j, $1 \leq j \leq 5$, are independent events. Show that

    (i) $(A_1 \cup A_2)A_3$ and $A_4^c \cup A_5^c$ are independent;
    (ii) $A_1 \cup A_2$, $A_3 \cap A_4$ and A_5^c are independent.

17. Suppose that in a certain casino there are three kinds of slot machines in equal numbers with pay-off frequencies $\frac{1}{3}, \frac{1}{2}, \frac{2}{3}$ respectively. One of these machines paid off twice in four cranks; what is the probability of a pay-off on the next crank?

18. A person takes four tests in succession. The probability of his passing the first test is p, that of his passing each succeeding test is p or $p/2$ according as he passes or fails the preceding one. He qualifies provided he passes at least three tests. What is his chance of qualifying?

19. 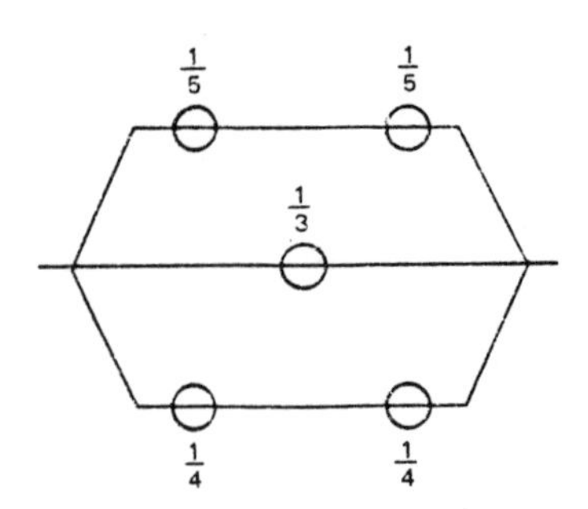

An electric circuit looks as in the figure where the numbers indicate the probabilities of failure for the various links, which are all independent. What is the probability that the circuit is in operation?

20. It rains half of the time in a certain city, and the weather forecast is correct 2/3 of the time. Mr. Milquetoast goes out every day and is much worried about rain. So he will take his umbrella if the forecast is rain, but he will also take it 1/3 of the time even if the forecast is no rain. Find

(a) the probability of his being caught in rain without an umbrella;
(b) the probability of his carrying an umbrella without rain.

These are the two kinds of errors defined by Neyman and Pearson in their statistical theory.* [Hint: compute the probability of "rain; forecast no rain; no umbrella," etc.]

21. Telegraphic signals "dot" and "dash" are sent in the proportion 3:4. Owing to conditions causing very erratic transmission, a dot becomes a dash with probability 1/4, whereas a dash becomes a dot with probability 1/3. If a dot is received what is the probability that it is sent as a dot?

22. *A* says *B* told him that *C* had lied. If each of these persons tells the truth with probability p, what is the probability that *C* indeed lied? [Believe it or not, this kind of question was taken seriously one time under the name of "credibility of the testimony of witnesses." In the popular phrasing given above it is grossly ambiguous, and takes a lot of words to explain the intended meaning. To cover one case in detail, suppose all three lied. Then *B* will tell *A* that *C* has told the truth, because *B* is supposed to know whether *C* has lied or not but decides to tell a lie himself; *A* will say that *B* told him that *C* had lied, since he wants to lie about what *B* told him, without knowing what *C* did. This is just one of the eight possible cases but the others can be similarly interpreted. A much clearer formulation is the model of transmission of signals used in No. 21. *C* transmits $-$ or $+$ according as he lies or not; then *B* transmits the message from *C* incorrectly or correctly according as he lies or not; then *A* transmits the message from *B* in a similar manner. There will be no semantic impasse even if we go on this way to any number

* The reader is supposed to translate the verbal descriptions so that answers are obtainable. In the real world, predictions and decisions are made on even vaguer grounds.

of witnesses. The question is: if "$-$" is received at the end of line, what is the probability that it is sent as such initially?]

23. A particle starts from the origin and moves on the line 1 unit to the right or left with probability 1/2 each, the successive movements being independent. Let Y_n denote its position after n moves. Find the following probabilities:
 (a) $P(Y_n \geq 0 \text{ for } 1 \leq n \leq 4)$;
 (b) $P(|Y_n| \leq 2 \text{ for } 1 \leq n \leq 4)$;
 (c) $P(Y_n \geq 0 \text{ for } 1 \leq n \leq 4 \mid Y_4 = 0)$.

24. In No. 23, show that if $j < k < n$, we have

$$P(Y_n = c \mid Y_j = a, Y_k = b) = P(Y_n = c \mid Y_k = b) = P(Y_{n-k} = c - b)$$

where a, b, c are any integers. Illustrate with $j = 4$, $k = 6$, $n = 10$, $a = 2$, $b = 4$, $c = 6$.

25. First throw an unbiased die, then throw as many unbiased coins as the point shown on the die.
 (a) What is the probability of obtaining k heads?
 (b) If 3 heads are obtained what is the probability that the die showed n?

26. In a nuclear reaction a certain particle may split into 2 or 3 particles, or not split at all. The probabilities for these possibilities are p_2, p_3 and p_1. The new particles behave in the same way and independently of each other as well as of the preceding reaction. Find the distribution of the total number of particles after two reactions.

27. An unbiased die is thrown n times; let M and m denote the maximum and minimum points obtained. Find $P(m = 2, M = 5)$. [Hint: begin with $P(m \geq 2, M \leq 5)$.]

28. Let X and Y be independent random variables with the same probability distribution $\{p_n, n \geq 1\}$. Find $P(X \leq Y)$ and $P(X = Y)$.

In Problems 29–32, consider two numbers picked at random in [0, 1].

29. If the smaller one is less than 1/4, what is the probability that the larger one is greater than 3/4?

30.* Given that the smaller one is less than x, find the distribution of the larger one. [Hint: consider $P(\min < x, \max < y)$ and the two cases $x \leq y$ and $x > y$.]

31.* The two points picked divide [0, 1] into three segments. What is the probability that these segments can be used to form a triangle? [Hint: this is the case if and only if the sum of lengths of any two is greater than the length of the third segment. Call the points X and Y and treat the case $X < Y$ first.]

32. Prove that the lengths of the three segments mentioned above have the same distributions. [Hint: consider the distribution of the smaller value picked, that of the difference between the two values, and use a

symmetrical argument for the difference between 1 and the larger value.]

33. In Pólya's urn scheme find:

    (a) $P(R_3 \mid B_1R_2)$;
    (b) $P(R_3 \mid R_1R_2)$;
    (c) $P(R_3 \mid R_2)$;
    (d) $P(R_1 \mid R_2R_3)$;
    (e) $P(R_1 \mid R_2)$;
    (f) $P(R_1 \mid R_3)$.

34. Consider two urns U_i containing r_i red and b_i black balls respectively, $i = 1, 2$. A ball is drawn at random from U_1 and put into U_2, then a ball is drawn at random from U_2 and put into urn U_1. After this what is the probability of drawing a red ball from U_1? Show that if $b_1 = r_1$, $b_2 = r_2$, then this probability is the same as if no transfers have been made.

35.* Assume that the a priori probabilities p in the sunrise problem (Example 9 of §5.2) can only take the values $\frac{k}{100}$, $1 \le k \le 100$, with probability $\frac{1}{100}$ each. Find $P(S^{n+1} \mid S^n)$. Replace 100 by N and let $N \to \infty$, what is the limit?

36.* Prove that the events $A_1, \dots, A_n$ are independent if and only if

$$P(\tilde{A}_1 \cdots \tilde{A}_n) = P(\tilde{A}_1) \cdots P(\tilde{A}_n)$$

where each $\tilde{A}_j$ may be A_j or A_j^c. [Hint: to deduce these equations from independence, use induction on n and also induction on k in $P(A_1^c \cdots A_k^c A_{k+1} \cdots A_n)$; the converse is easy by induction on n.]

37.* Spell out a proof of Theorem 2 in §5.3. [Hint: label all balls and show that any particular sequence of balls has the same probability of occupying any given positions if all balls are drawn in order.]

38. Verify Theorem 5 of §5.4 directly for the pairs (X_1, X_2), (X_1, X_3) and (X_2, X_3).

39. Assume that the three genotypes AA, Aa, aa are in the proportion $p^2{:}2pq{:}q^2$, where $p + q = 1$. If two parents chosen at random from the population have an offspring of type AA, what is the probability that another child of theirs is also of type AA? Same question with AA replaced by Aa.

40. Let X_1 and X_2 denote the genotype of a female parent and her child. Assuming that the unknown genotype of the male parent is distributed as in Problem No. 39 and using the notation of (5.6.3), find the nine conditional probabilities below:

$$P\{X_2 = k \mid X_1 = j\}, \quad j = 0, 1, 2;\ k = 0, 1, 2.$$

These are called the *transition probabilities* of a *Markov chain;* see §8.3.

41.* Prove that if the function φ defined on $[0, \infty)$ is nonincreasing and satisfies the *Cauchy functional equation*

$$\varphi(s + t) = \varphi(s)\varphi(t), \quad s \geq 0, t \geq 0;$$

then $\varphi(t) = e^{-\lambda t}$ for some $\lambda \geq 0$. Hence a positive random variable T has the property

$$P(T > s + t \mid T > s) = P(T > t), \quad s \geq 0, t \geq 0$$

if and only if it has an exponential distribution. [Hint: $\varphi(0) = 1$: $\varphi(1/n) = \alpha^{1/n}$ where $\alpha = \varphi(1)$, $\varphi(m/n) = \alpha^{m/n}$; if $m/n \leq t < (m + 1)/n$ then $\alpha^{(m+1)/n} \leq \varphi(t) \leq \alpha^{m/n}$; hence the general conclusion follows by letting $n \to \infty$.]

42. A needle of unit length is thrown onto a table which is marked with parallel lines at a fixed distance d from one another, where $d > 1$. Let the distance from the midpoint of the needle to the nearest line be x, and let the angle between the needle and the perpendicular from its midpoint to the nearest line be θ. It is assumed that x and θ are independent random variables, each of which is uniformly distributed over its range. What is the probability that the needle intersects a line? This is known as *Buffon's problem* and its solution suggests an empirical [Monte Carlo] method of determining the value of π.

Chapter 6

Mean, Variance and Transforms

6.1. Basic properties of expectation

The mathematical expectation of a random variable, defined in §4.3, is one of the foremost notions in probability theory. It will be seen to play the same role as integration in calculus—and we know "integral calculus" is at least half of all calculus. Recall its meaning as a probabilistically weighted average [in a countable sample space] and rewrite (4.3.11) more simply as:

$$E(X) = \sum_{\omega} X(\omega)P(\omega). \tag{6.1.1}$$

If we substitute $|X|$ for X above, we see that the proviso (4.3.12) may be written as

$$E(|X|) < \infty. \tag{6.1.2}$$

We shall say that the random variable X is *summable* when (6.1.2) is satisfied. In this case we say also that "X has a finite expectation (or mean)" or "its expectation exists." The last expression is actually a little vague because we generally allow $E(X)$ to be defined and equal to $+\infty$ when for instance $X \geq 0$ and the series in (6.1.1) diverges. See Exercises 27 and 28 of Chapter 4. We shall say so explicitly when this is the case.

It is clear that if X is bounded, namely when there exists a number M such that

$$|X(\omega)| \leq M \quad \text{for all } \omega \in \Omega,$$

then X is summable and in fact

$$E(|X|) = \sum_{\omega} |X(\omega)|\, P(\omega) \leq M \sum_{\omega} P(\omega) = M.$$

In particular if Ω is finite then every random variable is bounded (this does not mean all of them are bounded by the same number). Thus the class of random variables having a finite expectation is quite large. For this class the mapping

$$X \to E(X) \tag{6.1.3}$$

assigns a number to a random variable. For instance, if X is the height of students in a school, then $E(X)$ is their average height; if X is the income of wage earners, then $E(X)$ is their average income; if X is the number of vehicles passing through a toll bridge in a day, then $E(X)$ is the average daily traffic, etc.

If A is an event, then its indicator I_A (see §1.4) is a random variable, and we have

$$E(I_A) = P(A).$$

In this way the notion of mathematical expectation is seen to extend that of a probability measure.

Recall that if X and Y are random variables, then so is $X + Y$ (Proposition 1 of §4.2). If X and Y both have finite expectations, it is intuitively clear what the expectation of $X + Y$ should be. Thanks to the intrinsic nature of our definition, it is easy to prove the theorem.

Theorem 1. *If X and Y are summable, then so is $X + Y$ and we have*

$$E(X + Y) = E(X) + E(Y). \tag{6.1.4}$$

Proof: Applying the definition (6.1.1) to $X + Y$, we have

$$\begin{aligned} E(X + Y) &= \sum_\omega (X(\omega) + Y(\omega))P(\omega) \\ &= \sum_\omega X(\omega)P(\omega) + \sum_\omega Y(\omega)P(\omega) = E(X) + E(Y). \end{aligned}$$

This is the end of the matter. You may wonder wherever do we need the condition (6.1.2)? The answer is: we want the defining series for $E(X + Y)$ to converge absolutely, as explained in §4.3. This is indeed the case because

$$\begin{aligned} \sum_\omega |X(\omega) + Y(\omega)|\, P(\omega) &\le \sum_\omega (|X(\omega)| + |Y(\omega)|)P(\omega) \\ &= \sum_\omega |X(\omega)|\, P(\omega) + \sum_\omega |Y(\omega)|\, P(\omega) < \infty. \end{aligned}$$

Innocuous or ingenuous as Theorem 1 may appear, it embodies the most fundamental property of E. There is a pair of pale sisters as follows:

$$E(a) = a, \qquad E(aX) = aE(X) \tag{6.1.5}$$

for any constant a; and combining (6.1.4) and (6.1.5) we obtain

$$E(aX + bY) = aE(X) + bE(Y) \tag{6.1.6}$$

for any two constants a and b. This property makes the operation in (6.1.3) a "linear operator." This is a big name in mathematics; you may have heard of it in linear algebra or differential equations.

An easy extension of (6.1.4) by mathematical induction yields: if X_1, $X_2, \ldots, X_n$ are summable random variables, then so is their sum and

(6.1.7) $$E(X_1 + \cdots + X_n) = E(X_1) + \cdots + E(X_n).$$

Before we take up other properties of E, let us apply this to some interesting problems.

Example 1. A raffle lottery contains 100 tickets, of which there is one ticket bearing the prize \$10000, the rest being all zero. If I buy two tickets, what is my expected gain?

If I have only one ticket, my gain is represented by the random variable X which takes the value 10000 on exactly one ω and 0 on all the rest. The tickets are assumed to be equally likely to win the prize, hence

$$X = \begin{cases} 10000 & \text{with probability } \dfrac{1}{100} \\ 0 & \text{with probability } \dfrac{99}{100} \end{cases}$$

and

$$E(X) = 10000 \cdot \frac{1}{100} + 0 \cdot \frac{99}{100} = 100.$$

Thus my expected gain is \$100. This is trivial, but now if I have two tickets I know very well only one of them can possibly win, so there is definite interference [dependence] between the two random variables represented by the tickets. Will this affect my expectation? Thinking a bit more deeply: if I am not the first person to have bought the tickets, perhaps by the time I get mine someone else has already taken the prize, albeit unknown to all. Will it then make a difference whether I get the tickets early or late? Well, these questions have already been answered by the urn model discussed in §5.3. We need only assimilate the tickets to 100 balls of which exactly one is black. Then if we define the outcome of the nth drawing by X_n, we know from Theorem 1 there that X_n has the same probability distribution as the X shown above, and so also the same expectation. For $n = 2$ this was computed directly and easily without recourse to the general theorem. It follows that no matter what j and k are, namely for any two tickets drawn anytime, the expected value of both together is equal to

$$E(X_j + X_k) = E(X_j) + E(X_k) = 2E(X) = 200.$$

More generally, my expected gain is directly proportional to the number of tickets bought—a very fair answer, but is it so obvious in advance? In particular if I buy all 100 tickets I stand to win 100 $E(X) = 10000$ dollars. This may sound dumb but it checks out.

To go one step further, let us consider two lotteries of exactly the same kind. Instead of buying two tickets X and Y from the same lottery, I may choose to buy one from each lottery. Now I have a chance to win \$20000.

Does this make the scheme more advantageous to me? Yet Theorem 1 says that my expected gain is \$200 in either case. How is this accounted for? To answer this question you should figure out the distribution of $X + Y$ under each scheme and compute $E(X + Y)$ directly from it. You will learn a lot by comparing the results.

Example 2. [Coupon collecting problem.] There are N coupons marked 1 to N in a bag. We draw one coupon after another with replacement. Suppose we wish to collect r different coupons, what is the expected number of drawings to get them? This is the problem faced by school children who collect baseball star cards; or by housewives who can win a sewing machine if they have a complete set of coupons which come in some detergent boxes. In the latter case the coupons may well be stacked against them if certain crucial ones are made very scarce. Here of course we consider the fair case in which all coupons are equally likely and the successive drawings are independent.

The problem may be regarded as one of waiting time, namely: we wait for the rth *new* arrival. Let $X_1, X_2, \ldots$ denote the successive waiting times for a new coupon. Thus $X_1 = 1$ since the first is always new. Now X_2 is the waiting time for any coupon that is different from the first one drawn. Since at each drawing there are N coupons and all but one of them will be new, this reduces to Example 8 of §4.4 with success probability $p = \dfrac{N-1}{N}$; hence

$$E(X_2) = \frac{N}{N-1}.$$

After these two different coupons have been collected, the waiting time for the third new one is similar with success probability $p = \dfrac{N-2}{N}$; hence

$$E(X_3) = \frac{N}{N-2}.$$

Continuing this argument, we obtain for $1 \leq r \leq N$:

$$\begin{aligned} E(X_1 + \cdots + X_r) &= \frac{N}{N} + \frac{N}{N-1} + \cdots + \frac{N}{N-r+1} \\ &= N\left(\frac{1}{N-r+1} + \cdots + \frac{1}{N}\right). \end{aligned}$$

In particular if $r = N$, then

$$E(X_1 + \cdots + X_N) = N\left(1 + \frac{1}{2} + \cdots + \frac{1}{N}\right); \tag{6.1.8}$$

and if N is even and $r = N/2$,

(6.1.9) $$E(X_1 + \cdots + X_{N/2}) = N\left(\frac{1}{\frac{N}{2}+1} + \cdots + \frac{1}{N}\right).$$

Now there is a famous formula in mathematical analysis which says that

(6.1.10) $$1 + \frac{1}{2} + \cdots + \frac{1}{N} = \log N + C + \epsilon_N,$$

where the "log" is the natural logarithm to the base e, C is the Euler's constant $= .5772 \cdots$ [nobody in the world knows whether it is a rational number or not], and ϵ_N tends to zero as N goes to infinity. For most purposes, the more crude asymptotic formula is sufficient:

(6.1.11) $$\lim_{N\to\infty} \frac{1}{\log N}\left(1 + \frac{1}{2} + \cdots + \frac{1}{N}\right) = 1.$$

If we use this in (6.1.8) and (6.1.9), we see that for large values of N, the quantities there are roughly equal to $N \log N$ and $N \log 2 =$ about $\frac{7}{10} N$ respectively [how does one get log 2 in the second estimate?]. This means: whereas it takes somewhat more drawings than half the number of coupons to collect half of them, it takes "infinitely" more drawings to collect all of them. The last few items are the hardest to get even if the game is not rigged.

A terrifying though not so unrealistic application is to the effects of aerial bombing in warfare. The results of the strikes are pretty much randomized under certain circumstances such as camouflage, decoy, foul weather and intense enemy fire. Suppose there are 100 targets to be destroyed but each strike hits one of them at random, perhaps repeatedly. Then it takes "on the average" about 100 log 100 or about 460 strikes to hit all targets at least once. Thus if the enemy has a large number of retaliatory launching sites, it will be very hard to knock them all out without accurate military intelligence. The conclusion should serve as a mathematical deterrent to the preemptive strike theory.

6.2. The density case

To return to saner matters, we will extend Theorem 1 to random variables in an arbitrary sample space. Actually the result is true in general, provided the mathematical expectation is properly defined. An inkling of this may be given by writing it as an abstract integral as follows:

(6.2.1) $$E(X) = \int_\Omega X(\omega)\, d\omega,$$

where "$d\omega$" denotes the probability of the "element at ω," as is commonly done for an area or volume element in multi-dimensional calculus—the so-called "differential." In this form (6.1.4) becomes

$$\int_\Omega (X(\omega) + Y(\omega))\,d\omega = \int_\Omega X(\omega)\,d\omega + \int_\Omega Y(\omega)\,d\omega, \tag{6.2.2}$$

which is in complete analogy with the familiar formula in calculus:

$$\int_I (f(x) + g(x))\,dx = \int_I f(x)\,dx + \int_I g(x)\,dx,$$

where I is an interval, say $[0, 1]$. Do you remember anything of the proof of the last equation? It is established by going back to the definition of [Riemann] integrals through approximation by [Riemann] sums. For the probabilistic integral in (6.2.1) a similar procedure is followed. It is defined to be the limit of mathematical expectations of approximate discrete random variables [alluded to in §4.5]. These latter expectations are given by (6.1.1) and Theorem 1 is applicable to them. The general result (6.2.2) then follows by passing to the limit.

We cannot spell out the details of this proper approach in this text because it requires some measure theory, but there is a somewhat sneaky way to get Theorem 1 in the case when (X, Y) has a joint density as discussed in §4.6. Using the notation there, in particular (4.6.7), we have

$$E(X) = \int_{-\infty}^{\infty} uf(u, *)\,du, \quad E(Y) = \int_{-\infty}^{\infty} vf(*, v)\,dv. \tag{6.2.3}$$

On the other hand, if we substitute $\varphi(x, y) = x + y$ in (4.6.8), we have

$$E(X + Y) = \int_{-\infty}^{\infty}\int_{-\infty}^{\infty} (u + v)f(u, v)\,du\,dv. \tag{6.2.4}$$

Now this double integral can be split and evaluated by iterated integration:

$$\begin{aligned}\int_{-\infty}^{\infty} u\,du \left[\int_{-\infty}^{\infty} f(u, v)\,dv\right] &+ \int_{-\infty}^{\infty} v\,dv \left[\int_{-\infty}^{\infty} f(u, v)\,du\right] \\ &= \int_{-\infty}^{\infty} uf(u, *)\,du + \int_{-\infty}^{\infty} vf(*, v)\,dv.\end{aligned}$$

Comparison with (6.2.3) establishes (6.1.4).

The key to this method is the formula (6.2.4) whose proof was not given. The usual demonstration runs like this. "Now look here: if X takes the value u and Y takes the value v, then $X + Y$ takes the value $u + v$, and the probability that $X = u$ and $Y = v$ is $f(u, v)\,du\,dv$. See?" This kind of talk must be qualified as hand-waving or brow-beating. But it is a fact that applied scientists find such "demonstrations" quite convincing and one should go along until a second look becomes necessary, if ever. For the present the reader is advised to work out Exercise 40 below, which is the discrete analogue of the density argument above and is perfectly rigorous. These methods will be used again in the next section.

Example 3. Recall Example 5 in §4.2. The S_n's being the successive times when the claims arrive, let us put

$$S_1 = T_1, S_2 - S_1 = T_2, \ldots, S_n - S_{n-1} = T_n, \ldots.$$

Thus the T_n's are the *inter-arrival times.* They are significant not only for our example of insurance claims, but also for various other models such as the "idle-periods" for sporadically operated machinery, or "gaps" in a traffic pattern when the T's measure distance instead of time. In many applications it is these T's that are subject to statistical analysis. In the simplest case we may assume them to be exponentially distributed as in Example 12 of §4.5. If the density is $\lambda e^{-\lambda t}$ for all T_n, then $E(T_n) = 1/\lambda$. Since

$$S_n = T_1 + \cdots + T_n$$

we have by Theorem 1 in the density case:

$$E(S_n) = E(T_1) + \cdots + E(T_n) = \frac{n}{\lambda}.$$

Observe that there is no assumption about the independence of the T's, so that mutual influence between them is allowed. For example, several accidents may be due to the same cause such as a 20-car smash-up on the freeway. Furthermore, the T's may have different λ's due e.g., to diurnal or seasonal changes. If $E(T_n) = 1/\lambda_n$ then

$$E(S_n) = \frac{1}{\lambda_1} + \cdots + \frac{1}{\lambda_n}.$$

We conclude this section by solving a problem left over from §1.4 and Exercise 20 in Chapter 1; cf., also Problem 6 in §3.4.

Poincaré's formula. *For arbitrary events $A_1, \ldots, A_n$ we have*

$$(6.2.5)\qquad P\left(\bigcup_{j=1}^{n} A_j\right) = \sum_j P(A_j) - \sum_{j,k} P(A_jA_k) + \sum_{j,k,l} P(A_jA_kA_l) - + \cdots + (-1)^{n-1}P(A_1 \cdots A_n),$$

where the indices in each sum are distinct and range from 1 to n.

Proof: Let $\alpha_j = I_{A_j}$ be the indicator of A_j. Then the indicator of $A_1^c \cdots A_n^c$ is $\prod_{j=1}^{n} (1 - \alpha_j)$, hence that of its complement is given by

$$I_{A_1 \cup \cdots \cup A_n} = 1 - \prod_{j=1}^{n} (1 - \alpha_j) = \sum_j \alpha_j - \sum_{j,k} \alpha_j \alpha_k$$
$$+ \sum_{j,k,l} \alpha_j \alpha_k \alpha_l - + \cdots + (-1)^{n-1} \alpha_1 \cdots \alpha_n.$$

Now the expectation of an indicator random variable is just the probability of the corresponding event:

$$E(I_A) = P(A).$$

If we take the expectation of every term in the expansion above, and use (6.1.7) on the sums and differences, we obtain (6.2.5). [Henri Poincaré [1854–1912] was called the last universalist of mathematicians; his contributions to probability theory is largely philosophical and pedagogical. The formula above is a version of the "inclusion-exclusion principle" attributed to Sylvester.]

Example 4 [Matching problem or *problem of rencontre*]. Two sets of cards both numbered 1 to n are randomly matched. What is the probability of at least one match?

Solution. Let A_j be the event that the jth cards are matched, regardless of the others. There are $n!$ permutations of the second set against the first set, which may be considered as laid out in natural order. If the jth cards match, that leaves $(n - 1)!$ permutations for the remaining cards, hence

$$P(A_j) = \frac{(n-1)!}{n!} = \frac{1}{n}. \tag{6.2.6}$$

Similarly if the jth and kth cards are both matched, where $j \neq k$, that leaves $(n - 2)!$ permutations for the remaining cards, hence

$$P(A_j A_k) = \frac{(n-2)!}{n!} = \frac{1}{n(n-1)}; \tag{6.2.7}$$

next if j, k, l are all distinct, then

$$P(A_j A_k A_l) = \frac{(n-3)!}{n!} = \frac{1}{n(n-1)(n-2)},$$

and so on. Now there are $\binom{n}{1}$ terms in the first sum on the right side of (6.2.5), $\binom{n}{2}$ terms in the second, $\binom{n}{3}$ in the third, etc. Hence altogether the right side is equal to

$$\binom{n}{1}\frac{1}{n} - \binom{n}{2}\frac{1}{n(n-1)} + \binom{n}{3}\frac{1}{n(n-1)(n-2)} - + \cdots + (-1)^{n-1}\frac{1}{n!}$$
$$= 1 - \frac{1}{2!} + \frac{1}{3!} - + \cdots + (-1)^{n-1}\frac{1}{n!}.$$

Everybody knows (?) that

$$1 - e^{-1} = 1 - \frac{1}{2!} + \frac{1}{3!} - + \cdots + (-1)^{n-1}\frac{1}{n!} + \cdots = \sum_{n=1}^{\infty} \frac{(-1)^{n-1}}{n!}.$$

This series converges very rapidly, in fact it is easy to see that

$$\left|1 - e^{-1} - \left(1 - \frac{1}{2!} + \frac{1}{3!} - + \cdots + (-1)^{n-1}\frac{1}{n!}\right)\right| \le \frac{1}{(n+1)!}.$$

Hence as soon as $n \ge 4$, $\dfrac{1}{(n+1)!} \le \dfrac{1}{5!} = \dfrac{1}{120}$, and the probability of at least one match differs from $1 - e^{-1} \approx .63$ by less than .01. In other words, the probability of no match is about .63 for all $n \ge 4$.

What about the expected number of matches? The random number of matches is given by

$$N = I_{A_1} + \cdots + I_{A_n}. \tag{6.2.8}$$

[Why? Think this one through *thoroughly* and remember that the A's are neither disjoint nor independent.] Hence its expectation is, by another application of Theorem 1:

$$E(N) = \sum_{j=1}^{n} E(I_{A_j}) = \sum_{j=1}^{n} P(A_j) = n \cdot \frac{1}{n} = 1,$$

namely exactly 1 for all n. This is neat, but must be considered as a numerical accident.

6.3. Multiplication theorem; variance and covariance

We have indicated that Theorem 1 is really a general result in integration theory which is widely used in many branches of mathematics. In contrast, the next result requires stochastic independence and is special to probability theory.

Theorem 2. *If X and Y are independent summable random variables, then*

$$E(XY) = E(X)E(Y). \tag{6.3.1}$$

Proof: We will prove this first when Ω is countable. Then both X and Y have a countable range. Let $\{x_j\}$ denote all the *distinct* values taken by X, similarly $\{y_k\}$ for Y. Next, let

$$A_{jk} = \{\omega \mid X(\omega) = x_j,\ Y(\omega) = y_k\},$$

namely A_{jk} is the sample subset for which $X = x_j$ and $Y = y_k$. Then the sets A_{jk}, as (j, k) range over all pairs of indices are disjoint [why?] and their union is the whole space:

$$\Omega = \sum_j \sum_k A_{jk}.$$

The random variable XY takes the value $x_j y_k$ on the set A_{jk}, but some of these values may be equal, e.g., for $x_j = 2$, $y_k = 3$ and $x_j = 3$, $y_k = 2$. Nevertheless we get the expectation of XY by multiplying the probability of each A_{jk} with its value on the set, as follows:

$$E(XY) = \sum_j \sum_k x_j y_k P(A_{jk}). \tag{6.3.2}$$

This is a case of (4.3.15) and amounts merely to a grouping of terms in the defining series $\sum_\omega X(\omega)Y(\omega)P(\omega)$. Now by the assumption of independence,

$$P(A_{jk}) = P(X = x_j)P(Y = y_k).$$

Substituting this into (6.3.2), we see that the double sum splits into simple sums as follows:

$$\begin{aligned}\sum_j \sum_k & x_j y_k P(X = x_j)P(Y = y_k) \\ &= \Big\{\sum_j x_j P(X = x_j)\Big\}\Big\{\sum_k y_k P(Y = y_k)\Big\} = E(X)E(Y).\end{aligned}$$

Here again the reassembling is justified by absolute convergence of the double series in (6.3.2).

Next, we prove the theorem when (X, Y) has a joint density function f, by a method similar to that used in §6.2. Analogously to (6.2.4), we have

$$E(XY) = \int_{-\infty}^{\infty} \int_{-\infty}^{\infty} uvf(u, v)\,du\,dv.$$

Since we have by (5.5.9), using the notation of §4.6:

$$f(u, v) = f(u, *)f(*, v),$$

the double integral can be split as follows:

$$\int_{-\infty}^{\infty} uf(u, *)\,du \int_{-\infty}^{\infty} vf(*, v)\,dv = E(X)E(Y).$$

Strictly speaking, we should have applied the calculations first to $|X|$ and $|Y|$. These are also independent by Proposition 6 of §5.5, and we get

$$E(|XY|) = E(|X|)E(|Y|) < \infty.$$

Hence XY is summable and the manipulations above on the double series and double integral are valid. [These fussy details often distract from the main argument but are a necessary price to pay for mathematical rigor. The instructor as well as the reader is free to overlook some of these at his own discretion.]

The extension to any finite number of independent summable random variables is immediate:

$$E(X_1 \cdots X_n) = E(X_1) \cdots E(X_n). \tag{6.3.3}$$

This can be done directly or by induction. In the latter case we need that X_1X_2 is independent of X_3, etc. This is true and was mentioned in §5.5—another fussy detail.

In the particular case of Theorem 2 where each X_j is the indicator of an event A_j, (6.3.3) reduces to

$$P(A_1 \cdots A_n) = P(A_1) \cdots P(A_n).$$

This makes it crystal clear that Theorem 2 cannot hold without restriction on the dependence. Contrast this with the corresponding case of (6.1.7):

$$E(I_{A_1} + \cdots + I_{A_n}) = P(A_1) + \cdots + P(A_n).$$

Here there is no condition whatever on the events such as their being disjoint, and the left member is to be emphatically distinguished from $P(A_1 \cup \cdots \cup A_n)$ or any other probability. This is the kind of confusion which has pestered pioneers as well as beginners. It is known as *Cardano's paradox*. [Cardano (1501–1576) wrote the earliest book on games of chance.]

Example 5. Iron bars in the shape of slim cylinders are test-measured. Suppose the average length is 10 inches and average area of ends is 1 square inch. The average error made in the measurement of the length is .005 inch, that in the measurement of the area is .01 square inch. What is the average error made in estimating their weights?

Since weight is a constant times volume it is sufficient to consider the latter: $V = LA$ where L = length, A = area of ends. Let the errors be ΔL and ΔA respectively; then the error in V is given by

$$\Delta V = (L + \Delta L)(A + \Delta A) - LA = L\Delta A + A\Delta L + \Delta L \Delta A.$$

Assuming independence between the measurements, we have

$$E(\Delta V) = E(L)E(\Delta A) + E(A)E(\Delta L) + E(\Delta A)E(\Delta L)$$
$$= 10 \cdot \frac{1}{100} + 1 \cdot \frac{1}{200} + \frac{1}{100} \cdot \frac{1}{200}$$

$= .105$ cubic inch if the last term is ignored.

Definition of Moment. For positive integer r, the mathematical expectation $E(X^r)$ is called the rth *moment* [*moment of order* r] of X. Thus if X^r has a finite expectation, we say that X has a finite rth moment. For $r = 1$ of course the first moment is just the expectation or mean.

The case $r = 2$ is of special importance. Since $X^2 \geq 0$, we shall call $E(X^2)$ the second moment of X whether it is finite or equal to $+\infty$ according as the defining series [in a countable Ω] $\sum_\omega X^2(\omega)P(\omega)$ converges or diverges.

When the mean $E(X)$ is finite, it is often convenient to consider

$$X^0 = X - E(X) \tag{6.3.4}$$

instead of X because its first moment is equal to zero. We shall say X^0 is obtained from X by *centering*.

Definition of Variance and Standard Deviation. The second moment of X^0 is called the *variance* of X and denoted by $\sigma^2(X)$; its positive square root $\sigma(X)$ is called the *standard deviation* of X.

There is an important relation between $E(X)$, $E(X^2)$ and $\sigma^2(X)$, as follows.

Theorem 3. *If $E(X^2)$ is finite, then so is $E(|X|)$. We have then*

$$\sigma^2(X) = E(X^2) - E(X)^2; \tag{6.3.5}$$

consequently

$$E(|X|)^2 \leq E(X^2). \tag{6.3.6}$$

Proof: Since

$$X^2 - 2|X| + 1 = (|X| - 1)^2 \geq 0,$$

we must have (why?) $E(X^2 - 2|X| + 1) \geq 0$, and therefore $E(X^2) + 1 \geq 2E(|X|)$ by Theorem 1 and (6.1.5). This proves the first assertion of the theorem. Next we have

$$\sigma^2(X) = E\{(X - E(X))^2\} = E\{X^2 - 2E(X)X + E(X)^2\}$$
$$= E(X^2) - 2E(X)E(X) + E(X)^2 = E(X^2) - E(X)^2$$

Since $\sigma^2(X) \geq 0$ from the first equation above, we obtain (6.3.6) by substituting $|X|$ for X in (6.3.5).

What is the meaning of $\sigma(X)$? To begin with, X^0 is the deviation of X from its mean, and can take both positive and negative values. If we are only interested in its magnitude then the *mean absolute deviation* is $E(|X^0|) = E(|X - E(X)|)$. This can actually be used but it is difficult for calculations.

So we consider instead the *mean square deviation* $E(|X^0|^2)$ which is the variance. But then we should cancel out the squaring by extracting the root afterward, which gives us the standard deviation $+\sqrt{E(|X^0|^2)}$. This then is a gauge of the average deviation of a random variable [*sample value*] from its mean. The smaller it is, the better the random values cluster around its average and the population is well centered or concentrated. The true significance will be seen later when we discuss the convergence to a normal distribution in Chapter 7.

Observe that X and $X + c$ have the same variance for any constant c; in particular, this is the case for X and X^0. The next result resembles Theorem 1, but only in appearance.

Theorem 4. *If X and Y are independent and both have finite variances, then*

$$\sigma^2(X + Y) = \sigma^2(X) + \sigma^2(Y). \tag{6.3.7}$$

Proof: By the preceding remark, we may suppose that X and Y both have mean zero. Then $X + Y$ also has mean zero and the variances in (6.3.7) are the same as second moments. Now

$$E(XY) = E(X)E(Y) = 0$$

by Theorem 2, and

$$\begin{aligned} E\{(X + Y)^2\} &= E\{X^2 + 2XY + Y^2\} \\ &= E(X^2) + 2E(XY) + E(Y^2) = E(X^2) + E(Y^2) \end{aligned}$$

by Theorem 1, and this is the desired result.

The extension of Theorem 4 to any finite number of independent random variables is immediate. However, there is a general formula for the second moment without the assumption of independence, which is often useful. We begin with the algebraic identity:

$$(X_1 + \cdots + X_n)^2 = \sum_{j=1}^{n} X_j^2 + 2 \sum_{1 \le j < k \le n} X_j X_k.$$

Taking expectations of both sides and using Theorem 1, we obtain

$$E\{(X_1 + \cdots + X_n)^2\} = \sum_{j=1}^{n} E(X_j^2) + 2 \sum_{1 \le j < k \le n} E(X_j X_k). \tag{6.3.8}$$

When the X's are centered and assumed to be independent, then all the mixed terms in the second sum above vanish and the result is the extension of Theorem 4 already mentioned.

Let us introduce two *real indeterminants* [dummy variables] ξ and η and consider the identity:

$$E\{(X\xi + Y\eta)^2\} = E(X^2)\xi^2 + 2E(XY)\xi\eta + E(Y^2)\eta^2.$$

The right member is a quadratic form in (ξ, η) and it is never negative because the left member is the expectation of a random variable which does not take negative values. A well-known result in college algebra says that the coefficients of such a quadratic form $a\xi^2 + 2b\xi\eta + c\eta^2$ must satisfy the inequality $b^2 \leq ac$. Hence in the present case

$$E(XY)^2 \leq E(X^2)E(Y^2). \tag{6.3.9}$$

This is called the *Cauchy-Schwarz inequality*.

If X and Y both have finite variances, then the quantity

$$\begin{aligned} E(X^0Y^0) &= E\{(X - E(X))(Y - E(Y))\} \\ &= E\{XY - XE(Y) - YE(X) + E(X)E(Y)\} \\ &= E(XY) - E(X)E(Y) - E(Y)E(X) + E(X)E(Y) \\ &= E(XY) - E(X)E(Y) \end{aligned}$$

is called the *covariance* of X and Y and denoted by $\operatorname{Cov}(X, Y)$; the quantity

$$\rho(X, Y) = \frac{\operatorname{Cov}(X, Y)}{\sigma(X)\sigma(Y)}$$

is called the *coefficient of correlation* between X and Y, provided of course the denominator does not vanish. If it is equal to zero, then X and Y are said to be *uncorrelated*. This is implied by independence but is in general a weaker property. As a consequence of (6.3.9), we have always $-1 \leq \rho(X, Y) \leq 1$. The sign as well as the absolute value of ρ gives a sort of gauge of the mutual dependence between the random variables.† See also Exercise 30 of Chapter 7.

Example 6. The most classical application of the preceding results is to the case of Bernoullian random variables (see Example 9 of §4.4). These are independent with the same probability distribution as follows:

$$X = \begin{cases} 1 & \text{with probability } p; \\ 0 & \text{with probability } q = 1 - p. \end{cases} \tag{6.3.10}$$

We have encountered them in coin-tossing (Example 8 of §2.4), but the scheme can be used in any repeated trials in which there are only two outcomes: success ($X = 1$) and failure ($X = 0$). For instance, Example 1 in §6.1 is the case where $p = 1/100$ and the monetary unit is "ten grand." The chances of either "cure" or "death" in a major surgical operation is another illustration.

† The mathematician Emil Artin told me the following story in 1947. "Everybody knows that probability and statistics are the same thing, and statistics is nothing but correlation. Now the correlation is just the cosine of an angle. Thus, all is trivial."

The mean and variance of X are easy to compute:

$$E(X) = p, \quad \sigma^2(X) = p - p^2 = pq.$$

Let $\{X_n, n \geq 1\}$ denote Bernoullian random variables and write

$$S_n = X_1 + \cdots + X_n \tag{6.3.11}$$

for the nth partial sum. It represents the total number of successes in n trials. By Theorem 1,

$$E(S_n) = E(X_1) + \cdots + E(X_n) = np. \tag{6.3.12}$$

This would be true even without independence. Next by Theorem 3,

$$\sigma^2(S_n) = \sigma^2(X_1) + \cdots + \sigma^2(X_n) = npq. \tag{6.3.13}$$

The ease with which these results are obtained shows a great technical advance. Recall that (6.3.12) has been established in (4.4.16), via the binomial distribution of S_n and a tricky computation. A similar method is available for (6.3.13) and the reader is strongly advised to carry it out for practice and comparison. But how much simpler is our new approach, going from the mean and variance of the individual summands to those of the sum without the intervention of probability distributions. In more complicated cases the latter will be very hard if not impossible to get. That explains why we are devoting several sections to the discussion of mean and variance which often suffice for theoretical as well as practical purposes.

Example 7. Returning to the matching problem in §6.2, let us now compute the standard deviation of the number of matches. The I_{A_j}'s in (6.2.8) are not independent, but formula (6.3.8) is applicable and yields

$$E(N^2) = \sum_{j=1}^{n} E(I_{A_j}^2) + 2 \sum_{1 \leq j < k \leq n} E(I_{A_j} I_{A_k}).$$

Clearly,

$$E(I_{A_j}^2) = P(A_j) = \frac{1}{n}$$

$$E(I_{A_j} I_{A_k}) = P(A_j A_k) = \frac{1}{n(n-1)}$$

by (6.2.6) and (6.2.7). Substituting into the above, we obtain

$$E(N^2) = n \cdot \frac{1}{n} + 2 \binom{n}{2} \frac{1}{n(n-1)} = 1 + 1 = 2.$$

Hence

$$\sigma^2(N) = E(N^2) - E(N)^2 = 2 - 1 = 1.$$

Rarely an interesting general problem produces such simple numerical answers.

6.4. Multinomial distribution

A good illustration of the various notions and techniques developed in the preceding sections is the *multinomial distribution*. This is a natural generalization of the binomial distribution and serves as a model for repeated trials in which there are a number of possible outcomes instead of just "success or failure." We begin with the algebraic formula called the *multinomial theorem*:

$$(6.4.1) \qquad (x_1 + \cdots + x_r)^n = \sum \frac{n!}{n_1! \cdots n_r!} x_1^{n_1} \cdots x_r^{n_r}$$

where the sum ranges over all ordered r-tuples of integers $n_1, \ldots, n_r$ satisfying the following conditions:

$$(6.4.2) \qquad n_1 \geq 0, \ldots, n_r \geq 0, \quad n_1 + \cdots + n_r = n.$$

When $r = 2$ this reduces to the binomial theorem. For then there are $n + 1$ ordered couples

$$(0, n), (1, n - 1), \ldots, (k, n - k), \ldots, (n, 0)$$

with the corresponding coefficients

$$\frac{n!}{0!n!}, \frac{n!}{1!(n-1)!}, \ldots, \frac{n!}{k!(n-k)!}, \ldots, \frac{n!}{n!0!}$$

i.e.,

$$\binom{n}{0}, \binom{n}{1}, \ldots, \binom{n}{k}, \ldots, \binom{n}{n}.$$

Hence the sum can be written explicitly as

$$\binom{n}{0} x^0 y^n + \binom{n}{1} x^1 y^{n-1} + \cdots + \binom{n}{k} x^k y^{n-k} + \cdots + \binom{n}{n} x^n y^0$$
$$= \sum_{k=0}^{n} \binom{n}{k} x^k y^{n-k}.$$

In the general case the n identical factors $(x_1 + \cdots + x_r)$ on the left side of (6.4.1) are multiplied out and the terms collected on the right. Each term is of the form $x_1^{n_1} \cdots x_r^{n_r}$ with the exponents n_j satisfying (6.4.2). Such a term appears $n!/(n_1! \cdots n_r!)$ times because this is the number of ways of permuting n objects (the n factors) which belong to r different varieties (the x's), such that n_j of them belong to the jth variety. [You see some combinatorics are in the nature of things and cannot be avoided even if you have skipped most of Chapter 3.]

A concrete model for the multinomial distribution may be described as follows. An urn contains balls of r different colors in the proportions:

$$p_1 : \cdots : p_r, \text{ where } p_1 + \cdots + p_r = 1.$$

We draw n balls one after another with replacement. Assume independence of the successive drawings, which is simulated by a thorough shake-up of the urn after each drawing. What is the probability that so many of the balls drawn are of each color?

Let $X_1, \ldots, X_n$ be independent random variables all having the same distribution as the X below:

$$X = \begin{cases} 1 & \text{with probability } p_1, \\ 2 & \text{with probability } p_2, \\ \vdots & \\ r & \text{with probability } p_r. \end{cases} \tag{6.4.3}$$

What is the joint distribution of $(X_1, \ldots, X_n)$, namely

$$P(X_1 = x_1, \ldots, X_n = x_n) \tag{6.4.4}$$

for all possible choices of $x_1, \ldots, x_n$ from 1 to r? Here the numerical values correspond to labels for the colors. Such a quantification is not necessary but sometimes convenient. It also leads to questions which are not intended for the color scheme, such that the probability "$X_1 + \cdots + X_n = m$." But suppose we change the balls to lottery tickets bearing different monetary prizes, or to people having various ages or incomes, then the numerical formulation in (6.4.3) is pertinent. What about negative or fractional values for the X's? This can be accommodated by a linear transformation (cf. Example 14 in §4.5) provided all possible values are commensurable, say ordinary terminating decimals. For example if the values are in 3 decimal places and range from -10 up, then we can use

$$X' = 10000 + 1000X$$

instead of X. The value -9.995 becomes $10000 - 9995 = 5$ in the new scale. In a super-pragmatic sense, we might even argue that the multinomial distribution is all we need for sampling in independent trials. For in reality we shall never be able to distinguish between (say) $10^{10^{10}}$ different varieties of anything. But this kind of finitist attitude would destroy a lot of mathematics.

Let us evaluate (6.4.4). It is equal to $P(X_1 = x_1) \cdots P(X_n = x_n)$ by independence, and each factor is one of the p's in (6.4.3). To get an explicit expression we must know how many of the x_j's are 1 or 2 or $\cdots$ or r? Suppose

n_1 of them are equal to 1, n_2 of them equal to 2, . . . , n_r of them equal to r. Then these n_j's satisfy (6.4.2) and the probability in question is equal to $p_1^{n_1} \cdots p_r^{n_r}$. It is convenient to introduce new random variables N_j, $1 \le j \le r$, as follows:

$$N_j = \text{number of } X\text{'s among } (X_1, \ldots, X_n) \text{ that take the value } j.$$

Each N_j takes a value from 0 to n, but the random variables $N_1, \ldots, N_r$ cannot be independent since they are subject to the obvious restriction:

$$N_1 + \cdots + N_r = n. \tag{6.4.5}$$

However, their joint distribution can be written down:

$$P(N_1 = n_1, \ldots, N_r = n_r) = \frac{n!}{n_1! \cdots n_r!} p_1^{n_1} \cdots p_r^{n_r}. \tag{6.4.6}$$

The argument here is exactly the same as that given at the beginning of this section for (6.4.1), but we will repeat it. For any *particular*, or completely specified, sequence $(X_1, \ldots, X_n)$ satisfying the conditions $N_1 = n_1, \ldots, N_r = n_r$, we have just shown that the probability is given by $p_1^{n_1} \cdots p_r^{n_r}$. But there are $n!/(n_1! \cdots n_r!)$ different particular sequences satisfying the same conditions, obtained by permuting the n factors of which n_1 factors are p_1, n_2 factors are p_2, etc. To nail this down in a simple numerical example, let $n = 4$, $r = 3$, $n_1 = 2$, $n_2 = n_3 = 1$. This means in 4 drawings there are 2 balls of color 1, and 1 ball each of color 2 and 3. All the possible particular sequences are listed below:

(1, 1, 2, 3)	(1, 1, 3, 2)
(1, 2, 1, 3)	(1, 3, 1, 2)
(1, 2, 3, 1)	(1, 3, 2, 1)
(2, 1, 1, 3)	(3, 1, 1, 2)
(2, 1, 3, 1)	(3, 1, 2, 1)
(2, 3, 1, 1)	(3, 2, 1, 1)

Their number is $12 = \dfrac{4!}{2!1!1!}$ and the associated probability is $12p_1^2p_2p_3$.

Formula (6.4.6), in which the indices n_j range over all possible integer values subject to (6.4.2), is called the *multinomial distribution* for the random variables $(N_1, \ldots, N_r)$. Specifically, it may be denoted by $M(n; r; p_1, \ldots, p_{r-1}, p_r)$ subject to $p_1 + \cdots + p_r = 1$. For the binomial case $r = 2$, this is often written as $B(n; p)$, see Example 9 of §4.4.

If we divide (6.4.1) through by its left member, and put

$$p_j = \frac{x_j}{x_1 + \cdots + x_r}, \quad 1 \le j \le r,$$

we obtain

$$1 = \sum \frac{n!}{n_1! \cdots n_r!} p_1^{n_1} \cdots p_r^{n_r}. \tag{6.4.7}$$

This yields a check that the sum of all the terms of the multinomial distribution is indeed equal to 1. Conversely, since (6.4.7) is a consequence of its probabilistic interpretation, we can deduce (6.4.1) from it, at least when $x_j \geq 0$, by writing $(p_1 + \cdots + p_r)^n$ for the left member in (6.4.7). This is another illustration of the way probability theory can add a new meaning to an algebraic formula; cf. the last part of §3.3.

Marginal distributions (see §4.6) of $(N_1, \ldots, N_r)$ can be derived by a simple argument without computation. If we are interested in N_1 alone, then we can lump the $r - 1$ other varieties as "not 1" with probability $1 - p_1$. Thus the multinomial distribution collapses into a binomial one $B(n; p_1)$, namely:

$$P(N_1 = n_1) = \frac{n!}{n_1!(n - n_1)!} p_1^{n_1}(1 - p_1)^{n-n_1}.$$

From this we can deduce the mean and variance of N_1 as in Example 6 of §6.3. In general,

$$E(N_j) = np_j, \quad \sigma^2(N_j) = np_j(1 - p_j), \quad 1 \leq j \leq r. \tag{6.4.8}$$

Next, if we are interested in the pair (N_1, N_2), a similar lumping yields $M(n; 3; p_1, p_2, p_3)$, namely:

$$\begin{aligned} P(N_1 = n_1, N_2 = n_2) \\ &= \frac{n!}{n_1!n_2!(n - n_1 - n_2)!} p_1^{n_1} p_2^{n_2}(1 - p_1 - p_2)^{n-n_1-n_2}. \end{aligned} \tag{6.4.9}$$

We can now express $E(N_1N_2)$ by using (4.3.15) or (6.3.2) [without independence]:

$$\begin{aligned} E(N_1N_2) &= \sum n_1n_2P(N_1 = n_1, N_2 = n_2) \\ &= \sum \frac{n!}{n_1!n_2!n_3!} n_1n_2p_1^{n_1}p_2^{n_2}p_3^{n_3} \end{aligned}$$

where $n_3 = n - n_1 - n_2$, $p_3 = 1 - p_1 - p_2$ and the sum ranges as in (6.4.2) with $r = 3$. The multiple sum above can be evaluated by generalizing the device used in Example 9 of §4.4 for the binomial case. Take the indicated second partial derivative below:

$$\begin{aligned} \frac{\partial^2}{\partial x_1 \partial x_2}(x_1 + x_2 + x_3)^n &= n(n - 1)(x_1 + x_2 + x_3)^{n-2} \\ &= \sum \frac{n!}{n_1!n_2!n_3!} n_1n_2x_1^{n_1-1}x_2^{n_2-1}x_3^{n_3}. \end{aligned}$$

Multiply through by x_1x_2 and then put $x_1 = p_1$, $x_2 = p_2$, $x_3 = p_3$. The result is $n(n-1)p_1p_2$ on one side and the desired multiple sum on the other. Hence we have in general for $j \neq k$:

$$E(N_jN_k) = n(n-1)p_jp_k;$$

$$\text{(6.4.10)} \qquad \begin{aligned} \operatorname{Cov}(N_j, N_k) &= E(N_jN_k) - E(N_j)E(N_k) \\ &= n(n-1)p_jp_k - (np_j)(np_k) = -np_jp_k. \end{aligned}$$

It is fun to check out the formula (6.3.8), recalling (6.4.5):

$$\begin{aligned} n^2 &= E\{(N_1 + \cdots + N_r)^2\} \\ &= \sum_{j=1}^{r} \{n(n-1)p_j^2 + np_j\} + 2 \sum_{1 \le j < k \le r} n(n-1)p_jp_k \\ &= n(n-1)\left(\sum_{j=1}^{r} p_j\right)^2 + n \sum_{j=1}^{r} p_j = n(n-1) + n = n^2. \end{aligned}$$

There is another method of calculating $E(N_jN_k)$, similar to the first method in Example 6, §6.3. Let j be fixed and

$$\xi(x) = \begin{cases} 1 & \text{if } x = j, \\ 0 & \text{if } x \neq j. \end{cases}$$

As a function ξ of the real variable x, it is just the indicator of the singleton $\{j\}$. Now introduce the random variable

$$\text{(6.4.11)} \qquad \xi_\nu = \xi(X_\nu) = \begin{cases} 1 & \text{if } X_\nu = j, \\ 0 & \text{if } X_\nu \neq j. \end{cases}$$

namely, ξ_ν is the indicator for the event $\{X_\nu = j\}$. In other words the ξ_ν's count just those X_ν's taking the value j, so that $N_j = \xi_1 + \cdots + \xi_n$. Now we have

$$\begin{aligned} E(\xi_\nu) &= P(X_\nu = j) = p_j; \\ \sigma^2(\xi_\nu) &= E(\xi_\nu^2) - E(\xi_\nu)^2 = p_j - p_j^2 = p_j(1 - p_j). \end{aligned}$$

Finally, the random variables $\xi_1, \ldots, \xi_n$ are independent since $X_1, \ldots, X_n$ are, by Proposition 6 of §5.5. Hence by Theorems 1 and 4:

$$\text{(6.4.12)} \qquad \begin{aligned} E(N_j) &= E(\xi_1) + \cdots + E(\xi_n) = np_j, \\ \sigma^2(N_j) &= \sigma^2(\xi_1) + \cdots + \sigma^2(\xi_n) = np_j(1 - p_j). \end{aligned}$$

Next, let $k \neq j$ and define η and η_ν in the same way as ξ and ξ_ν are defined, but with k in place of j. Consider now for $1 \le \nu \le n$, $1 \le \nu' \le n$:

$$E(\xi_\nu \eta_{\nu'}) = P(X_\nu = j, X_{\nu'} = k) = \begin{cases} p_j p_k & \text{if } \nu \neq \nu', \\ 0 & \text{if } \nu = \nu'. \end{cases} \tag{6.4.13}$$

Finally we calculate

$$\begin{aligned} E(N_j N_k) &= E\left\{\left(\sum_{\nu=1}^{n} \xi_\nu\right)\left(\sum_{\nu'=1}^{n} \eta_{\nu'}\right)\right\} = E\left\{\sum_{\nu=1}^{n} \xi_\nu \eta_\nu + \sum_{1 \le \nu \neq \nu' \le n} \xi_\nu \eta_{\nu'}\right\} \\ &= \sum_{\nu=1}^{n} E(\xi_\nu \eta_\nu) + \sum_{1 \le \nu \neq \nu' \le n} E(\xi_\nu \eta_{\nu'}) \\ &= n(n-1) p_j p_k; \end{aligned}$$

by (6.4.13) because there are $(n)_2 = n(n-1)$ terms in the last written sum. This is of course the same result as in (6.4.10).

We conclude with a simple numerical illustration of a general problem mentioned above.

Example 8. Three identical dice are thrown. What is the probability of obtaining a total of 9? The dice are not supposed to be symmetrical and the probability of turning up face j is equal to p_j, $1 \le j \le 6$; same for all three dice.

Let us list the possible cases in terms of the X's and the N's respectively:

X_1	X_2	X_3	N_1	N_2	N_3	N_4	N_5	N_6	permutation number	probability
1	2	6	1	1				1	6	$6p_1p_3p_6$
1	3	5	1		1		1		6	$6p_1p_3p_5$
1	4	4	1			2			3	$3p_1p_4^2$
2	2	5		2			1		3	$3p_2^2p_5$
2	3	4		1	1	1			6	$6p_2p_3p_4$
3	3	3			3				1	p_3^3

Hence

$$P(X_1 + X_2 + X_3 = 9) = 6(p_1p_2p_6 + p_1p_3p_5 + p_2p_3p_4) + 3(p_1p_4^2 + p_2^2p_5) + p_3^3.$$

If all the p's are equal to $1/6$, then this is equal to

$$\frac{6 + 6 + 3 + 3 + 6 + 1}{6^3} = \frac{25}{216}.$$

The numerator 25 is equal to the sum

$$\sum \frac{3!}{n_1! \cdots n_6!}$$

where the n_j's satisfy the conditions

$$n_1 + n_2 + n_3 + n_4 + n_5 + n_6 = 3,$$
$$n_1 + 2n_2 + 3n_3 + 4n_4 + 5n_5 + 6n_6 = 9.$$

There are 6 solutions tabulated above as possible values of the N_j's.

In the general context of the X's discussed above, the probability $P(X_1 + \cdots + X_n = m)$ is obtained by summing the right side of (6.4.6) over all $(n_1, \ldots, n_r)$ satisfying both (6.4.2) and

$$\text{l}n_1 + 2n_2 + \cdots + rn_r = m.$$

It is obvious that we need a computing machine to handle such explicit formulas. Fortunately in most problems we are interested in cruder results such that

$$P(a_n \leq X_1 + \cdots + X_n \leq b_n)$$

for large values of n. The relevant asymptotic results and limit theorems will be the subject matter of Chapter 7. One kind of machinery needed for this purpose will be developed in the next section.

6.5. Generating function and the like

A powerful mathematical device, a true gimmick, is the generating function invented by the great prolific mathematician Euler [1707–1783] to study the partition problem in number theory. Let X be a random variable taking only nonnegative integer values with the probability distribution given by

$$P(X = j) = a_j, \quad j = 0, 1, 2, \ldots. \tag{6.5.1}$$

The idea is to put all the information contained above in a compact capsule. For this purpose a dummy variable z is introduced and the following power series in z set up:

$$g(z) = a_0 + a_1 z + a_2 z^2 + \cdots = \sum_{j=0}^{\infty} a_j z^j. \tag{6.5.2}$$

This is called the *generating function* associated with the sequence of numbers $\{a_j, j \geq 0\}$. In the present case we may also call it the generating function of the random variable X with the probability distribution (6.5.1). Thus g is a function of z which will be regarded as a real variable, although in some more advanced applications it is advantageous to consider z as a complex variable. Remembering that $\sum_j a_j = 1$, it is easy to see that the power series in (6.5.2) converges for $|z| \leq 1$. In fact it is *dominated* as follows:

$$|g(z)| \leq \sum_j |a_j|\,|z|^j \leq \sum_j a_j = 1, \quad \text{for } |z| \leq 1.$$

[It is hoped that your knowledge about power series goes beyond the "ratio test." The above estimate is more direct and says a lot more.] Now a theorem in calculus asserts that we can differentiate the series term by term to get the derivatives of g, so long as we restrict its domain of validity to $|z| < 1$;

$$(6.5.3)\qquad \begin{aligned} g'(z) &= a_1 + 2a_2z + 3a_3z^2 + \cdots = \sum_{n=1}^{\infty} na_nz^{n-1}, \\ g''(z) &= 2a_2 + 6a_3z + \cdots = \sum_{n=2}^{\infty} n(n-1)a_nz^{n-2}. \end{aligned}$$

In general we have

$$(6.5.4)\quad g^{(j)}(z) = \sum_{n=j}^{\infty} n(n-1)\cdots(n-j+1)a_nz^{n-j} = \sum_{n=j}^{\infty} \binom{n}{j} j!a_nz^{n-j}.$$

If we set $z = 0$ above, all the terms vanish except the constant term:

$$(6.5.5)\qquad g^{(j)}(0) = j!a_j \quad \text{or} \quad a_j = \frac{1}{j!}g^{(j)}(0).$$

This shows that we can recover all the a_j's from g. Therefore, not only does the probability distribution determine the generating function, but also *vice versa*. So there is no loss of information in the capsule. In particular, putting $z = 1$ in g' and g'' we get by (4.3.18)

$$g'(1) = \sum_{n=0}^{\infty} na_n = E(X), \quad g''(1) = \sum_{n=0}^{\infty} n^2a_n - \sum_{n=0}^{\infty} na_n = E(X^2) - E(X);$$

provided that the series converge, in which case (6.5.3) holds for $z = 1$†. Thus

$$(6.5.6)\qquad E(X) = g'(1), \quad E(X^2) = g'(1) + g''(1).$$

In practice, the following qualitative statement, which is a corollary of the above, is often sufficient.

Theorem 5. *The probability distribution of a nonnegative integer-valued random variable is uniquely determined by its generating function.*

Let Y be a random variable having the probability distribution $\{b_k, k \geq 0\}$ where $b_k = P(Y = k)$, and let h be its generating function:

$$h(z) = \sum_{k=0}^{\infty} b_kz^k.$$

Suppose that $g(z) = h(z)$ for all $|z| < 1$, then the theorem asserts that $a_k = b_k$ for all $k \geq 0$. Consequently X and Y have the same distribution, and this is what we mean by "unique determination." The explicit formula (6.5.4) of course implies this, but there is a simpler argument as follows. Since

† This is an Abelian theorem; cf. the discussion after (8.4.17).

$$\sum_{k=0}^{\infty} a_k z^k = \sum_{k=0}^{\infty} b_k z^k, \quad |z| < 1;$$

we get at once $a_0 = b_0$ by setting $z = 0$ in the equation. After removing these terms we can cancel a factor z on both sides, and then get $a_1 = b_1$ by again setting $z = 0$. Repetition of this process establishes the theorem. You ought to realize that we have just reproduced a terrible proof of a standard result which used to be given in some calculus text books! Can you tell what is wrong there and how to make it correct?

We proceed to discuss a salient property of generating functions when they are multiplied together. Using the notation above we have

$$g(z)h(z) = \left(\sum_{j=0}^{\infty} a_j z^j\right)\left(\sum_{k=0}^{\infty} b_k z^k\right) = \sum_j \sum_k a_j b_k z^{j+k}.$$

We will rearrange the terms of this double series into a power series in the usual form. Then

(6.5.7)
$$g(z)h(z) = \sum_{l=0}^{\infty} c_l z^l$$

where

(6.5.8)
$$c_l = \sum_{j+k=l} a_j b_k = \sum_{j=0}^{l} a_j b_{l-j}.$$

The sequence $\{c_j\}$ is called the *convolution* of the two sequences $\{a_j\}$ and $\{b_j\}$. What does c_l stand for? Suppose that the random variables X and Y are independent. Then we have by (6.5.8),

$$\begin{aligned} c_l &= \sum_{j=0}^{l} P(X = j)P(Y = l - j) \\ &= \sum_{j=0}^{l} P(X = j, Y = l - j) = P(X + Y = l). \end{aligned}$$

The last equation above is obtained by the rules in §5.2, as follows. Given that $X = j$, we have $X + Y = l$ if and only if $Y = l - j$, hence by Proposition 2 of §5.2 [cf. (5.2.4)]:

$$\begin{aligned} P(X + Y = l) &= \sum_{j=0}^{\infty} P(X = j)P(X + Y = l \mid X = j) \\ &= \sum_{j=0}^{\infty} P(X = j)P(Y = l - j \mid X = j) \\ &= \sum_{j=0}^{l} P(X = j)P(Y = l - j), \end{aligned}$$

because Y is independent of X and it does not take negative values. In other words, we have shown that for all $l \geq 0$,

$$P(X + Y = l) = c_l,$$

so that $\{c_l, l \geq 0\}$ is the probability distribution of the random variable $X + Y$. Therefore, by definition its generating function is given by the power series in (6.5.7) and so equal to the product of the generating functions of X and of Y. After an easy induction, we can state the result as follows.

Theorem 6. *If the random variables $X_1, \ldots, X_n$ are independent and have $g_1, \ldots, g_n$ as their generating functions, then the generating function of the sum $X_1 + \cdots + X_n$ is given by the product $g_1 \cdots g_n$.*

This theorem is of great importance since it gives a method to study sums of independent random variables via generating functions, as we shall see in Chapter 7. In some cases the product of the generating functions takes a simple form and then we can deduce its probability distribution by looking at its power series, or equivalently by using (6.5.5). The examples below will demonstrate this method.

For future reference let us take note that given a sequence of real numbers $\{a_j\}$, we can define the power series g as in (6.5.2). This will be called the generating function associated with the sequence. If the series has a nonzero radius of convergence, then the preceding analysis can be carried over to this case without the probability interpretations. In particular, the convolution of two such sequences can be defined as in (6.5.8), and (6.5.7) is still valid. In §8.4 below we shall use generating functions whose coefficients are probabilities, such that the series may diverge for $z = 1$.

Example 9. For the Bernoullian random variables $X_1, \ldots, X_n$ (Example 6 of §6.3), the common generating function is

$$g(z) = q + pz$$

since $a_0 = q$, $a_1 = p$ in (6.5.1). Hence the generating function of $S_n = X_1 + \cdots + X_n$, where the X's are independent, is given by the nth power of g:

$$g(z)^n = (q + pz)^n.$$

Its power series is therefore known from the binomial theorem, namely,

$$g(z)^n = \sum_{k=0}^{n} \binom{n}{k} q^{n-k} p^k z^k.$$

On the other hand, by definition of a generating function, we have

$$g(z)^n = \sum_{k=0}^{\infty} P(S_n = k)z^k.$$

Comparison of the last two expressions shows that

$$P(S_n = k) = \binom{n}{k} p^k q^{n-k}, \quad 0 \leqslant k \leqslant n; \quad P(S_n = k) = 0, \quad k > n.$$

This is the Bernoulli formula we learned sometime ago, but the derivation is new and it is machine-processed.

Example 10. For the waiting time distribution (§4.4), we have $p_j = q^{j-1}p$, $j \geq 1$; hence

$$g(z) = \sum_{j=1}^{\infty} q^{j-1}pz^j = \frac{p}{q}\sum_{j=1}^{\infty} (qz)^j = \frac{p}{q}\frac{qz}{1-qz} = \frac{pz}{1-qz}. \tag{6.5.9}$$

Let $S_n = T_1 + \cdots + T_n$ where the T's are independent and each has the g in (6.5.9) as generating function. Then S_n is the waiting time for the nth success. Its generating function is given by g^n, and this can be expanded into a power series by using the binomial series and (5.4.4):

$$\begin{aligned} g(z)^n &= \left(\frac{pz}{1-qz}\right)^n = p^n z^n \sum_{j=0}^{\infty} \binom{-n}{j}(-qz)^j \\ &= \sum_{j=0}^{\infty} \frac{n(n+1)\cdots(n+j-1)}{j!} p^n q^j z^{n+j} = \sum_{j=0}^{\infty} \binom{n+j-1}{n-1} p^n q^j z^{n+j} \\ &= \sum_{k=n}^{\infty} \binom{k-1}{n-1} p^n q^{k-n} z^k. \end{aligned}$$

Hence we obtain for $j \geq 0$,

$$P(S_n = n+j) = \binom{n+j-1}{j} p^n q^j = \binom{-n}{j} p^n (-q)^j.$$

The probability distribution given by $\left\{\binom{-n}{j} p^n(-q)^j, j \geq 0\right\}$ is called the *negative binomial distribution of order n*. The discussion above shows that its generating function is given by

$$\left(\frac{g(z)}{z}\right)^n = \left(\frac{p}{1-qz}\right)^n. \tag{6.5.10}$$

Now $g(z)/z$ is the generating function of $T_1 - 1$ (why?), which represents the number of failures before the first success. Hence the generating function in (6.5.10) is that of the random variable $S_n - n$, which is the total number of failures before the nth success.

Example 11. For the dice problem at the end of §6.4, we have $p_j = 1/6$ for $1 \leq j \leq 6$ if the dice are symmetrical. Hence the associated generating function is given by

$$g(z) = \frac{1}{6}(z + z^2 + z^3 + z^4 + z^5 + z^6) = \frac{z(1 - z^6)}{6(1 - z)}.$$

The generating function of the total points obtained by throwing 3 dice is just g^3. This can be expanded into a power series as follows:

$$(6.5.11)\quad \begin{aligned} g(z)^3 &= \frac{z^3}{6^3}\frac{(1 - z^6)^3}{(1 - z)^3} = \frac{z^3}{6^3}(1 - 3z^6 + 3z^{12} - z^{18})(1 - z)^{-3} \\ &= \frac{z^3}{6^3}(1 - 3z^6 + 3z^{12} - z^{18}) \sum_{k=0}^{\infty} \binom{k+2}{2} z^k. \end{aligned}$$

The coefficient of z^9 is easily found by inspection, since there are only two ways of forming it from the product above:

$$\frac{1}{6^3}\left\{1 \cdot \binom{6+2}{2} - 3 \cdot \binom{0+2}{2}\right\} = \frac{28 - 3}{6^3} = \frac{25}{6^3}.$$

You may not be overly impressed by the speed of this new method, as compared with a combinatorial counting done in §6.4, but you should observe how the machinery works:

Step 1°: Code the probabilities $\{P(X = j), j \geq 0\}$, into a generating function g;
Step 2°: Process the function by raising it to nth power g^n;
Step 3°: Decode the probabilities $\{P(S_n = k), k \geq 0\}$ from g^n by expanding it into a power series.

A characteristic feature of machine process is that parts of it can be performed mechanically such as the manipulations in (6.5.10). We need not keep track of what we are doing at every stage: plug something in, push a few buttons or crank some knobs, and out comes the product. To carry this gimmickry one step further, we will now exhibit the generating function of X in a form Euler would not have recognized [the concept of a random variable came late, not much before 1930]:

$$(6.5.12)\qquad g(z) = E(z^X)$$

namely the mathematical expectation of z^X. Let us first recall that for each z, the function $\omega \to z^{X(\omega)}$ is indeed a random variable. For countable Ω this is a special case of Proposition 2 in §4.2. When X takes the value j, z^X takes the value z^j, hence by (4.3.15) the expectation of z^X may be expressed as $\sum_{j=0}^{\infty} P(X = j)z^j$ which is $g(z)$.

An immediate pay-off is a new and smoother proof of Theorem 6 based on different principles. The generating function of $X_1 + \cdots + X_n$ is, by what has just been said, equal to

$$E(z^{X_1 + \cdots + X_n}) = E(z^{X_1}z^{X_2} \cdots z^{X_n})$$

by the law of exponentiation. Now the random variables $z^{X_1}, z^{X_2}, \ldots, z^{X_n}$ are independent by Proposition 6 of §5.5, hence by Theorem 2 of §6.3

$$E(z^{X_1}z^{X_2} \cdots z^{X_n}) = E(z^{X_1})E(z^{X_2}) \cdots E(z^{X_n}). \tag{6.5.13}$$

Since $E(z^{X_j}) = g_j(z)$ for each j this completes the proof of Theorem 6.

Another advantage of the expression $E(z^X)$ is that it leads to extensions. If X can take arbitrary real values this expression still has a meaning. For simplicity let us consider only $0 < z \leq 1$. Every such z can be represented as $e^{-\lambda}$ with $0 \leq \lambda < \infty$, in fact the correspondence $z = e^{-\lambda}$ is one-to-one; see Figure 25.

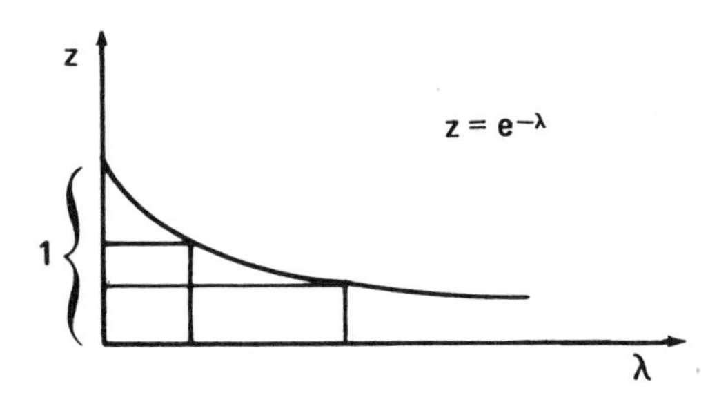

Figure 25

Now consider the new expression after such a change of variable:

$$E(e^{-\lambda X}), \quad 0 \leq \lambda < \infty. \tag{6.5.14}$$

If X has the probability distribution in (6.5.1), then

$$E(e^{-\lambda X}) = \sum_{j=0}^{\infty} a_j e^{-j\lambda}$$

which is of course just our previous $g(z)$ with $z = e^{-\lambda}$. More generally if X takes the values $\{x_j\}$ with probabilities $\{p_j\}$, then

$$E(e^{-\lambda X}) = \sum_{j} p_j e^{-\lambda x_j} \tag{6.5.15}$$

provided that the series converges absolutely. This is the case if all the values $x_j \geq 0$ because then $e^{-\lambda x_j} \leq 1$ and the series is dominated by $\sum_j p_j = 1$. Finally if X has the density function f, then by (4.5.6) with $\varphi(u) = e^{-\lambda u}$:

$$E(e^{-\lambda X}) = \int_{-\infty}^{\infty} e^{-\lambda u} f(u)\, du, \tag{6.5.16}$$

provided that the integral converges. This is the case if $f(u) = 0$ for $u < 0$, namely when X does not take negative values. We have therefore extended the notion of a generating function through (6.5.14) to a large class of random variables. This new gimmick is called the *Laplace transform* of X. In the analytic form given on the right side of (6.5.16) it is widely used in operational calculus, differential equations, and engineering applications.

If we replace the negative real parameter $-\lambda$ in (6.5.14) by the purely imaginary $i\theta$, where $i = \sqrt{-1}$ and θ is real, we get the *Fourier transform* $E(e^{i\theta X})$; in probability theory it is also known as the *characteristic function* of X. Let us recall De Moivre's formula (which used to be taught in high school trigonometry courses), for real u:

$$e^{iu} = \cos u + i \sin u;$$

and its consequence

$$|e^{iu}|^2 = (\cos u)^2 + (\sin u)^2 = 1.$$

This implies that for any real random variable X, we have $|e^{i\theta X}| = 1$; hence the function φ:

$$\varphi(\theta) = E(e^{i\theta X}), \quad -\infty < \theta < \infty, \tag{6.5.17}$$

is always defined, in fact $|\varphi(\theta)| \leq 1$ for all θ. Herein lies the superiority of this new transform over the others discussed above, which cannot be defined sometimes because the associated series or integral does not converge. On the other hand, we pay the price of having to deal with complex variables and functions which lie beyond the scope of an elementary text. Nevertheless, we will invoke both the Laplace and Fourier transforms in Chapter 7 and for future reference let us record the following theorem.

Theorem 7. *Theorems 5 and 6 remain true when the generating function is replaced by the Laplace transform (for non-negative random variables) or the Fourier transform (for arbitrary random variables).*

In the case of Theorem 6, this is immediate from (6.5.13) if the variable z there is replaced by $e^{-\lambda}$ or $e^{i\theta}$. For Theorem 5 the analogues lie deeper and require more advanced analysis (see [Chung 1; Chapter 6]). The reader is asked to accept their truth *by analogy* from the discussion above leading from $E(z^X)$ to $E(e^{-\lambda X})$ and $E(e^{i\theta X})$. After all, analogy is a time-honored method of learning.

Exercises

1. The Massachusetts state lottery has 1000000 tickets. There is one first prize of \$50000; 9 second prizes of \$2500 each; 90 third prizes of \$250 each; 900 fourth prizes of \$25 each. What is the expected value of one ticket? five tickets?
2. Suppose in the lottery above only 80% of the tickets are sold. What is the expected total to be paid out in prizes? If each ticket is sold at 50¢, what is the expected profit for the state?
3. Five residental blocks are polled for racial mixture. The number of houses having black or white owners are listed below

	1	2	3	4	5
Black	3	2	4	3	4
White	10	10	9	11	10

If two houses are picked at random from each block, what is the expected number of black-owned ones among them?

4. Six dice are thrown once. Find the mean and variance of the total points. Same question if the dice are thrown n times.
5. A lot of 1000 screws contain 1% with major defects and 5% with minor defects. If 50 screws are picked at random and inspected, what are the expected numbers of major and minor defectives?
6. In a bridge hand what is the expected number of spades? of different suits? [Hint: for the second question let $X_j = 1$ or 0 according as the jth suit is represented in the hand or not; consider $E(X_1 + X_2 + X_3 + X_4)$.]
7. An airport bus deposits 25 passengers at 7 stops. Assume that each passenger is as likely to get off at any stop as another and that they act independently. The bus stops only if someone wants to get off. What is the probability that nobody gets off at the third stop? What is the expected number of stops it will make? [Hint: let $X_j = 1$ or 0 according as someone gets off at the jth stop or not.]

8.* Given 500 persons picked at random, (a) What is the probability that more than one of them have January 1 as birthday? (b) What is the expected number among them who have this birthday? (c) What is the expected number of days of the year that are birthdays of at least one of these persons? (d) What is the expected number of days of the year that are the birthdays of more than one of these persons? Ignore leap years for simplicity. [Hint: for (b), (c), (d), proceed as in No. 7.]

9.* Problems 6, 7 and 8 are different versions of *occupancy problems* which may be formulated generally as follows. Put n unnumbered tokens into m numbered boxes (see §3.3). What is the expected number of boxes which get exactly [or at least] k tokens? One can also ask for instance: what is the expected number of tokens which do not share its box with any other token? Answer these questions and rephrase them in the language of Problem 6, 7 or 8.

10.* Using the occupancy model above, find the distribution of the tokens in the boxes, namely, the probabilities that exactly n_j tokens go into the jth box, $1 \leq j \leq m$. Describe this distribution in the language of Problem 6, 7 or 8.

11. An automatic machine produces a defective item with probability 2%. When this happens an adjustment is made. Find the average number of good items produced between adjustments.

12. Exactly one of six similar looking keys is known to open a certain door. If you try them one after another how many do you expect to have tried before the door is opened?

13. One hundred electric light bulbs are tested. If the probability of failure is p for each bulb, what is the mean and standard deviation of the number of failures? Assume stochastic independence of the bulbs.

14.* Fifty persons queue up for chest X-ray examinations. Suppose there are four "positive" cases among them. What is the expected number of "negative" cases before the first positive case is spotted? [Hint: think of the four as partitioning walls for the others. Thus the problem is equivalent to finding the expected number of tokens in the first box under (IV′) of §3.3.]

15. There are N coupons numbered 1 to N in a bag. Draw one after another with replacement. (a) What is the expected number of drawings until the first coupon drawn is drawn again? (b)* What is the expected number of drawings until the first time a duplication occurs? [Hint: for (b) compute first the probability of no duplication in n drawings.]

16.* In the problem above, what is the expected maximum coupon-number in n drawings? The same question if the coupons are drawn without replacement. [Hint: find $P(\text{maximum} \leq k)$.]

17. In Pólya's urn scheme with $c \geq -1$ (see §5.4),
    (a) What is the expected number of red balls in n drawings?
    (b) What is the expected number of red balls in the urn after the nth drawing (and putting back c balls)?

18. If $p_n \geq 0$, and $r_n = \sum_{k=n}^{\infty} p_k$ show that

$$\sum_{n=1}^{\infty} np_n = \sum_{n=1}^{\infty} r_n$$

whether both series converge or diverge to $+\infty$. Hence if X is a random variable taking nonnegative integer values, we have

$$E(X) = \sum_{n=1}^{\infty} P(X \geq n). \tag{6.6.1}$$

[Hint: Write $p_n = r_n - r_{n+1}$, rearrange the series (called Abel's method of summation in some calculus textbooks).]

19. Apply the formula (6.6.1) to compute the mean waiting time discussed in Example 8 of §4.4. Note that $P(X \geq n) = q^{n-1}$, $n \geq 1$.
20. Let $X_1, \ldots, X_m$ be independent nonnegative integer-valued random variables all having the same distribution $\{p_n, n \geq 0\}$; and $r_n = \sum_{k=n}^{\infty} p_k$. Show that

$$E\{\min(X_1, \ldots, X_m)\} = \sum_{n=1}^{\infty} r_n^m.$$

[Hint: use No. 18.]
21. Let X be a nonnegative random variable with density function f. Show that if $r(u) = \int_u^\infty f(t)\,dt$, then

$$E(X) = \int_0^\infty P(X \geq u)\,du = \int_0^\infty r(u)\,du. \tag{6.6.2}$$

[Hint: this is the analogue of No. 18. Calculation with integrals is smoother than with sums.]
22. Apply formula (6.6.2) to an X with the exponential density $\lambda e^{-\lambda t}$.
23. The duration T of a certain type of telephone call is found to satisfy the relation

$$P(T > t) = ae^{-\lambda t} + (1 - a)e^{-\mu t}, \quad t \geq 0;$$

where $0 \leq a \leq 1$, $\lambda > 0$, $\mu > 0$ are constants determined statistically. Find the mean and variance of T. [Hint: for the mean a quick method is to use No. 21.]
24. Suppose that the "life" of an electronic device has the exponential density $\lambda e^{-\lambda t}$ in hours. Knowing that it has been in use for n hours, how much longer can it be expected to last? Compare this with its initial life expectancy. Do you see any contradiction?
25. Let five devices described above be tested simultaneously. (a) How long can you expect before one of them fails? (b) How long can you expect before all of them fail?
26. The average error committed in measuring the diameter of a circular disk is .2% and the area of the disk is computed from this measurement. What is the average percentage error in the area if we ignore the square of the percentage error in the diameter?
27. Express the mean and variance of $aX + b$ in terms of those of X, where a and b are two constants. Apply this to the conversion of temperature from Centigrade to Fahrenheit:

$$F = \frac{9}{5}C + 32.$$

28. A gambler figures that he can always beat the house by doubling his bet each time to re-coup any losses. Namely he will quit as soon as he wins, otherwise he will keep doubling his ante until he wins. The only drawback to this winning system is that he may be forced to quit when he runs out of funds. Suppose that he has a capital of \$150 and begins with a dollar bet, and suppose he has an even chance to win each time. What is the probability that he will quit winning, and how much will he have won? What is the probability that he will quit because he does not have enough left to double his last ante, and how much will he have lost in this case? What is his overall mathematical expectation by using this system? The same questions if he will bet all his remaining capital when he can no longer double.
29. Pick n points at random in $[0, 1]$. Find the expected value of the maximum, minimum, and range (= maximum minus minimum).
30. Consider n independent events A_j with $P(A_j) = p_j$, $1 \leq j \leq n$. Let N denote the (random) number of occurrences among them. Find the generating function of N and compute $E(N)$ from it.
31. Let $\{p_j, j \geq 0\}$ be a probability distribution and

$$u_k = \sum_{j=0}^{k} p_j,$$

$$g(z) = \sum_{k=0}^{\infty} u_k z^k.$$

Show that the power series converges for $|z| < 1$. As an example let S_n be as in Example 9 of §6.5, and $p_j = P\{S_n = j\}$. What is the meaning of u_k? Find its generating function g.
32. It is also possible to define the generating function of a random variable which takes positive and negative values. To take a simple case, if

$$P(X = k) = p_k, \quad k = 0, \pm 1, \pm 2, \ldots, \pm N,$$

then

$$g(z) = \sum_{k=-N}^{+N} p_k z^k$$

is a *rational function* of z, namely the quotient of two polynomials. Find g when $p_k = \dfrac{1}{2N+1}$ above, which corresponds to the uniform distribution over the set of integers $\{-N, -(N-1), \ldots, -1, 0, +1, \ldots, N-1, N\}$. Compute the mean from g' for a check.
33. Let $\{X_j, 1 \leq j \leq n\}$ be independent random variables such that

$$X_j = \begin{cases} 1 & \text{with probability } \frac{1}{4}, \\ 0 & \text{with probability } \frac{1}{2}, \\ -1 & \text{with probability } \frac{1}{4}; \end{cases}$$

and $S_n = \sum_{j=1}^{n} X_j$. Find the generating function of S_n in the sense of No. 32, and compute $P(S_n = 0)$ from it. As a concrete application, suppose A and B toss an unbiased coin n times each. What is the probability that they score the same number of heads? [This problem can also be solved without using generating function, by using formula (3.3.9).]

34.* In the coupon collecting problem of No. 15, let T denote the number of drawings until a complete set of coupons is collected. Find the generating function of T. Compute the mean from it for a beautiful check with (6.1.8). [Hint: Let T_j be the waiting time between collecting the $(j - 1)$th and the jth new card; then it has a geometric distribution with $p_j = (N - j + 1)/N$. The T_j's are independent.]

35. Let X and g be as in (6.5.1) and (6.5.2). Derive explicit formulas for the first four moments of X in terms of g and its derivatives.

36. Denote the Laplace transform of X in (6.5.16) by $L(\lambda)$. Express the nth moment of X in terms of L and its derivatives.

37. Find the Laplace transform corresponding to the density function f given below.

(a) $f(u) = \frac{1}{c}$ in $(0, c)$, $c > 0$.

(b) $f(u) = \frac{2u}{c^2}$ in $(0, c)$, $c > 0$.

(c) $f(u) = \frac{\lambda^n u^{n-1}}{(n-1)!} e^{-\lambda u}$ in $[0, \infty)$, $\lambda > 0$, $n \geq 1$. [First verify that this is a density function! The corresponding distribution is called the *gamma distribution* $\Gamma(n; \lambda)$.]

38. Let $S_n = T_1 + \cdots + T_n$ where the T_j's are independent random variables all having the density $\lambda e^{-\lambda t}$. Find the Laplace transform of S_n. Compare with the result in No. 37(c). We can now use Theorem 7 to identify the distribution of S_n.

39.* Consider a population of N taxpayers paying various amounts of taxes, of which the mean is m and variance is σ^2. If n of these are selected at random, show that the mean and variance of their total taxes are equal to

$$nm \quad \text{and} \quad \frac{N-n}{N-1} n\sigma^2$$

respectively. [Hint: denote the amounts by $X_1, \ldots, X_n$ and use (6.3.8). Some algebra may be saved by noting that $E(X_j X_k)$ does not depend on n, so it can be determined when $n = N$, but this trick is by no means necessary.]

40. Prove Theorem 1 by the method used in the proof of Theorem 2. Do the density case as well.
41. Let $a(\cdot)$ and $b(\cdot)$ be two probability density functions and define their *convolution* $c(\cdot)$ as follows:

$$c(v) = \int_{-\infty}^{\infty} a(u)b(v-u)\,du, \quad -\infty < v < \infty;$$

cf. (6.5.8). Show that $c(\cdot)$ is also a probability density function, often denoted by $a * b$.

42.* If $a(u) = \lambda e^{-\lambda u}$ for $u \geq 0$, find the convolution of $a(\cdot)$ with itself. Find by induction the n-fold convolution $\underbrace{a * a * \cdots * a}_{n \text{ times}}$. [Hint: the result is given in No. 37(c).]
43. Prove Theorem 4 for nonnegative integer-valued random variables by using generating functions. [Hint: express the variance by generating functions as in No. 35 and then use Theorem 6.]
44. Prove the analogues of Theorem 6 for Laplace and Fourier transforms.
45. Consider a sequence of independent trials each having probability p for success and q for failure. Show that the probability that the nth success is preceded by exactly j failures is equal to

$$\binom{n+j-1}{j} p^n q^j.$$

46.* Prove the formula

$$\sum_{j+k=l} \binom{m+j-1}{j}\binom{n+k-1}{k} = \binom{m+n+l-1}{l}$$

where the sum ranges over all $j \geq 0$ and $k \geq 0$ such that $j + k = l$. [Hint: this may be more recognizable in the form

$$\sum_{j+k=l} \binom{-m}{j}\binom{-n}{k} = \binom{-m-n}{l};$$

cf. (3.3.9). Use $(1-z)^{-m}(1-z)^{-n} = (1-z)^{-m-n}$.]

47.* The general case of the problem of points (Example 6 of §2.2) is as follows. Two players play a series of independent games in which A has probability p, B has probability $q = 1 - p$ of winning each game. Suppose that A needs m and B needs n more games to win the series. Show that the probability that A will win is given by either one of the expressions below:

(i) $$\sum_{k=m}^{m+n-1} \binom{m+n-1}{k} p^k q^{m+n-1-k};$$

(ii) $$\sum_{k=0}^{n-1} \binom{m+k-1}{k} p^m q^k.$$

The solutions were first given by Montmort (1678–1719). [Hint: solution (i) follows at once from Bernoulli's formula by an obvious interpretation. This is based on the idea (see Example 6 of §2.2) to complete $m + n - 1$ games even if A wins before the end. Solution (ii) is based on the more natural idea of terminating the series as soon as A wins m games before B wins n games. Suppose this happens after exactly $m + k$ games, then A must win the last game and also $m - 1$ among the first $m + k - 1$ games, and $k \leq n - 1$.]

48.* Prove directly that the two expressions (i) and (ii) given in No. 47 are equal. [Hint: one can do this by induction on n, for fixed m; but a more interesting method is suggested by comparison of the two ideas involved in the solutions. This leads to the expansion of (ii) into

$$\begin{aligned}\sum_{k=0}^{n-1} \binom{m+k-1}{k} p^m q^k (p+q)^{n-1-k} &= \sum_{k=0}^{n-1} \binom{m+k-1}{k} p^m q^k \sum_{j=0}^{n-k-1} \binom{n-1-k}{j} p^{n-1-k-j} q^j \\ &= \sum_{l=0}^{n-1} p^{m+n-1-l} q^l \sum_{j+k=l} \binom{m+k-1}{k} \binom{n-k-1}{j};\end{aligned}$$

now use No. 46. Note that the equality relates a binomial distribution to a negative binomial distribution.]

Chapter 7

Poisson and Normal Distributions

7.1. Models for Poisson distribution

The Poisson distribution is of great importance in theory and in practice. It has the added virtue of being a simple mathematical object. We could have introduced it at an earlier stage in the book, and the reader was alerted to this in §4.4. However, the belated entrance will give it more prominence, as well as a more thorough discussion than would be possible without the benefit of the last two chapters.

Fix a real positive number α and consider the probability distribution $\{a_k, k \in N^0\}$, where N^0 is the set of all nonnegative integers, given by

$$a_k = \frac{e^{-\alpha}}{k!}\,\alpha^k. \tag{7.1.1}$$

We must first verify that

$$\sum_{k=0}^{\infty} a_k = e^{-\alpha} \sum_{k=0}^{\infty} \frac{\alpha^k}{k!} = e^{-\alpha} \cdot e^{\alpha} = 1$$

where we have used the Taylor series of e^{α}. Let us compute its mean as well:

$$\sum_{k=0}^{\infty} ka_k = e^{-\alpha} \sum_{k=0}^{\infty} k\frac{\alpha^k}{k!} = e^{-\alpha}\alpha \sum_{k=1}^{\infty} \frac{\alpha^{k-1}}{(k-1)!}$$
$$= e^{-\alpha}\alpha \sum_{k=0}^{\infty} \frac{\alpha^k}{k!} = e^{-\alpha}\alpha e^{+\alpha} = \alpha.$$

[This little summation has been spelled out since I have found that students often do not learn such problems of "infinite series" from their calculus course.] Thus the parameter α has a very specific meaning indeed. We shall call the distribution in (7.1.1) the *Poisson distribution with parameter* α. It will be denoted by $\pi(\alpha)$, and the term with subscript k by $\pi_k(\alpha)$. Thus if X is a random variable having this distribution, then

$$P(X = k) = \pi_k(\alpha) = \frac{e^{-\alpha}}{k!}\,\alpha^k,\ k \in N^0;$$

and

$$E(X) = \alpha. \tag{7.1.2}$$

Next, let us find the generating function g as defined in §6.5. We have, using Taylor's series for $e^{\alpha z}$ this time:

$$g(z) = \sum_{k=0}^{\infty} a_k z^k = e^{-\alpha} \sum_{k=0}^{\infty} \frac{\alpha^k}{k!} z^k = e^{-\alpha} e^{\alpha z} = e^{\alpha(z-1)}. \tag{7.1.3}$$

This is a simple function and can be put to good use in calculations. If we differentiate it twice, we get

$$g'(z) = \alpha e^{\alpha(z-1)}, \quad g''(z) = \alpha^2 e^{\alpha(z-1)}.$$

Hence by (6.5.6),

$$\begin{aligned} E(X) &= g'(1) = \alpha \\ E(X^2) &= g'(1) + g''(1) = \alpha + \alpha^2, \\ \sigma^2(X) &= \alpha. \end{aligned} \tag{7.1.4}$$

So the variance as well as the mean is equal to the parameter α (see below for an explanation).

Mathematically, the Poisson distribution can be derived in a number of significant ways. One of these is a limiting scheme via the binomial distribution. This is known historically as Poisson's limit law, and will be discussed first. Another way, that of adding exponentially distributed random variables, is the main topic of the next section.

Recall the binomial distribution $B(n; p)$ in §4.4 and write

$$B_k(n; p) = \binom{n}{k} p^k (1 - p)^{n-k}, \quad 0 \le k \le n. \tag{7.1.5}$$

We shall allow p to vary with n; this means only that we put $p = p_n$ in the above. Specifically, we take

$$p_n = \frac{\alpha}{n}, \quad n \ge 1. \tag{7.1.6}$$

We are therefore considering the sequence of binomial distributions $B(n; \alpha/n)$, a typical term of which is given by

$$B_k\left(n; \frac{\alpha}{n}\right) = \binom{n}{k} \left(\frac{\alpha}{n}\right)^k \left(1 - \frac{\alpha}{n}\right)^{n-k}, \quad 0 \le k \le n. \tag{7.1.7}$$

For brevity let us denote this by $b_k(n)$. Now fix k and let n go to infinity. It turns out that $b_k(n)$ converges for every k, and can be calculated as follows. To begin at the beginning, take $k = 0$: then we have

$$\lim_{n\to\infty} b_0(n) = \lim_{n\to\infty} \left(1 - \frac{\alpha}{n}\right)^n = e^{-\alpha}. \tag{7.1.8}$$

This is one of the fundamental formulas for the exponential function which you ought to remember from calculus. An easy way to see it is to take natural logarithm and use the Taylor series $\log(1 - x) = -\sum_{n=1}^{\infty} x^n/n$:

(7.1.9) $$\log\left(1 - \frac{\alpha}{n}\right)^n = n \log\left(1 - \frac{\alpha}{n}\right) = n\left\{-\frac{\alpha}{n} - \frac{\alpha^2}{2n^2} - \cdots\right\}$$
$$= -\alpha - \frac{\alpha^2}{2n} - \cdots.$$

When $n \to \infty$ the last-written quantity converges to $-\alpha$ which is $\log e^{-\alpha}$. Hence (7.1.8) may be verified by taking logarithms and expanding into power series, a method very much in use in applied mathematics. A rigorous proof must show that the three dots at the end of (7.1.9) above can indeed be overlooked; see Exercise 18 below.

To proceed, we take the ratio of consecutive terms in (7.1.7):

$$\frac{b_{k+1}(n)}{b_k(n)} = \frac{n-k}{k+1}\left(\frac{\alpha}{n}\right)\left(1 - \frac{\alpha}{n}\right)^{-1} = \frac{\alpha}{k+1}\left[\left(\frac{n-k}{n}\right)\left(1 - \frac{\alpha}{n}\right)^{-1}\right].$$

The two factors within the square brackets above both converge to 1 as $n \to \infty$, hence

(7.1.10) $$\lim_{n\to\infty} \frac{b_{k+1}(n)}{b_k(n)} = \frac{\alpha}{k+1}.$$

Starting with (7.1.8), and using (7.1.10) for $k = 0, 1, 2, \ldots$, we obtain

$$\lim_{n\to\infty} b_1(n) = \frac{\alpha}{1} \lim_{n\to\infty} b_0(n) = \alpha e^{-\alpha},$$

$$\lim_{n\to\infty} b_2(n) = \frac{\alpha}{2} \lim_{n\to\infty} b_1(n) = \frac{\alpha^2}{1\cdot 2} e^{-\alpha},$$

$$\cdots$$

$$\lim_{n\to\infty} b_k(n) = \frac{\alpha}{k} \lim_{n\to\infty} b_{k-1}(n) = \frac{\alpha^k}{1\cdot 2\cdots k} e^{-\alpha}.$$

$$\cdots$$

These limit values are the successive terms of $\pi(\alpha)$. Therefore we have proved Poisson's theorem in its simplest form as follows.

Poisson's limit law:

$$\lim_{n\to\infty} B_k\left(n; \frac{\alpha}{n}\right) = \pi_k(\alpha), \quad k \in N^0.$$

This result remains true if the α/n on the left side above is replaced by α_n/n, where $\lim_n \alpha_n = \alpha$. In other words, instead of taking $p_n = \alpha/n$ as we did in (7.1.6), so that $np_n = \alpha$, we may take $p_n = \alpha_n/n$, so that $np_n = \alpha_n$ and

$$\lim_{n \to \infty} np_n = \lim_n \alpha_n = \alpha. \tag{7.1.11}$$

The derivation is similar to the above except that (7.1.8) is replaced by the stronger result below: if $\lim_{n \to \infty} \alpha_n = \alpha$, then

$$\lim_{n \to \infty} \left(1 - \frac{\alpha_n}{n}\right)^n = e^{-\alpha}. \tag{7.1.12}$$

With this improvement, we can now enunciate the theorem in a more pragmatic form as follows. A binomial probability $B_k(n; p)$, when n is large compared with np which is nearly α, may be approximated by $\pi_k(\alpha)$, for modest values of k. Recall that np is the mean of $B(n; p)$ (see §4.4); it is no surprise that its approximate value α should also be the mean of the approximate Poisson distribution, as we have seen under (7.1.2). Similarly, the variance of $B(n; p)$ is $npq = n\frac{\alpha}{n}\left(1 - \frac{\alpha}{n}\right)$ for $p = \frac{\alpha}{n}$; as $n \to \infty$ the limit is also α as remarked under (7.1.4).

The mathematical introduction of the Poisson distribution is thus done. The limiting passage from the binomial scheme is quite elementary, in contrast to what will be done in §7.3 below. But does the condition (7.1.6), or the more relaxed (7.1.11), make sense in any real situation? The astonishing thing here is that a great variety of natural and man-made random phenomena are found to fit the pattern nicely. We give four examples to illustrate the ways in which the scheme works to a greater or lesser degree.

Example 1. Consider a *rare* event, namely one with small probability p of occurrence. For instance if one bets on a single number at roulette, the probability of winning is equal to $1/37 \approx .027$, assuming that the 36 numbers and one "zero" are equally likely. [The roulette wheels in Monte Carlo have a single "zero," but those in Las Vegas have "double zeros."] If one does this 37 times, he can "expect" to win once. (Which theorem says this?) But we can also compute the probabilities that he wins no time, once, twice, etc. The exact answers are of course given by the first three terms of $B(37; 1/37)$:

$$\left(1 - \frac{1}{37}\right)^{37},$$

$$\binom{37}{1}\left(1 - \frac{1}{37}\right)^{36} \frac{1}{37} = \left(1 - \frac{1}{37}\right)^{36},$$

$$\binom{37}{2}\left(1 - \frac{1}{37}\right)^{35} \frac{1}{37^2} = \frac{36}{2 \times 37}\left(1 - \frac{1}{37}\right)^{35}.$$

If we set

$$c = \left(1 - \frac{1}{37}\right)^{37} \approx .363,$$

then the three numbers above are:

$$c, \quad \frac{37}{36}c, \quad \frac{37}{36} \times \frac{1}{2}c.$$

Hence if we use the approximation $e^{-1} \approx .368$ for c, committing thereby an error of 1.5%; and furthermore "confound" $\frac{37}{36}$ with 1, committing another error of 3%, but in the opposite direction; we get the first three terms of $\pi(1)$, namely:

$$e^{-1}, \; e^{-1}, \; \frac{1}{2}e^{-1}.$$

Further errors will be compounded if we go on, but some may balance others. We may also choose to bet, say, one hundred eleven times ($111 = 37 \times 3$) on a single number, and vary it from time to time as gamblers usually do at a roulette table. The same sort of approximation will then yield

$$\left(1 - \frac{1}{37}\right)^{111} = c^3 \approx e^{-3},$$

$$\frac{111}{37}\left(1 - \frac{1}{37}\right)^{110} = \frac{37}{36} \times 3c^3 \approx 3e^{-3},$$

$$\frac{111 \times 110}{2}\left(1 - \frac{1}{37}\right)^{109} = \frac{111 \times 110}{36 \times 36}\,\frac{1}{2}c^3 \approx \frac{9}{2}e^{-3},$$

etc. Here of course c^3 is a worse approximation of e^{-3} than c is of e^{-1}. Anyway it should be clear that we are simply engaged in more or less crude but handy numerical approximations, without going to any limit. For no matter how small p is, so long as it is fixed as in this example, np will of course go to infinity with n, and the limiting scheme discussed above will be wide of the mark when n is large enough. Nevertheless a reasonably good approximation can be obtained for values of n and p such that np is relatively small compared with n. It is just a case of pure and simple numerical approximation, but many such applications have been made to various rare events. In fact the Poisson law was very popular at one time under the name of "the law of small numbers." Well kept statistical data such as the number of Prussian cavalry men killed each year by a kick from a horse, or the number of child suicides in Prussia, were cited as typical examples of this remarkable distribution (see [Keynes]).

Example 2. Consider the card-matching problem in §6.2. If a person who claims ESP (extrasensory perception) is a fake and is merely trying to match the cards at random, will his average score be better or worse when the number of cards is increased? Intuitively, two opposite effects are apparent. On one hand, there will be more cards to score; on the other, it will be harder to score each. As it turns out (see §6.2) these two effects balance each other so nicely that the expected number is equal to 1 irrespective of the number of cards! Here is an ideal setting for (7.1.6) with $\alpha = 1$. In fact, we can make it conform exactly to the previous scheme by allowing duplication in the guessing. That is, if we think of a deck of n cards laid face down on the table, we are allowed to guess them one by one with total forgetfulness. Then we can guess each card to be any one of the n cards, with equal probability $1/n$, and independently of all other guesses. The probability of exactly k matches is then given by (7.1.7) with $\alpha = 1$, and so the Poisson approximations $\pi_k(1)$ applies if n is large.

This kind of matching corresponds to sampling with replacement. It is not a realistic model when two decks of cards are matched against each other. There is then mutual dependence between the various guesses and the binomial distribution above of course does not apply. But it can be shown that when n is large the effect of dependence is small, as follows. Let the probability of "no match" be q_n when there are n cards to be matched. We see in Example 4 of §6.2 that

$$q_n \approx e^{-1}$$

is an excellent approximation even for moderate values of n. Now an easy combinatorial argument (Exercise 19) shows that the probability of exactly k matches is equal to

$$\binom{n}{k} \frac{1}{(n)_k} q_{n-k} = \frac{1}{k!} q_{n-k}. \tag{7.1.13}$$

Hence for fixed k, this converges to $\frac{1}{k!} e^{-1} = \pi_k(1)$.

Example 3. The Poisson law in a spatial distribution is typified by the counting of particles in a sort of "homogeneous chaos." For instance we may count the number of virus particles with a square grid under the microscope. Suppose that the average number per small square is μ and that there are N squares in the grid. The virus moves freely about in such a way that its distribution over the grid may be approximated by the "tokens in boxes" model described under (I') in §3.3. Namely, there are μN particles to be placed into the N squares, and each particle can go into any of the squares with probability $1/N$, independently of each other. Then the probability of finding exactly k particles in a given square is given by the binomial distribution:

$$B_k\left(\mu N; \frac{1}{N}\right) = \binom{\mu N}{k}\left(\frac{1}{N}\right)^k\left(1 - \frac{1}{N}\right)^{\mu N - k}.$$

Now we should imagine that the virus specimen under examination is part of a much larger specimen with the same average spatial proportion μ. In practice, this assumption is reasonably correct when for example a little blood is drawn from a sick body. It is then legitimate to approximate the above probability by $\pi_k(\mu)$ when N is large. The point here is that the small squares in which the counts are made remain fixed in size, but the homogeneity of space permits a limiting passage when the number of such squares is multiplied.

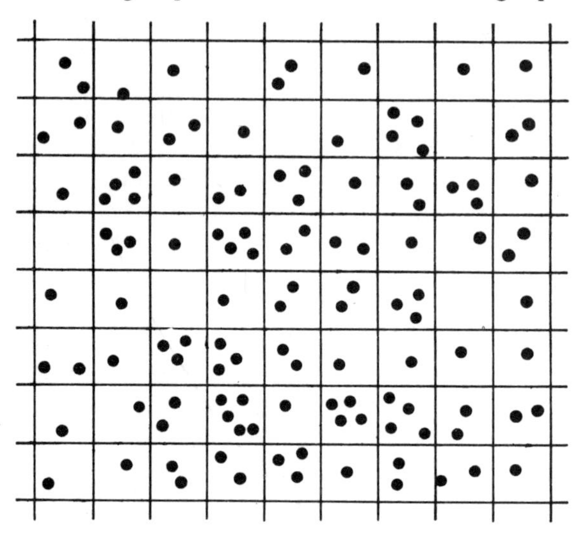

Figure 26

A grim example of the spatial scheme is furnished by the counting of flying-bomb hits on the south of London during World War II. The area was divided into $N = 576$ squares each of 1/4 square mile, and μ was found statistically to be about .930. The table below shows the actual counts N_k and the Poisson approximations $\pi_k(\mu)$ with $\mu = .9323$. The close fit in this case might be explained by the deliberate randomness of the attacks which justified the binomial model above.

k	0	1	2	3	4	≥ 5
N_k	229	211	93	35	7	1
$N\pi_k$	226.74	211.39	98.54	30.62	7.14	1.59

Example 4. In a large class of applications, time plays the role of space in the preceding example. If random occurrences are distributed over a period of time in such a way that their number per unit time may be supposed to be fairly constant over the period, then the Poisson scheme will operate with time acting as the medium for the homogeneous chaos. One could repeat the multiplication argument in Example 3 with time substituting for space, but here it is perhaps more plausible to subdivide the time. Suppose for example some cosmic ray impacts are registered on a geiger counter at the average rate of α per second. Then the probability of a register in a small time interval

δ is given by $\alpha\delta + o(\delta)$, where the "little-o" term represents an error term which is of smaller order of magnitude than δ, or roughly, "very small." Now divide the time interval $[0, t]$ into N equal parts, so that the probability of a counter register in each subinterval is

$$\frac{\alpha t}{N} + o\left(\frac{t}{N}\right)$$

with $\delta = t/N$ above. Of course $\alpha t/N$ is much smaller than 1 when t is fixed and N is large. Let us first assume that for large enough values of N, the probability of more than one register in any small subinterval may be neglected, so that we may suppose that the number of impacts received in each of the N subintervals is either 0 or 1. These numbers can then be treated as Bernoullian random variables taking the values 0 and 1 with probabilities $1 - \frac{\alpha t}{N}$ and $\frac{\alpha t}{N}$ respectively. Finally we assume that they are independent of each other. This assumption can be justified on empirical grounds; for a deeper analysis in terms of the Poisson process, see the next section. Under these assumptions it is now clear that the probability of receiving exactly k impacts in the entire period $[0, t]$ is given by the binomial $B_k\left(N; \frac{\alpha t}{N}\right)$; in fact the total number registered in $[0, t]$ is just the sum of N independent Bernoullian random variables described above. (See Example 9 of §4.4.) Since N is at our disposal and may be made arbitrarily large, in the limit we get $\pi_k(\alpha t)$. Thus in this case the validity of the Poisson scheme may be attributed to the infinite subdivisibility of time. The basic assumption concerning the independence of actions in disjoint subintervals will be justified in Theorem 2 of the following section.

7.2.* Poisson process

For a deeper understanding of the Poisson distribution we will construct a model in which it takes its proper place. The model is known as *Poisson process* and is a fundamental stochastic process.

Consider a sequence of independent positive random variables all of which have the exponential density $\alpha e^{-\alpha t}$, $\alpha > 0$; see Example 3 of §6.2. Let them be denoted by $T_1, T_2, \ldots$ so that for each j,

(7.2.1) $$P(T_j \leq t) = 1 - e^{-\alpha t}, \quad P(T_j > t) = e^{-\alpha t}, \; t \geq 0.$$

Since they are independent, we have for any nonnegative $t_1, \ldots, t_n$:

$$\begin{aligned} P(T_1 > t_1, \ldots, T_n > t_n) &= P(T_1 > t_1) \cdots P(T_n > t_n) \\ &= e^{-\alpha(t_1 + \cdots + t_n)}. \end{aligned}$$

This determines the joint distribution of the T_j's, although we have given the "tail probabilities" for obvious simplicity. Examples of such random variables have been discussed before. For instance, they may be the *inter-arrival* times between vehicles in a traffic flow, or between claims received by an insurance company (see Example 5 of §4.2). They can also be the durations of successive telephone calls, or sojourn times of atoms at a specific energy level. Since

$$E(T_j) = \frac{1}{\alpha}, \tag{7.2.2}$$

it is clear that the smaller α is, the longer the average *inter-arrival*, or *waiting*, or *holding* time. For instance if T is the inter-arrival time between automobiles at a check point, then the corresponding α must be much larger on a Los Angeles freeway than in a Nevada desert. In this particular case α is also known as the *intensity of the flow*, in the sense that heavier traffic means a higher intensity, as every driver knows from his nerves.

Now let us put $S_0 = 0$ and for $n \geq 1$:

$$S_n = T_1 + \cdots + T_n. \tag{7.2.3}$$

Then by definition S_n is the waiting time till the nth arrival; and the event $\{S_n \leq t\}$ means that the nth arrival occurs before time t. [We shall use the preposition "before" loosely to mean "before or at" (time t). The difference can often be overlooked in continuous time models but must be observed in discrete time.] Equivalently, this means "the total number of arrivals in the time interval $[0, t]$" is at least n. This kind of dual point of view is very useful, so we will denote the number just introduced by $N(t)$. We can then record the assertion as follows:

$$\{N(t) \geq n\} = \{S_n \leq t\}. \tag{7.2.4}$$

Like S_n, $N(t)$ is also a random variable: $N(t, \omega)$ with the ω omitted from the notation as in $T_j(\omega)$. If you still remember our general discussion of random variables as functions of a sample point ω, now is a good time to review the situation. What is ω here? Just as in the examples of §4.2, each ω may be regarded as a possible record of the traffic flow or insurance claims or telephone service or nuclear transition. More precisely, $N(t)$ is determined by the whole sequence $\{T_j, j \geq 1\}$, and depends on ω through the T_j's. In fact, taking differences of both sides in the equations (7.2.4) for n and $n + 1$, we obtain

$$\{N(t) = n\} = \{S_n \leq t\} - \{S_{n+1} \leq t\} = \{S_n \leq t < S_{n+1}\}. \tag{7.2.5}$$

The meaning of this new equation is clear from a direct interpretation: there are exactly n arrivals in $[0, t]$ if and only if the nth arrival occurs before t but the $(n + 1)$st occurs after t. For each value of t, the probability distribution of the random variable $N(t)$ is therefore given by

(7.2.6) $$P\{N(t) = n\} = P\{S_n \le t\} - P\{S_{n+1} \le t\}, \quad n \in N^0.$$

Observe the use of our convention $S_0 = 0$ in the above. We proceed to show that this is the Poisson distribution $\pi(\alpha t)$.

We shall calculate the probability $P\{S_n \le t\}$ via the Laplace transform of S_n (see §6.5). The first step is to find the Laplace transform $L(\lambda)$ of each T_j, which is defined since $T_j \ge 0$. By (6.5.16) with $f(u) = \alpha e^{-\alpha u}$, we have

(7.2.7) $$L(\lambda) = \int_0^\infty e^{-\lambda u} \alpha e^{-\alpha u}\, du = \frac{\alpha}{\alpha + \lambda}.$$

Since the T_j's are independent, an application of Theorem 7 of §6.5 yields the Laplace transform of S_n:

(7.2.8) $$L(\lambda)^n = \left(\frac{\alpha}{\alpha + \lambda}\right)^n.$$

To get the distribution or density function of S_n from its Laplace transform is called an inversion problem; and there are tables of common Laplace transforms from which you can look up the *inverse*, namely the distribution or density associated with it. In the present case the answer has been indicated in Exercise 38 of Chapter 6. However here is a trick which leads to it quickly. The basic formula is

(7.2.9) $$\int_0^\infty e^{-xt}\, dt = \frac{1}{x}, \quad x > 0.$$

Differentiating both sides $n - 1$ times, which is easy to do, we obtain

$$\int_0^\infty (-t)^{n-1} e^{-xt}\, dt = \frac{(-1)^{n-1}(n-1)!}{x^n},$$

or

(7.2.10) $$\frac{1}{(n-1)!} \int_0^\infty t^{n-1} e^{-xt}\, dt = \frac{1}{x^n}.$$

Substituting $\alpha + \lambda$ for x in the above and multiplying both sides by α^n, we deduce

$$\int_0^\infty \frac{\alpha^n}{(n-1)!} u^{n-1} e^{-\alpha u} e^{-\lambda u}\, du = \left(\frac{\alpha}{\alpha + \lambda}\right)^n.$$

Thus if we put

(7.2.11) $$f_n(u) = \frac{\alpha^n}{(n-1)!} u^{n-1} e^{-\alpha u},$$

we see that f_n is the density function for the Laplace transform in (7.2.8), namely that of S_n.† Hence we can rewrite the right side of (7.2.6) explicitly as

(7.2.12) $$\int_0^t f_n(u)\, du - \int_0^t f_{n+1}(u)\, du.$$

To simplify this we integrate the first integral by parts as indicated below:

$$\frac{\alpha^n}{(n-1)!}\int_0^t e^{-\alpha u}u^{n-1}\, du = \frac{\alpha^n}{(n-1)!}\left\{\frac{u^n}{n}e^{-\alpha u}\Big|_0^t + \int_0^t \frac{u^n}{n}e^{-\alpha u}\alpha\, du\right\}$$
$$= \frac{\alpha^n}{n!}t^n e^{-\alpha t} + \frac{\alpha^{n+1}}{n!}\int_0^t u^n e^{-\alpha u}\, du.$$

But the last-written integral is just the second integral in (7.2.12); hence the difference there is precisely $\frac{\alpha^n}{n!}t^n e^{-\alpha t} = \pi_n(\alpha t)$. For fixed n and α, this is the density function of the gamma distribution $\Gamma(n; \alpha)$; see p. 189. Let us record this as a theorem.

Theorem 1. *The total number of arrivals in a time interval of length t has the Poisson distribution $\pi(\alpha t)$, for each $t > 0$.*

The reader should observe that the theorem asserts more than has been proved. For in our formulation above we have implicitly chosen an initial instant from which time is measured, namely the zero-time for the first arrival time T_1. Thus the result was proved only for the total number of arrivals in the interval $[0, t]$. Now let us denote the number of arrivals in an arbitrary time interval $[s, s+t]$ by $N(s, s+t)$. Then it is obvious that

$$N(s, s+t) = N(s+t) - N(s)$$

in our previous notation, and $N(0) = 0$. But we have yet to show that the distribution of $N(s, s+t)$ is the same as $N(0, t)$. The question becomes: if we start counting arrivals from time s on, will the same pattern of flow hold as from time 0 on? The answer is "yes" but it involves an essential property of the exponential distribution of the T_j's. Intuitively speaking, if a waiting time such as T_j is broken somewhere in between, its duration after the break follows the original exponential distribution regardless how long it has already endured before the break. This property is sometimes referred to as "lack of memory," and can be put in symbols: for any $s \geq 0$ and $t \geq 0$, we have

(7.2.13) $$P(T > t + s \mid T > s) = P(T > t) = e^{-\alpha t};$$

† Another derivation of this is contained in Exercise 42 of Chapter 6.

see Example 4 of §5.1. There is a converse: if a nonnegative random variable T satisfies the first equation in (7.2.13), then it must have an exponential distribution; see Exercise 41 of Chapter 5. Thus the lack of memory is characteristic of an exponential inter-arrival time.

We can now argue that the pattern of flow from time s on is the same as from 0 on. For the given instant s breaks one of the inter-arrival times, say T_k, into two stretches as shown below:

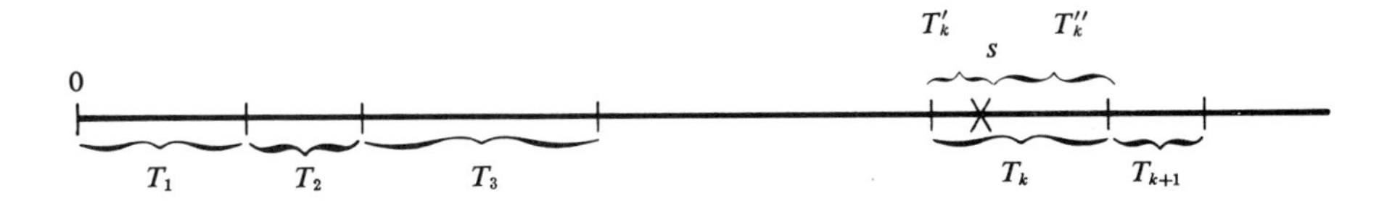

Figure 27

According to the above, the second stretch T_k'' of the broken T_k has the same distribution as T_k, and it is *clearly* independent of all the succeeding T_{k+1}, $T_{k+2}, \ldots$. [The clarity is intuitive enough, but a formal proof takes some doing and is omitted.] Hence the new shifted inter-arrival times from s onward:

$$T_k'', T_{k+1}, T_{k+2}, \ldots \tag{7.2.14}$$

follow the same probability pattern as the original inter-arrival times beginning at 0:

$$T_1, T_2, T_3, \ldots. \tag{7.2.15}$$

Therefore our previous analysis applies to the shifted flow as well as the original one. In particular the number of arrivals in $[s, s + t]$ must have the same distribution as that in $[0, t]$. This is the assertion of Theorem 1.

The fact that $N(s, s + t)$ has the same distribution for all s is referred to as the *time-homogeneity* of the flow. Let us remember that this is shown under the assumption that the intensity α is constant for all time. In practice such an assumption is tenable only over specified periods of time. For example in the case of traffic flow on a given highway, it may be assumed for the rush hour or from 2 a.m. to 3 a.m. with different values of α. However, for longer periods of time such as one day, an average value of α over 24 hours may be used. This may again vary from year to year, even week to week.

So far we have studied the number of arrivals in one period of time, of arbitrary length and origin. For a more complete analysis of the flow we must consider several such periods and their mutual dependence. In other words, we want to find the joint distribution of

(7.2.16) $$N(s_1, s_1 + t_1),\ N(s_2, s_2 + t_2),\ N(s_3, s_3 + t_3), \ldots$$

etc. The answer is given in the next theorem.

Theorem 2. *If the intervals* $(s_1, s_1 + t_1)$, $(s_2, s_2 + t_2)$, . . . *are disjoint, then the random variables in* (7.2.16) *are independent and have the Poisson distributions* $\pi(\alpha t_1)$, $\pi(\alpha t_2)$, ...

It is reasonable and correct to think that if we know the joint action of N over any arbitrary finite set of disjoint time intervals, then we know all about it in principle. Hence with Theorem 2 we shall be in full control of the process in question.

The proof of Theorem 2 depends again on the lack-of-memory property of the T_j's. We will indicate the main idea here without going into formal details. Going back to the sequence in (7.2.14), where we put $s = s_2$, we now make the further observation that all the random variables there are not only independent of one another, but also of all those which precede s, namely:

(7.2.17) $$T_1, \ldots, T_{k-1}, T'_k.$$

The fact that the two broken stretches T'_k and T''_k are independent is a consequence of (7.2.13), whereas the independence of all the rest should be intuitively obvious because they have not been disturbed by the break at s. [Again, it takes some work to justify the intuition.] Now the "past history" of the flow up to time s is determined by the sequence in (7.2.17), while its "future development" after s is determined by the sequence in (7.2.14). Therefore relative to the "present" s, past and future are independent. In particular, $N(s_1, s_1 + t_1)$ which is part of the past, must be independent of $N(s_2, s_2 + t_2)$, $N(s_3, s_3 + t_3)$, . . . , which are all part of the future. Repeating this argument for $s = s_3, s_4, \ldots$, the assertion of Theorem 2 follows.

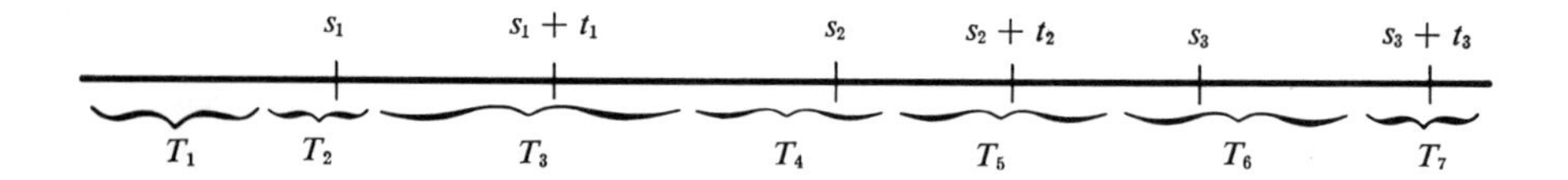

Figure 28

We are now ready to give a general definition for the "flow" we have been discussing all along.

Definition of Poisson Process. A family of random variables $\{X(t)\}$, indexed by the continuous variable t ranging over $[0, \infty)$, is called a *Poisson process with parameter* (or *mean*) α iff it satisfies the following conditions:

(i) $X(0) = 0$;
(ii) the increments $X(s_i + t_i) - X(s_i)$, over an arbitrary finite set of disjoint intervals $(s_i, s_i + t_i)$, are independent random variables;
(iii) for each $s \geq 0$, $t \geq 0$, $X(s + t) - X(s)$ has the Poisson distribution $\pi(\alpha t)$.

According to Theorems 1 and 2 above, the family $\{N(t), t \geq 0\}$ satisfies these conditions and therefore forms a Poisson process. Conversely, it can be shown that every Poisson process is representable as the $N(t)$ above.

The concept of a stochastic process has already been mentioned in §§5.3–5.4, in connection with Pólya's urn model. The sequence $\{X_n, n \geq 1\}$ in Theorem 5 of §5.4 may well be called a Pólya process. In principle a stochastic process is just any family of random variables; but this is putting matters in an esoteric way. What is involved here goes back to the foundations of probability theory discussed in Chapters 2, 4 and 5. There is a sample space Ω with points ω, a probability measure P defined for certain sets of ω, a family of functions $\omega \to X_t(\omega)$ called random variables, and the process is concerned with the joint action or behavior of this family: the marginal and joint distributions, the conditional probabilities, the expectations, and so forth. Everything we have discussed (and are going to discuss) may be regarded as questions in stochastic processes, for in its full generality the term encompasses any random variable or sample set (via its indicator). But in its customary usage we mean a rather numerous and well organized family governed by significant and useful laws. The preceding characterization of a Poisson process is a good example of this description.

As defined, $\omega \to N(t, \omega)$ is a random variable for each t, with the Poisson distribution $\pi(\alpha t)$. There is a dual point of view which is equally important in the study of a process, and that is the function $t \to N(t, \omega)$ for each ω. Such a function is called a *sample function* (*path* or *trajectory*). For example in the case of telephone calls, to choose a sample point ω may mean to pick a day's record of the actual counts at a switch-board over a 24-hour period. This of course varies from day to day so the function $t \to N(t, \omega)$ gives only a *sample* (denoted by ω) of the telephone service. Its graph may look like Fig. 29. The points of jumps are the successive arrival times $S_n(\omega)$, each jump being equal to 1, and the horizontal stretches indicate the inter-arrival times. So the sample function is a monotonically nondecreasing function which increases only by jumps of size one and is flat between jumps. Such a graph is typical of the sample function of a Poisson process. If the flow is intense then the points of jumps are crowded together.

The sequence $\{S_n, n \geq 1\}$ defined in (7.2.3) is also a stochastic process, indexed by n. A sample function $n \to S_n(\omega)$ for this process is an increasing sequence of positive numbers $\{S_1(\omega), S_2(\omega), \ldots, S_n(\omega), \ldots\}$. Hence it is often called a *sample sequence*. There is a reciprocal relation between this and the sample function $N(t, \omega)$ above. If we interchange the two coordinate axes, which can be done by turning the page 90°, and look at Figure 29

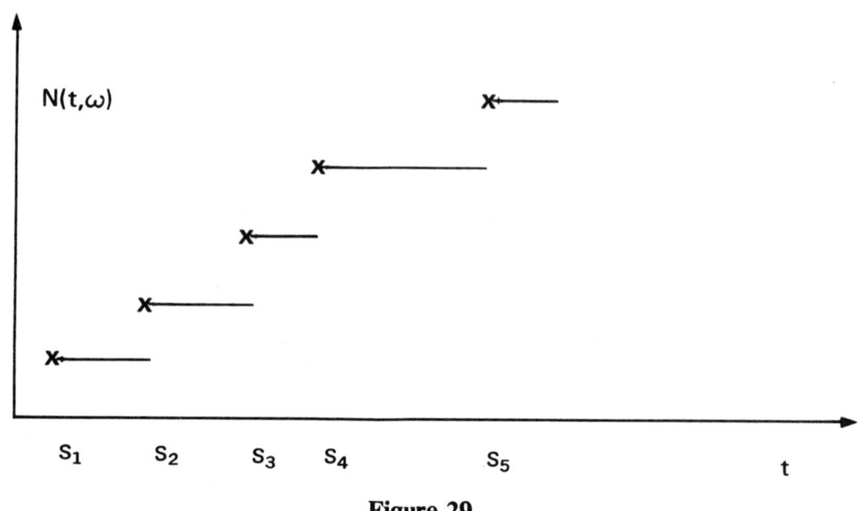

Figure 29

through the light from the back, we get the graph of $n \to S_n(\omega)$. Ignore the now vertical stretches except the lower end points which indicate the values of S_n.

The following examples illustrate some of the properties of the Poisson distribution and process.

Example 5. Consider the number of arrivals in two disjoint time intervals: $X_1 = N(s_1, s_1 + t_1)$ and $X_2 = N(s_2, s_2 + t_2)$ as in (7.2.16). What is the probability that the total number $X_1 + X_2$ is equal to n?

By Theorem 2, X_1 and X_2 are independent random variables with the distributions $\pi(\alpha t_1)$ and $\pi(\alpha t_2)$ respectively. Hence

$$\begin{aligned} P(X_1 + X_2 = n) &= \sum_{j+k=n} P(X_1 = j)P(X_2 = k) \\ &= \sum_{j+k=n} \frac{e^{-\alpha t_1}(\alpha t_1)^j}{j!} \frac{e^{-\alpha t_2}(\alpha t_2)^k}{k!} \\ &= \frac{e^{-\alpha(t_1+t_2)}}{n!} \sum_{j=0}^{n} \binom{n}{j} (\alpha t_1)^j (\alpha t_2)^{n-j} \\ &= \frac{e^{-\alpha(t_1+t_2)}}{n!} (\alpha t_1 + \alpha t_2)^n = \pi_n(\alpha t_1 + \alpha t_2). \end{aligned}$$

Namely, $X_1 + X_2$ is also Poissonian with parameter $\alpha t_1 + \alpha t_2$. The general proposition is as follows.

Theorem 3. *Let X_j be independent random variables with Poisson distributions $\pi(\alpha_j)$, $1 \le j \le n$. Then $X_1 + \cdots + X_n$ has Poisson distribution $\pi(\alpha_1 + \cdots + \alpha_n)$.*

This follows from an easy induction, but we can also make speedy use of generating functions. If we denote the generating function of X_i by g_{X_i}, then by Theorem 6 of 6.5:

$$\begin{aligned} g_{X_1+\cdots+X_n}(z) &= g_{X_1}(z)g_{X_2}(z)\cdots g_{X_n}(z) \\ &= e^{\alpha_1(z-1)}e^{\alpha_2(z-1)}\cdots e^{\alpha_n(z-1)} \\ &= e^{(\alpha_1+\cdots+\alpha_n)(z-1)}. \end{aligned}$$

Thus $X_1 + \cdots + X_n$ has the generating function associated with $\pi(\alpha_1 + \cdots + \alpha_n)$, and so by the uniqueness stated in Theorem 7 of §6.5 it has the latter as distribution.

Example 6. At a crossroad of America we watch cars zooming by bearing license plates of various states. Assume that the arrival process is Poissonian with intensity α, and that the probabilities of each car being from the states of California, Nevada and Arizona, are respectively $p_1 = 1/25$, $p_2 = 1/100$, $p_3 = 1/80$. In a unit period of time what are the number of cars counted with these license plates?

We are assuming that if n cars are counted the distribution of various license plates follows a multinomial distribution $M(n; 50; p_1, \ldots, p_{50})$ where the first three p's are given. Now the number of cars passing in the period of time is a random variable N such that

$$P(N = n) = \frac{e^{-\alpha}}{n!}\alpha^n, \quad n = 0, 1, 2, \ldots.$$

Among these N cars, the number bearing the kth state license is also a random variable N_k; of course

$$N_1 + N_2 + \cdots + N_{50} = N.$$

The problem is to compute

$$P(N_1 = n_1, N_2 = n_2, N_3 = n_3)$$

for arbitrary n_1, n_2, n_3. Let $q = 1 - p_1 - p_2 - p_3$; this is the probability of a license plate not being from one of the three indicated states. For $n \geq n_1 + n_2 + n_3$, the conditional probability under the hypothesis that $N = n$ is given by the multinomial; hence

$$P(N_1 = n_1, N_2 = n_2, N_3 = n_3 | N = n) = \frac{n!p_1^{n_1}p_2^{n_2}p_3^{n_3}q^k}{n_1!n_2!n_3!k!}$$

where $k = n - n_1 - n_2 - n_3$. Using the formula for "total probability" (5.2.3), we get

$$P(N_1 = n_1, N_2 = n_2, N_3 = n_3)$$

$$= \sum_n P(N = n)P(N_1 = n_1, N_2 = n_2, N_3 = n_3 | N = n)$$

$$= \sum_{k=0}^{\infty} \frac{e^{-\alpha}}{n!} \alpha^n \frac{n!}{n_1!n_2!n_3!} \frac{p_1^{n_1}p_2^{n_2}p_3^{n_3}q^k}{k!}.$$

Since $n_1 + n_2 + n_3$ is fixed and $n \geq n_1 + n_2 + n_3$ the summation above reduces to that with k ranging over all nonnegative integers. Now write in the above

$$e^{-\alpha} = e^{-\alpha(p_1+p_2+p_3)}e^{-\alpha q}, \quad \alpha^n = \alpha^{n_1+n_2+n_3}\alpha^k$$

and take out the factors which do not involve the index of summation k. The result is

$$\frac{e^{-\alpha(p_1+p_2+p_3)}}{n_1!n_2!n_3!} (\alpha p_1)^{n_1}(\alpha p_2)^{n_2}(\alpha p_3)^{n_3} \sum_{k=0}^{\infty} \frac{e^{-\alpha q}}{k!} (\alpha q)^k = \pi_{n_1}(\alpha p_1)\pi_{n_2}(\alpha p_2)\pi_{n_3}(\alpha p_3)$$

since the last-written sum equals one. Thus the random variables N_1, N_2, N_3 are independent (why?) and have the Poisson distributions $\pi(\alpha p_1)$, $\pi(\alpha p_2)$, $\pi(\alpha p_3)$.

The substitution of the "fixed number n" in the multinomial $M(n; r; p_1, \ldots, p_r)$ by the "random number N" having a Poisson distribution is called in statistical methodology "randomized sampling." In the example here the difference is illustrated by either counting a fixed number of cars, or counting whatever number of cars in a chosen time interval, or by some other selection method which allows a chance variation of the number. Which way of counting is more appropriate will in general depend on the circumstances and the information sought.

There is of course a general proposition behind the example above which may be stated as follows.

Theorem 4. *Under randomized sampling from a multinomial population $M(n; r; p_1, \ldots, p_r)$ where the total number sampled is a Poisson random variable N with mean α, the numbers $N_1, \ldots, N_r$ of the various varieties obtained by the sampling become independent Poisson variables with means $\alpha p_1, \ldots, \alpha p_r$.*

As an illustration of the finer structure of the Poisson process we will derive a result concerning the location of the jumps of its sample functions. Let us begin with the remark that although (almost) all sample functions have infinitely many jumps in $(0, \infty)$, the probability that a jump occurs at any prescribed instant of time is equal to zero. For if $t > 0$ is fixed, then as $\delta \downarrow 0$ we have

$$P\{N(t + \delta) - N(t - \delta) \geq 1\} = 1 - \pi_0(\alpha, 2\delta) = 1 - e^{-2\alpha\delta} \to 0.$$

In particular, the number of jumps in an interval (t_1, t_2) has the same distribution whether the end-points t_1 and t_2 are included or not. As before let us write $N(t_1, t_2)$ for this number. Now suppose $N(0, t) = n$ for a given t, where $n \geq 1$, and consider an arbitrary subinterval (t_1, t_2) of $(0, t)$. We have for $0 \leq j \leq n$:

$$P\{N(t_1, t_2) = j;\ N(0, t) = n\} = P\{N(t_1, t_2) = j;\ N(0, t_1) + N(t_2, t) = n - j\}.$$

Let $t_2 - t_1 = s$, then the sum of the lengths of the two intervals $(0, t_1)$ and (t_2, t) is equal to $t - s$. By property (ii) and Theorem 3 the random variable $N(0, t_1) + N(t_2, t)$ has the distribution $\pi(\alpha(t - s))$, and is independent of $N(t_1, t_2)$. Hence the probability above is equal to

$$e^{-\alpha s} \frac{(\alpha s)^j}{j!} e^{-\alpha(t-s)} \frac{(\alpha t - \alpha s)^{n-j}}{(n-j)!}.$$

Dividing this by $P\{N(0, t) = n\} = e^{-\alpha t} \dfrac{(\alpha t)^n}{n!}$, we obtain the conditional probability:

$$P\{N(t_1, t_2) = j \mid N(0, t) = n\} = \binom{n}{j} \left(\frac{s}{t}\right)^j \left(1 - \frac{s}{t}\right)^{n-j}. \tag{7.2.18}$$

This is the binomial probability $B_j(n; s/t)$.

Now consider an arbitrary partition of $(0, t)$ into a finite number of subintervals $I_1, \ldots, I_l$ of lengths $s_1, \ldots, s_l$ so that $s_1 + \cdots + s_l = t$. Let $n_1, \ldots, n_l$ be arbitrary nonnegative integers with $n_1 + \cdots + n_l = n$. If we denote by $N(I_k)$ the number of jumps of the process in the interval I_k, then we have by a calculation similar to the above:

$$\begin{aligned} &P\{N(I_k) = n_k, 1 \leq k \leq l \mid N(0, t) = n\} \\ &= \prod_{k=1}^{l} \frac{e^{-\alpha s_k}(\alpha s_k)^{n_k}}{n_k!} \left(e^{-\alpha t} \frac{(\alpha t)^n}{n!}\right)^{-1} \qquad (7.2.19) \\ &= \frac{n!}{n_1! \cdots n_l!} \prod_{k=1}^{l} \left(\frac{s_k}{t}\right)^{n_k}. \end{aligned}$$

This is the multinomial distribution discussed in §6.4.

Let us pick n points at random in $(0, t)$ and arrange them in nondecreasing order $0 < \xi_1 \leq \xi_2 \leq \cdots \leq \xi_n < t$. Using the notation above let $\tilde{N}(I_k)$ denote the number of these points lying in I_k. It is not hard to see that the n-dimensional distribution of $(\xi_1, \ldots, \xi_n)$ is uniquely determined by the distribution of $(\tilde{N}(I_1), \ldots, \tilde{N}(I_l))$ for all possible partitions of $(0, t)$; for a rigorous proof of this fact see Exercise 26 below. In particular, if the n points are picked independently of one another and each is uniformly distributed in $(0, t)$, then it follows from the discussion in §6.4 that the probability $P\{\tilde{N}(I_k) = n_k, 1 \leq k \leq l\}$

is given by the last term in (7.2.19). Therefore, under the hypothesis that there are exactly n jumps of the Poisson process in $(0, t)$, the *conditional distribution* of the n points of jump is the same as if they are picked in the manner just described. This has been described as a sort of "homogeneous chaos."

7.3. From binomial to normal

From the point of view of approximating the binomial distribution $B(n; p)$ for large values of n, the case discussed in §7.1 leading to the Poisson distribution is *abnormal*, because p has to be so small that np remains constant, or nearly so. The fact that many random phenomena follow this law rather nicely was not known in the early history of probability. One must remember that not only radioactivity had yet to be discovered, but neither the telephone nor automobile traffic existed as modern problems. On the other hand counting heads by tossing coins or points by rolling dice, and the measurement of all kinds of physical and biological quantities were already done extensively. These led to binomial and multinomial distributions, and since computing machines were not available it became imperative to find manageable formulas for the probabilities. The *normal* way to approximate the binomial distribution:

$$B_k(n; p) = \binom{n}{k} p^k(1 - p)^{n-k}, \quad 0 \leq k \leq n; \tag{7.3.1}$$

is for a fixed value of p and large values of n. To illustrate by the simplest kind of example, suppose an unbiased coin is tossed 100 times; what is the probability of obtaining exactly 50 heads? The answer

$$\binom{100}{50} \frac{1}{2^{100}} = \frac{100!}{50!50!} \frac{1}{2^{100}}$$

gives little satisfaction as we have no idea of the magnitude of this probability. Without some advanced mathematics (which will be developed presently), who can guess whether this is near $\frac{1}{2}$ or $\frac{1}{10}$ or $\frac{1}{50}$?

Now it is evident that the *key* to such combinatorial formulas is the factorial $n!$ which just crops up everywhere. Take a look back at Chapter 3. So the problem is to find a handy formula: another function $\chi(n)$ of n which is a good approximation for $n!$ but of a simpler structure for computations. But what is "good"? Since $n!$ increases very rapidly with n (see the short table in §3.2) it would be hopeless to make the difference $|n! - \chi(n)|$ small. [Does it really make a difference to have a million dollars or a million and three?] What counts is the ratio $n!/\chi(n)$ which should be close to 1. For two positive functions ψ and χ of the integer variable n, there is a standard notation

(7.3.2) $$\psi(n) \sim \chi(n) \quad \text{which means} \quad \lim_{n\to\infty} \frac{\psi(n)}{\chi(n)} = 1.$$

We say also that $\psi(n)$ and $\chi(n)$ are *asymptotically equal* (or *equivalent*) as $n \to \infty$. If so we have also

$$\lim_{n\to\infty} \frac{|\psi(n) - \chi(n)|}{\chi(n)} = 0$$

provided $\chi(n) > 0$ for large n; thus the difference $|\psi(n) - \chi(n)|$ is negligible in comparison with $\chi(n)$ or $\psi(n)$, though it may be large indeed in absolute terms. Here is a trivial example which you should have retained from a calculus course (under the misleading heading "indeterminate form"):

$$\psi(n) = 2n^2 + 10n - 100, \quad \chi(n) = 2n^2.$$

More generally, a polynomial in n is asymptotically equal to its highest term. Here, of course, we are dealing with something far more difficult: to find a simple enough $\chi(n)$ such that

$$\lim_{n\to\infty} \frac{n!}{\chi(n)} = 1.$$

Such a χ is given by *Stirling's formula* (see Appendix 2):

(7.3.3) $$\chi(n) = \left(\frac{n}{e}\right)^n \sqrt{2\pi n} = n^{n+(1/2)}e^{-n}\sqrt{2\pi}$$

or more precisely

(7.3.4) $$n! = \left(\frac{n}{e}\right)^n \sqrt{2\pi n}\, e^{\omega(n)}, \text{ where } \frac{1}{12\left(n + \frac{1}{2}\right)} < \omega(n) < \frac{1}{12n}.$$

You may think $\chi(n)$ is uglier looking than $n!$, but it is much easier to compute because powers are easy to compute. Here we will apply it at once to the little problem above. It does not pay to get involved in numericals at the beginning so we will consider

(7.3.5) $$\binom{2n}{n}\frac{1}{2^{2n}} = \frac{(2n)!}{n!n!}\frac{1}{2^{2n}}.$$

Substituting $\chi(n)$ and $\chi(2n)$ for $n!$ and $(2n)!$ respectively, we see that this is asymptotically equal to

$$\frac{\left(\frac{2n}{e}\right)^{2n}\sqrt{4\pi n}}{\left(\frac{n}{e}\right)^{2n}2\pi n}\frac{1}{2^{2n}} = \frac{1}{\sqrt{\pi n}}.$$

In particular for $n = 50$ we get the desired answer $1/\sqrt{50\pi} = .08$ approximately. Try to do this by using logarithms on

$$\frac{(100)_{50}}{50!}\frac{1}{2^{100}}$$

and you will appreciate Stirling's formula more. We proceed at once to the slightly more general

$$\binom{2n}{n+k}\frac{1}{2^{2n}} = \frac{(2n)!}{(n+k)!(n-k)!}\frac{1}{2^{2n}} \tag{7.3.6}$$

where k is fixed. A similar application of (7.3.3) yields

$$\frac{\left(\frac{2n}{e}\right)^{2n}\sqrt{4\pi n}\cdot\frac{1}{2^{2n}}}{\left(\frac{n+k}{e}\right)^{n+k}\sqrt{2\pi(n+k)}\left(\frac{n-k}{e}\right)^{n-k}\sqrt{2\pi(n-k)}}$$

$$= \frac{n^{2n}}{(n+k)^{n+k}(n-k)^{n-k}}\sqrt{\frac{n}{\pi(n+k)(n-k)}}$$

$$= \left(\frac{n}{n+k}\right)^{n+k}\left(\frac{n}{n-k}\right)^{n-k}\sqrt{\frac{n}{\pi(n^2-k^2)}}.$$

Clearly the last-written factor is asymptotically equal to $1/\sqrt{\pi n}$. As for the two preceding ones, it follows from (7.1.8) that

$$\begin{aligned}\lim_{n\to\infty}\left(\frac{n}{n+k}\right)^{n+k} &= \lim_{n\to\infty}\left(1-\frac{k}{n+k}\right)^{n+k} = e^{-k},\\ \lim_{n\to\infty}\left(\frac{n}{n-k}\right)^{n-k} &= \lim_{n\to\infty}\left(1+\frac{k}{n-k}\right)^{n-k} = e^{k}.\end{aligned} \tag{7.3.7}$$

Hence the asymptotic value of (7.3.6) is

$$e^{-k}e^{k}\frac{1}{\sqrt{\pi n}} = \frac{1}{\sqrt{\pi n}},$$

exactly as in (7.3.5) which is the particular case $k = 0$.

As a consequence, for any fixed number l, we have

(7.3.8)
$$\lim_{n\to\infty} \sum_{k=-l}^{l} \binom{2n}{n+k} \frac{1}{2^{2n}} = 0$$

because each term in the sum has limit 0 as $n \to \infty$ as just shown, and there are only a fixed number of terms. Now if we remember Pascal's triangle (3.3.5), the binomial coefficients $\binom{2n}{n+k}$, $-n \le k \le n$, assume their maximum value $\binom{2n}{n}$ for the middle term $k = 0$ and decreases as $|k|$ increases (see Exercise 6 below). According to (7.3.8), the sum of a fixed number of terms centered around the middle term approaches zero, hence *a fortiori* the sum of any fixed number of terms will also approach zero, namely for any fixed a and b with $a < b$, we have

$$\lim_{n\to\infty} \sum_{j=a}^{b} \binom{2n}{j} \frac{1}{2^{2n}} = 0.$$

Finally this result remains true if we replace $2n$ by $2n + 1$ above, because the ratio of corresponding terms

$$\binom{2n+1}{j} \frac{1}{2^{2n+1}} \bigg/ \binom{2n}{j} \frac{1}{2^{2n}} = \frac{2n+1}{2n+1-j} \cdot \frac{1}{2}$$

approaches 1/2 which does not affect the zero limit. Now let us return to the probability meaning of the terms, and denote as usual by S_n the number of heads obtained in n tosses of the coin. The result then asserts that for any fixed numbers a and b, we have

(7.3.9)
$$\lim_{n\to\infty} P(a \le S_n \le b) = 0.$$

Observe that there are $n + 1$ possible values for S_n, whereas if the range $[a, b]$ is fixed irrespective of n, it will constitute a negligible fraction of n when n is large. Thus the result (7.3.9) is hardly surprising, though certainly disappointing.

It is clear that in order to "catch" a sufficient number of possible values of S_n to yield a non-zero limit probability, the range allowed must increase to infinity with n. Since we saw that the terms near the middle are of the order of magnitude $1/\sqrt{n}$, it is plausible that the number of terms needed will be of the order of magnitude $\sqrt{n}$. More precisely, it turns out that for each l,

(7.3.10)
$$P\left(\frac{n}{2} - l\sqrt{n} \le S_n \le \frac{n}{2} + l\sqrt{n}\right) = \sum_{\left|j-\frac{n}{2}\right| \le l\sqrt{n}} \binom{n}{j} \frac{1}{2^n}$$

will have a limit strictly between 0 and 1 as $n \to \infty$. Here the range for S_n is centered around the middle value $n/2$ and contains about $2l\sqrt{n}$ terms. When

n is large this is still only a very small fraction of n, but it increases just rapidly enough to serve our purpose. The choice of $\sqrt{n}$ rather than say $n^{1/3}$ or $n^{3/5}$ is crucial and is determined by a rather deep bit of mathematical analysis which we proceed to explain.

Up to here we have considered the case $p = 1/2$ in (7.3.1), in order to bring out the essential features in the simplest case. However, this simplification would obscure the role of np and npq in the general formula below. The reader is advised to carry out the following calculations by himself in the easier case $p = q = 1/2$ to obtain some practice and confidence in such calculations.

Theorem 5. *Suppose* $0 < p < 1$; *put* $q = 1 - p$, *and*

$$x_{nk} = \frac{k - np}{\sqrt{npq}}, \quad 0 \leq k \leq n. \tag{7.3.11}$$

Clearly x_{nk} depends on both n and k, but it will be written as x_k below.

Let A be an arbitrary but fixed positive constant. Then in the range of k such that

$$|x_k| \leq A, \tag{7.3.12}$$

we have

$$\binom{n}{k} p^k q^{n-k} \sim \frac{1}{\sqrt{2\pi npq}} e^{-x_k^2/2}. \tag{7.3.13}$$

The convergence is uniform with respect to k in the range specified above.

Proof: We have from (7.3.11),

$$k = np + \sqrt{npq}\, x_k, \quad n - k = nq - \sqrt{npq}\, x_k. \tag{7.3.14}$$

Hence in the range indicated in (7.3.12),

$$k \sim np, \quad n - k \sim nq. \tag{7.3.15}$$

Using Stirling's formula (7.3.3) we may write the left member of (7.3.13) as

$$\frac{\left(\frac{n}{e}\right)^n \sqrt{2\pi n}\, p^k q^{n-k}}{\left(\frac{k}{e}\right)^k \sqrt{2\pi k} \left(\frac{n-k}{e}\right)^{n-k} \sqrt{2\pi(n-k)}} \tag{7.3.16}$$

$$= \sqrt{\frac{n}{2\pi k(n-k)}}\, \varphi(n, k) \sim \frac{1}{\sqrt{2\pi npq}}\, \varphi(n, k)$$

by (7.3.15), where

$$\varphi(n, k) = \left(\frac{np}{k}\right)^k \left(\frac{nq}{n-k}\right)^{n-k}.$$

Taking logarithms and using the Taylor series

$$\log(1 + x) = x - \frac{x^2}{2} + \cdots + (-1)^{n-1}\frac{x^n}{n} + \cdots, \quad |x| < 1;$$

we have by (7.3.14)

$$\begin{aligned}
\log\left(\frac{np}{k}\right)^k &= k \log\left(1 - \frac{\sqrt{npq}\, x_k}{k}\right) \\
&= k\left(-\frac{\sqrt{npq}\, x_k}{k} - \frac{npqx_k^2}{2k^2} - \cdots\right); \\
\log\left(\frac{nq}{n-k}\right)^{n-k} &= (n-k)\log\left(1 + \frac{\sqrt{npq}\, x_k}{n-k}\right) \\
&= (n-k)\left(\frac{\sqrt{npq}\, x_k}{n-k} - \frac{npqx_k^2}{2(n-k)^2} + \cdots\right);
\end{aligned} \tag{7.3.17}$$

provided that

$$\left|\frac{\sqrt{npq}\, x_k}{k}\right| < 1 \quad \text{and} \quad \left|\frac{\sqrt{npq}\, x_k}{n-k}\right| < 1. \tag{7.3.17'}$$

These conditions are satisfied for sufficiently large value of n, in view of (7.3.12) and (7.3.15). Adding the two series expansions above whereupon the first terms cancel out each other obligingly, ignoring the dots but using "$\sim$" instead of "$=$," we obtain

$$\log \varphi(n, k) \sim -\frac{npqx_k^2}{2k} - \frac{npqx_k^2}{2(n-k)} = -\frac{n^2pqx_k^2}{2k(n-k)}.$$

In Appendix 2 we will give a rigorous demonstration of this relation. Using (7.3.15) again, we see that

$$\log \varphi(n, k) \sim -\frac{n^2pqx_k^2}{2npnq} = -\frac{x_k^2}{2}. \tag{7.3.18}$$

In view of (7.3.12) [why do we need this reminder?], this is equivalent to

$$\varphi(n, k) \sim e^{-x_k^2/2}.$$

Going back to (7.3.16) we obtain (7.3.13).

Theorem 6 (*De Moivre-Laplace Theorem*). *For any two constants a and b, $-\infty < a < b < +\infty$, we have*

(7.3.19) $$\lim_{n\to\infty} P\left(a < \frac{S_n - np}{\sqrt{npq}} \leq b\right) = \frac{1}{\sqrt{2\pi}} \int_a^b e^{-x^2/2}\, dx.$$

Proof: Let k denote a possible value of S_n so that $S_n = k$ means $(S_n - np)/\sqrt{npq} = x_k$ by the transformation (7.3.11). Hence the probability on the left side of (7.3.19) is just

$$\sum_{a < x_k \leq b} P(S_n = k) = \sum_{a < x_k \leq b} \binom{n}{k} p^k q^{n-k}.$$

Substituting for each term its asymptotic value given in (7.3.13), and observing from (7.3.11) that

$$x_{k+1} - x_k = \frac{1}{\sqrt{npq}},$$

we obtain

(7.3.20) $$\frac{1}{\sqrt{2\pi}} \sum_{a < x_k \leq b} e^{-x_k^2/2}(x_{k+1} - x_k).$$

The correspondence between k and x_k is one-to-one and when k varies from 0 to n, x_k varies in the interval $[-\sqrt{np/q}, \sqrt{nq/p}]$, not continuously but by an increment $x_{k+1} - x_k = 1/\sqrt{npq}$. For large enough n the interval contains the given $(a, b]$ and the points x_k falling inside $(a, b]$ form a partition of it into equal subintervals of length $1/\sqrt{npq}$. Suppose the smallest and greatest values of k satisfying the condition $a < x_k \leq b$ are j and l, then we have

$$x_{j-1} \leq a < x_j < x_{j+1} < \cdots < x_{l-1} < x_l \leq b < x_{l+1}$$

and the sum in (7.3.20) may be written as follows:

(7.3.21) $$\sum_{k=j}^{l} \varphi(x_k)(x_{k+1} - x_k); \quad \text{where } \varphi(x) = \frac{1}{\sqrt{2\pi}} e^{-x^2/2}.$$

This is a Riemann sum for the definite integral $\int_a^b \varphi(x)\, dx$, although in standard textbook treatments of Riemann integration the endpoints a and b are usually included as points of partition. But this makes no difference as $n \to \infty$ and the partition becomes finer, so the sum above converges to the integral as shown in (7.3.19).

The result in (7.3.19) is called the *De Moivre-Laplace Theorem* [Abraham De Moivre (1667–1754), considered as successor to Newton, gave this result in his *Doctrine of Chances* (1714). Apparently he had priority over Stirling (1692–1770) for the formula named after the latter. Laplace extended it and realized its importance in his monumental *Théorie Analytique des Probabilités* (1812)]. It was the first known particular case of *The Central Limit Theorem* to be discussed in the next section. It solves the problem of approximation stated at the beginning of the section. The right member of (7.3.20) involves a new probability distribution to be discussed in the next section. Simple examples of application will be given at the end of §7.5 and among the exercises.

7.4. Normal distribution

The probability distribution with the φ in (7.3.21) as density function will now be formally introduced:

$$\Phi(x) = \frac{1}{\sqrt{2\pi}} \int_{-\infty}^{x} e^{-u^2/2}\, du, \quad \varphi(x) = \frac{1}{\sqrt{2\pi}} e^{-x^2/2}.$$

It is called the *normal distribution*, also the *Laplace-Gauss distribution;* and sometimes the prefix *unit* is attached to distinguish it from a whole family of normal distributions derived by a linear transformation of the variable x; see below. But we have yet to show that φ is a true probability density as defined in §4.5, namely that

$$\int_{-\infty}^{\infty} \varphi(x)\, dx = 1. \tag{7.4.1}$$

A heuristic proof of this fact may be obtained by setting $a = -\infty$, $b = +\infty$ in (7.3.19), whereupon the probability on the left side certainly becomes 1. Why is this not rigorous? Because two (or three) passages to limit are involved here which are not necessarily interchangeable. Actually the argument can be justified (see Appendix 2); but it may be more important that you should convince yourself that a justification is needed at all. This is an instance where advanced mathematics separates from the elementary kind we are doing mostly in this book.

A direct proof of (7.4.1) is also very instructive; although it is given in most calculus texts we will reproduce it for its sheer ingenuity. The trick is to consider the square of the integral in (7.4.1) and convert it to a double integral:

$$\left(\int_{-\infty}^{\infty} \varphi(x)\, dx\right)\left(\int_{-\infty}^{\infty} \varphi(y)\, dy\right)$$
$$= \int_{-\infty}^{\infty}\int_{-\infty}^{\infty} \varphi(x)\varphi(y)\, dx\, dy = \frac{1}{2\pi}\int_{-\infty}^{\infty}\int_{-\infty}^{\infty} \exp\left(-\frac{1}{2}(x^2 + y^2)\right) dx\, dy.$$

We can then use polar coordinates:

$$\rho^2 = x^2 + y^2, \quad dx\, dy = \rho\, d\rho\, d\theta$$

to evaluate it:

$$\frac{1}{2\pi}\int_0^{2\pi}\int_0^{\infty}\exp\left(-\frac{1}{2}\rho^2\right)\rho\,d\rho\,d\theta = \frac{1}{2\pi}\int_0^{2\pi} -\exp\left(-\frac{1}{2}\rho^2\right)\Big|_0^{\infty}\,d\theta$$

$$= \frac{1}{2\pi}\int_0^{2\pi} 1\,d\theta = 1.$$

This establishes (7.4.1) if we take the positive square root.

The normal density φ has many remarkable analytical properties, in fact Gauss determined it by selecting a few of them as characteristics of a "law of errors." [Carl Friedrich Gauss (1777–1855) ranked as one of the greatest of all mathematicians, also did fundamental work in physics, astronomy and geodesy. His major contribution to probability was through his theory of errors of observations, known as the method of least squares.] Let us observe first that it is a symmetric function of x, namely $\varphi(x) = \varphi(-x)$, from which the convenient formula follows:

$$\text{(7.4.2)} \qquad \int_{-x}^{x}\varphi(u)\,du = \Phi(x) - \Phi(-x) = 2\Phi(x) - 1.$$

Next, φ has derivatives of all orders, and each derivative is the product of φ by a polynomial called a *Hermite polynomial.* The existence of all derivatives makes the curve $x \to \varphi(x)$ very smooth, and it is usually described as "bell-shaped."† Furthermore as $|x| \to \infty$, $\varphi(x)$ decreases to 0 very rapidly. The following estimate of the tail of Φ is often useful:

$$1 - \Phi(x) = \int_x^{\infty}\varphi(u)\,du \le \frac{\varphi(x)}{x} = \frac{e^{-x^2/2}}{\sqrt{2\pi}\,x}, \qquad x > 0.$$

To see this, note that $-\varphi'(u) = u\varphi(u)$, hence

$$\int_x^{\infty} 1\cdot\varphi(u)\,du \le \int_x^{\infty}\frac{u}{x}\varphi(u)\,du = \frac{-1}{x}\int_x^{\infty}\varphi'(u)\,du = \frac{-1}{x}\varphi(u)\Big|_x^{\infty} = \frac{\varphi(x)}{x},$$

another neat trick. It follows that not only Φ has moments of all orders, but even the integral

$$\text{(7.4.3)} \qquad M(\theta) = \int_{-\infty}^{\infty} e^{\theta x}\varphi(x)dx = \int_{-\infty}^{\infty}\exp\left(\theta x - \frac{x^2}{2}\right)dx$$

is finite for every real θ, because $e^{-x^2/2}$ decreases much faster than $e^{|\theta x|}$ increases as $|x| \to \infty$. As a function of θ, M is called the *moment generating function* of φ or Φ. Note that if we replace θ by the purely imaginary $i\theta$, then

† See the graph attached to the Table of $\Phi(x)$ on p. 320.

$M(i\theta)$ becomes the characteristic function or Fourier transform of Φ [see (6.5.17)]. The reason why we did not introduce the moment generating function in §6.5 is because the integral in (7.4.3) rarely exists if φ is replaced by an arbitrary density function, but for the normal φ, $M(\theta)$ is cleaner than $M(i\theta)$ and serves as well. Let us now calculate $M(\theta)$. This is done by completing a square in the exponent in (7.4.3):

$$\theta x - \frac{x^2}{2} = \frac{\theta^2}{2} - \frac{(x-\theta)^2}{2}.$$

Now we have

$$M(\theta) = e^{\theta^2/2} \int_{-\infty}^{\infty} \varphi(x-\theta)\,dx = e^{\theta^2/2}. \tag{7.4.4}$$

From this we can derive all the moments of Φ by successive differentiation of M with respect to θ, as in the case of a generating function discussed in §6.5. More directly, we may expand the $e^{\theta x}$ in (7.4.3) into its Taylor series in θ and compare the result with the Taylor series of $e^{\theta^2/2}$ in (7.4.4):

$$\int_{-\infty}^{\infty} \left\{1 + \theta x + \frac{(\theta x)^2}{2!} + \cdots + \frac{(\theta x)^n}{n!} + \cdots\right\} \varphi(x)\,dx$$
$$= 1 + \frac{\theta^2}{2} + \frac{1}{2!}\left(\frac{\theta^2}{2}\right)^2 + \cdots + \frac{1}{n!}\left(\frac{\theta^2}{2}\right)^n + \cdots.$$

If we denote the nth moment by $m^{(n)}$:

$$m^{(n)} = \int_{-\infty}^{\infty} x^n \varphi(x)\,dx,$$

the above equation becomes

$$\sum_{n=0}^{\infty} \frac{m^{(n)}}{n!}\theta^n = \sum_{n=0}^{\infty} \frac{1}{2^n n!}\theta^{2n}.$$

It follows from the uniqueness of power series expansion (cf. §6.5) that the corresponding coefficients on both sides must be equal: thus for $n \geq 1$:

$$m^{(2n-1)} = 0, \qquad m^{(2n)} = \frac{(2n)!}{2^n n!}. \tag{7.4.5}$$

Of course the vanishing of all moments of odd order is an immediate consequence of the symmetry of φ.

In general, for any real m and $\sigma^2 > 0$, a random variable X is said to have a *normal distribution* $N(m, \sigma^2)$ iff the *reduced variable*

$$X^* = \frac{X - m}{\sigma}$$

has Φ as its distribution function. In particular for $m = 0$ and $\sigma^2 = 1$, $N(0, 1)$ is just the unit normal Φ. The density function of $N(m, \sigma^2)$ is

$$\frac{1}{\sqrt{2\pi}\,\sigma} \exp\left(-\frac{(x - m)^2}{2\sigma^2}\right) = \frac{1}{\sigma}\,\varphi\left(\frac{x - m}{\sigma}\right). \tag{7.4.6}$$

This follows from a general proposition (see Exercise 13 of Chapter 4). The moment-generating function M_X of X is most conveniently calculated through that of X^* as follows:

$$\begin{aligned} M_X(\theta) &= E(e^{\theta(m+\sigma X^*)}) = e^{m\theta}E(e^{(\sigma\theta)X^*}) \\ &= e^{m\theta}M(\sigma\theta) = e^{m\theta+\sigma^2\theta^2/2} \end{aligned} \tag{7.4.7}$$

A basic property of the *normal family* is given below. Cf. the analogous Theorem 3 in §7.2 for the Poisson family.

Theorem 7. *Let X_j be independent random variables with normal distributions $N(m_j, \sigma_j^2)$, $1 \leq j \leq n$. Then $X_1 + \cdots + X_n$ has the normal distribution* $N\left(\sum_{j=1}^n m_j, \sum_{j=1}^n \sigma_j^2\right)$.

Proof: It is sufficient to prove this for $n = 2$, since the general case follows by induction. This is easily done by means of the moment-generating function. We have by the product theorem as in Theorem 6 of §6.5:

$$\begin{aligned} M_{X_1+X_2}(\theta) = M_{X_1}(\theta)M_{X_2}(\theta) &= e^{m_1\theta+\sigma_1{}^2\theta^2/2}e^{m_2\theta+\sigma_2{}^2\theta^2/2} \\ &= e^{(m_1+m_2)\theta+(\sigma_1{}^2+\sigma_2{}^2)\theta^2/2}, \end{aligned}$$

which is the moment-generating function of $N(m_1 + m_2, \sigma_1^2 + \sigma_2^2)$ by (7.4.7). Hence $X_1 + X_2$ has this normal distribution since it is uniquely determined by the moment-generating function. [We did not prove this assertion but see the end of §6.5.]

7.5.* Central limit theorem

We will now return to the De Moivre-Laplace Theorem 6 and give it a more general formulation. Recall that

$$S_n = X_1 + \cdots + X_n, \quad n \geq 1, \tag{7.5.1}$$

where the X_j's are independent Bernoullian random variables. We know that for every j:

$$E(X_j) = p, \quad \sigma^2(X_j) = pq;$$

and for every n:

$$E(S_n) = np, \quad \sigma^2(S_n) = npq;$$

see Example 6 of §6.3. Put

$$X_j^* = \frac{X_j - E(X_j)}{\sigma(X_j)}; \quad S_n^* = \frac{S_n - E(S_n)}{\sigma(S_n)} = \frac{1}{\sqrt{n}} \sum_{j=1}^{n} X_j^*. \tag{7.5.2}$$

The S_n^*'s are the random variables appearing in the left member of (7.3.19), and are sometimes called the *normalized* or *normed sums*. We have for every j and n:

$$\begin{aligned} E(X_j^*) &= 0, \quad \sigma^2(X_j^*) = 1; \\ E(S_n^*) &= 0, \quad \sigma^2(S_n^*) = 1. \end{aligned} \tag{7.5.3}$$

The linear transformation from X_j to X_j^* or S_n to S_n^* amounts to a change of origin and scale in the measurement of a random quantity in order to reduce its mean to zero and variance to one as shown in (7.5.3). Each S_n^* is a random variable taking the set of values

$$x_{n,k} = \frac{k - np}{\sqrt{npq}}, \quad k = 0, 1, \ldots, n.$$

This is just the x_{nk} in (7.3.11). The probability distribution of S_n^* is given by

$$P(S_n^* = x_{n,k}) = \binom{n}{k} p^k q^{n-k}, \quad 0 \le k \le n.$$

It is more convenient to use the corresponding distribution function; call it F_n so that

$$P(S_n^* \le x) = F_n(x), \quad -\infty < x < \infty.$$

Finally, if I is the finite interval $(a, b]$, and F is any distribution function, we shall write

$$F(I) = F(b) - F(a).$$

[By now you should understand why we used $(a, b]$ rather than (a, b) or $[a, b]$. It makes no difference if F is continuous, but the F_n's above are not continuous. Of course in the limit the difference disappears in the present case, but it cannot be ignored generally.] After these elaborate preparations, we can rewrite the De Moivre-Laplace formula in the elegant form below: for any finite interval I,

$$\lim_{n\to\infty} F_n(I) = \Phi(I). \tag{7.5.4}$$

Thus we see that we are dealing with the convergence of a sequence of distribution functions to a given distribution function in a certain sense.

In this formulation the subject is capable of a tremendous generalization. The sequence of distribution functions need not be those of normalized sums, the given limit need not be the normal distribution nor even specified in advance, and the sense of convergence need not be that specified above. For example, the Poisson limit theorem discussed in §7.1 can be viewed as a particular instance. The subject matter has been intensively studied in the last forty years and is still undergoing further evolutions. [For some reference books in English see [Feller 2], [Chung 1].] Here we must limit ourselves to one such generalization, the so-called Central Limit Theorem in its classical setting, which is about the simplest kind of extension of Theorem 6 in §7.3. Even so we shall need a powerful tool from more advanced theory which we can use but not fully explain. This extension consists in replacing the Bernoullian variables above by rather arbitrary ones, as we proceed to describe.

Let $\{X_j, j \geq 1\}$ be a sequence of *independent and identically distributed* random variables. The phrase "identically distributed" means they have a common distribution, which need not be specified. But it is assumed that the mean and variance of each X_j are finite and denoted by m and σ^2 respectively, where $0 < \sigma^2 < \infty$. Define S_n and S_n^* exactly as before, then

$$E(S_n) = nm, \quad \sigma^2(S_n) = n\sigma^2 \tag{7.5.5}$$

and (7.5.3) holds as before. Again let F_n denote the distribution of the normalized sum S_n^*. Then Theorem 8 below asserts that (7.5.4) remains true under the liberalized conditions for the X_j's. To mention just some simple cases, each X_j may now be a "die-rolling" instead of a "coin-tossing" random variable to which Theorem 6 is applicable; or it may be uniformly distributed ("point-picking" variable); or again it may be exponentially distributed ("telephone-ringing" variable). Think of some other varieties if you wish.

Theorem 8. *For the sums S_n under the generalized conditions spelled out above, we have for any $a < b$:*

$$\lim_{n\to\infty} P\left(a < \frac{S_n - nm}{\sqrt{n}\,\sigma} \leq b\right) = \frac{1}{\sqrt{2\pi}} \int_a^b e^{-x^2/2}\, dx. \tag{7.5.6}$$

Proof: The powerful tool alluded to earlier is that of the characteristic function discussed in §6.5. [We could not have used the moment-generating function since it may not exist for S_n.] For the unit normal distribution Φ, its characteristic function g can be obtained by substituting $i\theta$ for θ in (7.4.4):

$$g(\theta) = e^{-\theta^2/2}. \tag{7.5.7}$$

With each arbitrary distribution F_n there is also associated its characteristic function g_n which is in general expressible by means of F_n as a *Stieltjes integral.* This is beyond the scope of this book but luckily we can by-pass it in the following treatment by using the associated random variables. [Evidently we will leave the reader to find out what may be concealed!] We can now state the following result.

Theorem 9. *If we have for every θ,*

$$\lim_{n\to\infty} g_n(\theta) = g(\theta) = e^{-\theta^2/2}, \tag{7.5.8}$$

then we have for every x:

$$\lim_{n\to\infty} F_n(x) = \Phi(x) = \frac{1}{\sqrt{2\pi}} \int_{-\infty}^{x} e^{-u^2/2}\, du; \tag{7.5.9}$$

in particular (7.5.4) is true.

Although we shall not prove this (see [Chung 1; Chapter 6], let us at least probe its significance. According to Theorem 7 of §6.5, each g_n uniquely determines F_n, and g determines Φ. The present theorem carries this correspondence between distribution function and its transform (characteristic function) one step further; for it says that the *limit* of the sequence $\{g_n\}$ also determines the *limit* of the sequence $\{F_n\}$. Hence it has been called the "continuity theorem" for the transform. In the case of the normal Φ above the result is due to Pólya; the general case is due to Paul Lévy (1886–1972) and Harald Cramér (1893–), both pioneers of modern probability theory.

Next we need a little lemma about characteristic functions.

Lemma. *If X has mean 0 and variance 1, then its characteristic function h has the following Taylor expansion at $\theta = 0$:*

$$h(\theta) = 1 - \frac{\theta^2}{2}(1 + \epsilon(\theta)) \tag{7.5.10}$$

where ϵ is a function depending on h such that $\lim_{\theta\to 0} \epsilon(\theta) = 0$.

Proof: According to a useful form of Taylor's theorem (look it up in your calculus book): if h has a second derivative at $\theta = 0$, then we have

$$h(\theta) = h(0) + h'(0)\theta + \frac{h''(0)}{2}\theta^2(1 + \epsilon(\theta)). \tag{7.5.11}$$

From

$$h(\theta) = E(e^{i\theta X})$$

we obtain by formal differentiation:

$$h'(\theta) = E(e^{i\theta X} iX), \quad h''(\theta) = E(e^{i\theta X}(iX)^2);$$

hence

$$h'(0) = E(iX) = 0, \quad h''(0) = E(-X^2) = -1.$$

Substituting into (7.5.11) we get (7.5.10).

Theorem 8 can now be proved by a routine calculation. Consider the characteristic function of S_n^*:

$$E(e^{i\theta S_n^*}) = E(e^{i\theta(X_1^* + \cdots + X_n^*)/\sqrt{n}})$$

Since the X_j^*'s are independent and identically distributed as well as the X_j's, by the analogue of Theorem 6 of §6.5 the right member above is equal to

$$E(e^{i\theta X_1^*/\sqrt{n}})^n = h\left(\frac{\theta}{\sqrt{n}}\right)^n \tag{7.5.12}$$

where h denotes the characteristic function of X_1^*. It follows from the Lemma that

$$h\left(\frac{\theta}{\sqrt{n}}\right) = 1 - \frac{\theta^2}{2n}\left(1 + \epsilon\left(\frac{\theta}{\sqrt{n}}\right)\right) \tag{7.5.13}$$

where θ is fixed and $n \to \infty$. Consequently we have

$$\begin{aligned} \lim_{n\to\infty} E(e^{i\theta S_n^*}) &= \lim_{n\to\infty} \left[1 - \frac{\theta^2}{2n}\left(1 + \epsilon\left(\frac{\theta}{\sqrt{n}}\right)\right)\right]^n \\ &= e^{-\theta^2/2} \end{aligned}$$

by an application of (7.1.12). This means the characteristic functions of S_n^* converge to that of the unit normal, therefore by Theorem 9, the distribution F_n converges to Φ in the sense of (7.5.9), from which (7.5.6) follows.

The name "central limit theorem" is used generally to designate a convergence theorem in which the normal distribution appears as the limit. More particularly it applies to sums of random variables as in Theorem 8. Historically these variables arose as errors of observations of chance fluctuations, so that the result is the all-embracing assertion that under "normal" conditions they all obey the same *normal* law, also known as the "error function." For this reason it had been regarded by some as a law of nature! Even in this narrow context Theorem 8 can be generalized in several directions: the assumptions of a finite second moment, of a common distribution, and of strict independence can all be relaxed. Finally, if the normal conditions are radically altered, then the central limit theorem will no longer apply, and random phenomena abound in which the limit distribution is no longer normal. The Poisson case discussed in §7.1 may be considered as one such example but there are other laws closely related to the normal which are called "stable" and "infinitely divisible" laws. See [Chung 1; Chapter 7] for a discussion of the various possibilities mentioned here.

It should be stressed that the central limit theorem as stated in Theorems 6 and 8 is of the form (7.5.4), without giving an estimate of the "error" $F_n(I) - \Phi(I)$. In other words, it asserts convergence without indicating any "speed of convergence." This renders the result useless in accurate numerical computations. However, under specified conditions it is possible to obtain bounds for the error. For example in the De Moivre-Laplace case (7.3.19) we can show that the error does not exceed $C/\sqrt{n}$ where C is a numerical constant involving p but not a or b; see [Chung 1; §7.4] for a more general result. In crude, quick-and-dirty applications the error is simply ignored, as will be done below.

In contrast to the mathematical developments, simple practical applications which form the backbone of "large sample theory" in statistics are usually of the cook-book variety. The great limit theorem embodied in (7.5.6) is turned into a rough approximate formula which may be written as follows:

$$P(x_1\sigma\sqrt{n} < S_n - mn < x_2\sigma\sqrt{n}) \approx \Phi(x_2) - \Phi(x_1).$$

In many situations we are interested in a symmetric spread around the mean, i.e., $x_1 = -x_2$, then the above becomes by (7.4.2):

$$P(|S_n - mn| < x\,\sigma\sqrt{n}) \approx 2\Phi(x) - 1. \tag{7.5.14}$$

Extensive tabulations of the values of Φ and its inverse function Φ^{-1} are available; a short table is appended at the end of the book. The following example illustrates the routine applications of the central limit theorem.

Example 7. A physical quantity is measured many times for accuracy. Each measurement is subject to a random error. It is judged reasonable to assume that it is uniformly distributed between -1 and $+1$ in a conveniently chosen unit. Now if we take the arithmetical mean [average] of n measurements, what is the probability that it differs from the true value by less than a fraction δ of the unit?

Let the true value be denoted by m and the actual measurements obtained by X_j, $1 \le j \le n$. Then the hypothesis says that

$$X_j = m + \xi_j,$$

where ξ_j is a random variable which has the uniform distribution in $[-1, +1]$. Thus

$$E(\xi_j) = \int_{-1}^{+1} \frac{x}{2}\, dx = 0, \quad \sigma^2(\xi_j) = E(\xi_j^2) = \int_{-1}^{+1} \frac{1}{2} x^2\, dx = \frac{1}{3};$$

$$E(X_j) = m, \qquad \sigma^2(X_j) = \frac{1}{3}.$$

In our notation above, we want to compute the approximate value of $P\{|S_n - mn| < \delta n\}$. This probability must be put into the form given in (7.5.6), and the limit relation there becomes by (7.5.14):

$$P\left\{\left|\frac{S_n - mn}{\sqrt{n/3}}\right| < \delta\sqrt{3n}\right\} \approx 2\Phi(\delta\sqrt{3n}) - 1.$$

For instance, if $n = 25$ and $\delta = 1/5$, then the result is equal to

$$2\Phi(\sqrt{3}) - 1 \approx 2\Phi(1.73) - 1 \approx .92,$$

from the Table on p. 321. Thus, if 25 measurements are taken, then we are 92% sure that their average is within one fifth of a unit from the true value.

Often the question is turned around: how many measurements should we take in order that the probability will exceed α (the "significance level") that the average will differ from the true value by at most δ? This means we must first find the value x_α such that

$$2\Phi(x_\alpha) - 1 = \alpha, \quad \text{or} \quad \Phi(x_\alpha) = \frac{1 + \alpha}{2};$$

and then choose n to make

$$\delta\sqrt{3n} > x_\alpha.$$

For instance, if $\alpha = .95$ and $\delta = 1/5$, then the Table shows that $x_\alpha \approx 1.96$; hence

$$n > \frac{x_\alpha^2}{3\delta^2} \approx 32.$$

Thus, seven or eight more measurements should increase our degree of confidence from 92% to 95%. Whether this is worthwhile may depend on the cost of doing the additional work as well as the significance of the enhanced probability.

It is clear that there are three variables involved in questions of this kind, namely: δ, α and n. If two of them are fixed, we can solve for the third. Thus if $n = 25$ is fixed because the measurements are found in recorded data and not repeatable, and our credulity demands a high degree of confidence α, say

99%, then we must compromise on the coefficient of accuracy δ. We leave this as an exercise.

Admittedly these practical applications of the great theorem are dull stuff, but so are e.g. Newton's laws of motion on the quotidian level.

7.6. Law of large numbers

In this section we collect two results related to the central limit theorem: the law of large numbers and Chebyshev's inequality.

The celebrated Law of Large Numbers can be deduced from Theorem 8 as an easy consequence.

Theorem 10. *Under the same conditions as in Theorem* 8, *we have for a fixed but arbitrary constant* $c > 0$:

$$\lim_{n\to\infty} P\left(\left|\frac{S_n}{n} - m\right| < c\right) = 1. \tag{7.6.1}$$

Proof: Since c is fixed, for any positive constant l, we have

$$l\sigma\sqrt{n} < cn \tag{7.6.2}$$

for all sufficiently large values of n. Hence the event

$$\left\{\left|\frac{S_n - mn}{\sigma\sqrt{n}}\right| < l\right\} \text{ certainly implies } \left\{\left|\frac{S_n - mn}{n}\right| < c\right\}$$

and so

$$P\left(\left|\frac{S_n - mn}{\sigma\sqrt{n}}\right| < l\right) \leq P\left(\left|\frac{S_n - mn}{n}\right| < c\right) \tag{7.6.3}$$

for large n. According to (7.5.6) with $a = -l$, $b = +l$, the left member above converges to

$$\frac{1}{\sqrt{2\pi}} \int_{-l}^{l} e^{-x^2/2}\, dx$$

as $n \to \infty$. Given any $\delta > 0$ we can first choose l so large that the value of the integral above exceeds $1 - \delta$, then choose n so large that (7.6.3) holds. It follows that

$$P\left(\left|\frac{S_n}{n} - m\right| < c\right) > 1 - \delta \tag{7.6.4}$$

for all sufficiently large n, and this is what (7.6.1) says.

Briefly stated, the law of large numbers is a corollary to the central limit theorem because any large multiple of $\sqrt{n}$ is negligible in comparison with any small multiple of n.

In the Bernoullian case the result was first proved by Jakob Bernoulli as a crowning achievement. [*Jakob* or *Jacques Bernoulli* (1654–1705), Swiss mathematician and physicist, author of the first treatise on probability: *Ars conjectandi* (1713) which contained this theorem.] His proof depended on direct calculations with binomial coefficients without of course the benefit of such formulas as Stirling's. In a sense the DeMoivre-Laplace Theorem 6 was a sequel to it. By presenting it in reverse to the historical development it is made to look like a trivial corollary. As a matter of fact, the law of large numbers is a more fundamental but also more primitive limit theorem. It holds true under much broader conditions than the central limit theorem. For instance, in the setting of Theorem 8, it is sufficient to assume that the common mean of X_j is finite, without any assumption on the second moment. Since the assertion of the law concerns only the mean such an extension is significant and was first proved by A. Ya. Khintchine [1894–1959, one of the most important of the school of Russian probabilists]. In fact, it can be proved by the method used in the proof of Theorem 8 above, except that it requires an essential extension of Theorem 9 which will take us out of our depth here. (See Theorem 6.4.3 of [Chung 1].) Instead we will give an extension of Theorem 10 in another direction, when the random variables $\{X_j\}$ are not necessarily identically distributed. This is easy via another celebrated but simple result known as Chebyshev's inequality. [P. L. Chebyshev (1821–1894), together with A. A. Markov (1856–1922) and A. M. Ljapunov (1857–1918), were founders of the Russian school of probability.]

Theorem 11. *Suppose the random variable X has a finite second moment, then for any constant $c > 0$ we have*

$$P(|X| \geq c) \leq \frac{E(X^2)}{c^2}. \tag{7.6.5}$$

Proof: We will carry out the proof for a countably valued X and leave the analogous proof for the density case as an exercise. The idea of the proof is the same for a general random variable.

Suppose that X takes the values v_i with probabilities p_i, as in §4.3. Then we have

$$E(X^2) = \sum_j p_j v_j^2. \tag{7.6.6}$$

If we consider only those values v_j satisfying the inequality $|v_j| \geq c$ and denote by A the corresponding set of indices j, namely $A = \{j \mid |v_j| \geq c\}$, then of course $v_j^2 \geq c^2$ for $j \in A$, whereas

$$P(|X| \geq c) = \sum_{j \in A} p_j.$$

Hence if we sum the index j only over the partial set A, we have

$$E(X^2) \geq \sum_{j\in A} p_j v_j^2 \geq \sum_{j\in A} p_j c^2 = c^2 \sum_{j\in A} p_j = c^2 P(|X| \geq c),$$

which is (7.6.5).

We can now state an extended form of the law of large numbers as follows.

Theorem 12. *Let $\{X_j, j \geq 1\}$ be a sequence of independent random variables such that for each j:*

$$E(X_j) = m_j, \quad \sigma^2(X_j) = \sigma_j^2; \tag{7.6.7}$$

and furthermore suppose there exists a constant $M < \infty$ such that for all j:

$$\sigma_j^2 \leq M. \tag{7.6.8}$$

Then we have for each fixed $c > 0$:

$$\lim_{n\to\infty} P\left(\left|\frac{X_1 + \cdots + X_n}{n} - \frac{m_1 + \cdots + m_n}{n}\right| < c\right) = 1. \tag{7.6.9}$$

Proof: If we write $X_j^0 = X_j - m_j$, $S_n^0 = \sum_{j=1}^{n} X_j^0$, then the expression between the bars above is just S_n^0/n. Of course $E(S_n^0) = 0$, whereas

$$E((S_n^0)^2) = \sigma^2(S_n^0) = \sum_{j=1}^{n} \sigma^2(X_j^0) = \sum_{j=1}^{n} \sigma_j^2.$$

This string of equalities follows easily from the properties of variances and you ought to have no trouble recognizing them now at a glance. [If you still do then you should look up the places in preceding chapters where they are discussed.] Now the condition in (7.6.8) implies that

$$E((S_n^0)^2) \leq Mn, \quad E\left(\left(\frac{S_n^0}{n}\right)^2\right) \leq \frac{M}{n}. \tag{7.6.10}$$

It remains to apply Theorem 11 to $X = S_n^0/n$ to obtain

$$P\left(\left|\frac{S_n^0}{n}\right| \geq c\right) \leq \frac{E((S_n^0/n)^2)}{c^2} \leq \frac{M}{c^2 n}. \tag{7.6.11}$$

Hence the probability above converges to zero as $n \to \infty$, which is equivalent to the assertion in (7.6.9).

Actually the proof yields more: it gives an estimate on the "speed of convergence." Namely, given M, c and δ we can tell how large n must be in order that the probability in (7.6.9) exceeds $1 - \delta$. Note also that Theorem 10 is a particular case of Theorem 12 because there all the σ_j^2's are equal and we may take $M = \sigma_1^2$.

Perhaps the reader will agree that the above derivations of Theorems 11 and 12 are relatively simple doings compared with the fireworks in §§7.3–7.4. Looking back, we may find it surprising that it took two centuries before the *right* proof of Bernoulli's theorem was discovered by Chebyshev. It is an instance of the triumph of an *idea*, a new way of thinking, but even Chebyshev himself buried his inequality among laborious and unnecessary details. The cleaning up as shown above was done by later authors. Let us observe that the method of proof is applicable to any sum S_n of random variables, whether they are independent or not, provided that the crucial estimates in (7.6.10) are valid.

We turn now to the meaning of the law of large numbers. This is best explained in the simplest Bernoullian scheme where each X_j takes the values 1 and 0 with probabilities p and $q = 1 - p$, as in Theorems 5 and 6 above. In this case $S_n^0 = S_n - np$ and $E((S_n^0)^2) = \sigma^2(S_n) = npq$, so that (7.6.11) becomes,

$$(7.6.12) \qquad P\left(\left|\frac{S_n}{n} - p\right| \geq c\right) \leq \frac{pq}{c^2n} \leq \frac{1}{4c^2n};$$

note that $p(1 - p) \leq 1/4$ for $0 \leq p \leq 1$. In terms of coin-tossing with p as the probability of a head, S_n/n represents the relative frequency of heads in n independent tosses; cf. Example 3 of §2.1. This is of course a random number varying from one experiment to another. It will clarify things if we reinstate the long-absent sample point ω and write explicitly:

$$\frac{S_n(\omega)}{n} = \begin{array}{l}\text{relative frequency of heads in } n \text{ tosses associated}\\ \text{with the experiment denoted by } \omega.\end{array}$$

Thus each ω is conceived as the record of the outcomes of a sequence of coin-tossing called briefly an *experiment*, so that different ω's correspond to different experiments. The sequence of tossing should be considered as *infinite* [indefinite, unending] in order to allow arbitrarily large values of n in computing the frequency [but we are not necessarily assuming the existence of a limiting frequency; see below]. Symbolically, ω may be regarded as a concise representation of the sequence of successive outcomes:

$$\omega = \{X_1(\omega), X_2(\omega), \ldots, X_n(\omega), \ldots\},$$

see the discussion in §4.1. We can talk about the probability of certain sets of ω; indeed we have written the probability of the set

$$\Lambda_n(c) = \left\{\omega \,\middle|\, \left|\frac{S_n(\omega)}{n} - p\right| \geq c\right\}$$

in (7.6.12); see also e.g. (7.3.10) and (7.3.19) where similar probabilities are evaluated. In the precise form given in (7.6.12), if the value c as well as any given positive number ϵ is assigned, we can determine how large n need be to make $P(\Lambda_n(c)) \leq \epsilon$, namely:

$$\frac{1}{4c^2n} \leq \epsilon \quad \text{or} \quad n \geq \frac{1}{4c^2\epsilon}. \tag{7.6.13}$$

This bound for n is larger than necessary because Chebyshev's inequality is a crude estimate in general. A sharper bound can be culled from Theorem 6 as follows. Rewrite the probability in (7.6.12) and use the approximation given in (7.3.19) simplified by (7.4.2):

$$P\left(\left|\frac{S_n - np}{\sqrt{npq}}\right| \geq \sqrt{\frac{n}{pq}}\,c\right) \approx 2\left[1 - \Phi\left(\sqrt{\frac{n}{pq}}\,c\right)\right]. \tag{7.6.14}$$

This is *not* an asymptotic relation as it pretends to be, because the constants a and b must be fixed in Theorem 6, whereas here $\pm c\sqrt{n}/\sqrt{pq}$ vary with n (and increase too rapidly with n). It can be shown that the difference of the two sides above is of the form $A/\sqrt{n}$ where A is a numerical constant depending on p, and this error term should not be ignored. But we do so below. Now put

$$\eta = \sqrt{\frac{n}{pq}}\,c;$$

our problem is to find the value of η to make

$$2[1 - \Phi(\eta)] \leq \epsilon \quad \text{or} \quad \Phi(\eta) \geq 1 - \frac{\epsilon}{2}; \tag{7.6.15}$$

then solve for n from η. This can be done by looking up a table of values of Φ; a short one is appended at the end of this book.

Example 8. Suppose $c = 2\%$ and $\epsilon = 5\%$. Then (7.6.15) becomes

$$\Phi(\eta) \geq 1 - \frac{5}{200} = .975.$$

From the table we see that this is satisfied if $\eta \geq 1.96$. Thus

$$n \geq \frac{(1.96)^2pq}{c^2} = \frac{(1.96)^2 \times 10000}{4}\,pq.$$

The last term depends on p, but $p(1 - p) \leq 1/4$ for all p, as already noted, and so $n \geq 10000 \cdot \frac{1}{4} = 2500$ will do. For comparison, the bound given in

(7.6.13) requires $n \geq 12500$; but that estimate has been rigorously established whereas the normal approximation is a rough-and-dirty one. We conclude that if the coin is tossed more than 2500 times, then we can be 95% sure that relative frequency of heads computed from the actual experiment will differ from the true p by no more than 2%.

Such a result can be applied in two ways (both envisioned by Bernoulli): (i) if we consider p as known then we can make a prediction on the outcome of the experiment; (ii) if we regard p as unknown then we can make an estimate of its value by performing an actual experiment. The second application has been called a problem of "inverse probability" and is the origin of the so-called Monte-Carlo method. Here is a numerical example. In an actual experiment 10000 tosses were made and the total number of heads obtained is 4979; see [Feller 1; p. 21] for details. The computation above shows that we can be 95% sure that

$$\left|p - \frac{4979}{10000}\right| \leq \frac{2}{100} \quad \text{or} \quad .4779 \leq p \leq .5179.$$

Returning to the general situation in Theorem 10, we will state the law of large numbers in the following form reminiscent of the definition of an ordinary limit. For any $\epsilon > 0$, there exists an $n_0(\epsilon)$ such that for all $n \geq n_0(\epsilon)$ we have

$$P\left(\left|\frac{S_n}{n} - m\right| < \epsilon\right) > 1 - \epsilon. \tag{7.6.16}$$

We have taken both c and δ in (7.6.4) to be ϵ without loss of generality [see Exercise 22 below]. If we interpret this as in the preceding example as an assertion concerning the proximity of the theoretical mean m to the empirical average S_n/n, the double hedge [margin of error] implied by the two ϵ's in (7.6.16) seems inevitable. For in any experiment one can neither be 100% sure nor 100% accurate, otherwise the phenomenon would not be a random one. Nevertheless mathematicians are idealists and long for perfection. What cannot be realized in the empirical world may be achieved in a purely mathematical scheme. Such a possibility was uncovered by Borel in 1909, who created a new chapter in probability by his discovery described below. In the Bernoullian case, his famous result may be stated as follows:

$$P\left(\lim_{n\to\infty} \frac{S_n}{n} = p\right) = 1.\dagger \tag{7.6.17}$$

This is known as a "strong law of large numbers," which is an essential improvement on Bernoulli's "weak law of large numbers." It asserts the existence of a limiting frequency equal to the theoretical probability p, for all sample points ω except possibly a set of probability zero (but not necessarily an empty set). Thus the limit in (2.1.10) indeed exists, but only *for almost all*

† For a discussion of Borel's theorem and related topics, see Chapter 5 of Chung [1].

ω, so that the empirical theory of frequencies beloved by the applied scientist is justifiable through a sophisticated theorem. The difference between this and Bernoulli's weaker theorem:

$$\forall \epsilon > 0: \lim_{n \to \infty} P\left(\left|\frac{S_n}{n} - p\right| < \epsilon\right) = 1,$$

is subtle and cannot be adequately explained without measure theory. The astute reader may observe that although we claim 100% certainty and accuracy in (7.6.17), the limiting frequency is not an empirically observable thing—so that the cynic might say that what we are sure of is only an *ideal*, whereas the sophist could retort that we shall never be caught wanting! Even so a probabilistic certainty does not mean absolute certainty in the deterministic sense. There is an analogue of this distinction in the Second Law of Thermodynamics (which comes from statistical mechanics). According to that law, e.g., when a hot body is in contact with a cold body, it is *logically possible* that heat will flow from the cold to the hot, but the probability of this happening is zero. A similar exception is permitted in Borel's theorem. For instance, if a coin is tossed indefinitely, it is logically possible that it's heads every single time. Such an event constitutes an exception to the assertion in (7.6.17), but its probability is equal to $\lim_{n \to \infty} p^n = 0$.

The strong law of large numbers is the foundation of a mathematical theory of probability based on the concept of frequency; see §2.1. It makes better sense than the weak one and is indispensable for certain theoretical investigations. [In statistical mechanics it is known in an extended form under the name *Ergodic Theorem*.] But the dyed-in-the-wool empiricist, as well as a radical school of logicians called intuitionists, may regard it as an idealistic fiction. It is amusing to quote two eminent authors on the subject:

Feller: "[the weak law of large numbers] is of very limited interest and should be replaced by the more precise and more useful strong law of large numbers" (p. 152 of [Feller 1]).

van der Waerden: "[the strong law of large numbers] scarcely plays a role in mathematical statistics" (p. 98 of *Mathematische Statistik*, 3rd ed., Springer-Verlag, 1971).

Let us end this discussion by keeping in mind the gap between observable phenomena in the real world and the theoretical models used to study them; see Einstein's remark on p. 123. The law of large numbers, weak or strong, is a mathematical theorem deduced from axioms. Its applicability to true-life experiences such as the tossing of a penny or nickel is necessarily limited and imperfect. The various examples given above to interpret and illustrate the theorems should be viewed with this basic understanding.

Exercises

1. Suppose that a book of 300 pages contains 200 misprints. Use Poisson approximation to write down the probability that there is more than one misprint on a particular page.

2. In a school where 4% of the children write with their left hands, what is the probability that there are no left-handed children in a class of 25?
3. Six dice are thrown 200 times by the players. Estimate the probability of obtaining "six different faces" k times, where $k = 0, 1, 2, 3, 4, 5$.
4. A home bakery made 100 loaves of raisin bread using 2000 raisins. Write down the probability that the loaf you bought contains 20 to 30 raisins.
5. It is estimated that on a certain island of 15 square miles there are 20 giant tortoises of one species and 30 of another species left. An ecological survey team spotted 2 of them in an area of 1 square mile, but neglected to record which species. Use Poisson distribution to find the probabilities of the various possibilities.
6. Find the maximum term or terms in the binomial distribution $B_k(n; p)$, $0 \leq k \leq n$. Show that the terms increase up to the maximum and then decrease. [Hint: take ratios of consecutive terms.]
7. Find the maximum term or terms in the Poisson distribution $\pi_k(\alpha)$, $0 \leq k < \infty$. Show the same behavior of the terms as in No. 6.
8. Let X be a random variable such that $P(X = c + kh) = \pi_k(\alpha)$ where c is a real and h is a positive number. Find the Laplace transform of X.
9. Find the convolution of two sequences given by Poisson distributions $\{\pi_k(\alpha)\}$ and $\{\pi_k(\beta)\}$.

10.* If X_α has the Poisson distribution $\pi(\alpha)$, then

$$\lim_{\alpha \to \infty} P\left\{\frac{X_\alpha - \alpha}{\sqrt{\alpha}} \leq u\right\} = \Phi(u)$$

for every u. [Hint: use the Laplace transform $E(e^{-\lambda(X_\alpha - \alpha)/\sqrt{\alpha}})$, show that as $\alpha \to \infty$ it converges to $e^{\lambda^2/2}$, and invoke the analogue of Theorem 9 of §7.5.]

11. Assume that the distance between cars going in one direction on a certain highway is exponentially distributed with mean value 100 meters. What is the probability that in a stretch of 5 kilometers there are between 50 to 60 cars?
12. On a certain highway the flow of traffic may be assumed to be Poissonian with intensity equal to 30 cars per minute. Write down the probability that it takes more than N seconds for n consecutive cars to pass by an observation post. [Hint: use (7.2.11).]
13. A perfect die is rolled 100 times. Find the probability that the sum of all points obtained is between 330 and 380.
14. It is desired to find the probability p that a certain thumb tack will fall on its flat head when tossed. How many trials are needed in order that we may be 95% sure that the observed relative frequency differs from p by less than $p/10$? [Hint: try it a number of times to get a rough bound for p.]
15. Two movie theatres compete for 1000 customers. Suppose that each

customer chooses one of the two with "total indifference" and independently of other customers. How many seats should each theatre have so that the probability of turning away any customer for lack of seats is less than 1%?

16. A sufficient number of voters are polled to determine the percentage in favor of a certain candidate. Assume that an unknown proportion p of the voters favor him and they act independently of one another, how many should be polled to predict the value of p within 4.5% with 95% confidence? [This is the so-called "four percent margin of error in predicting elections, presumably because $<.045$ becomes $\leq .04$ by the rule of rounding decimals.]
17. Write $\Phi((a, b))$ for $\Phi(b) - \Phi(a)$ where $a < b$ and Φ is the unit normal distribution. Show that $\Phi((0, 2)) > \Phi((1, 3))$ and generalize to any two intervals of the same length. [Hint: $e^{-x^2/2}$ decreases as $|x|$ increases.]
18. Complete the proof of (7.1.8) and then use the same method to prove (7.1.12). [Hint: $|\log(1 - x) + x| \leq \frac{1}{2} \sum_{n=2}^{\infty} |x|^n = \frac{1}{2} \frac{x^2}{1 - |x|}$; hence if $|x| \leq \frac{1}{2}$ this is bounded by x^2.]
19. Prove (7.1.13).
20. Prove Chebyshev's inequality when X has a density. [Hint: $\sigma^2(X) = \int_{-\infty}^{\infty} (x - m)^2 f(x)\, dx \geq \int_{|x-m|>c} (x - m)^2 f(x)\, dx$.]
21. Prove the following analogue of Chebyshev's inequality where the absolute first moment is used in place of the second moment:

$$P(|X - m| > c) \leq \frac{1}{c} E(|X - m|).$$

22.* Show that $\lim_{n\to\infty} P(|X_n| > \epsilon) = 0$ for every ϵ if and only if given any ϵ, there exists $n_0(\epsilon)$ such that

$$P(|X_n| > \epsilon) < \epsilon \quad \text{for} \quad n > n_0(\epsilon).$$

This is also equivalent to: given any δ and ϵ, there exists $n_0(\delta, \epsilon)$ such that

$$P(|X_n| > \epsilon) < \delta \quad \text{for} \quad n > n_0(\delta, \epsilon).$$

[Hint: consider $\epsilon' = \delta \wedge \epsilon$ and apply the first form.]

23. If X has the distribution Φ, show that $|X|$ has the distribution Ψ, where $\Psi = 2\Phi - 1$; Ψ is called the "positive normal distribution."
24. If X has the distribution Φ, find the density function of X^2 and the corresponding distribution. This is known as the "chi-square distribution" in statistics. [Hint: differentiate $P(X^2 < x) = 2/\sqrt{2\pi} \int_0^{\sqrt{x}} e^{-u^2/2}\, du$.]

25.* Use No. 24 to show that

$$\int_0^\infty x^{-1/2}e^{-x}\,dx = \sqrt{\pi}.$$

The integral is equal to $\Gamma\left(\frac{1}{2}\right)$ where Γ is the *gamma function* defined by $\Gamma(\alpha) = \int_0^\infty x^{\alpha-1}e^{-x}\,dx$ for $\alpha > 0$. [Hint: consider $E(X^2)$ in No. 24.]

26.* Let $\{\xi_k, 1 \leq k \leq n\}$ be n random variables satisfying $0 < \xi_1 \leq \xi_2 \leq \cdots \leq \xi_n \leq t$; let $(0, t] = \bigcup_{k=1}^{l} I_k$ be an arbitrary partition of $(0, t]$ into subintervals $I_k = (x_{k-1}, x_k]$ where $x_0 = 0$; and $\tilde{N}(I_k)$ denote the number of ξ's belonging to I_k. How can we express the event $\{\xi_k \leq x_k; 1 \leq k \leq l\}$ by means of $\tilde{N}(I_k)$, $1 \leq k \leq l$? Here of course $0 < x_1 < x_2 < \cdots < x_n \leq t$. Now suppose that x_k, $1 \leq k \leq l$, are arbitrary and answer the question again. [Hint: try $n = 2$ and 3 to see what is going on; relabel the x_k in the second part.]

27.* Let $\{X(t), t \geq 0\}$ be a Poisson process with parameter α. For a fixed $t > 0$ define $\delta(t)$ to be the distance from t to the last jump before t if there is one, and to be t otherwise. Define $\delta'(t)$ to be the distance from t to the next jump after t. Find the distributions of $\delta(t)$ and $\delta'(t)$. [Hint: if $u < t$, $P\{\delta(t) > u\} = P\{N(t-u, t) = 0\}$; for all $u > 0$, $P\{\delta'(t) > u\} = P\{N(t, t+u) = 0\}$.]

28.* Let $\tau(t) = \delta(t) + \delta'(t)$ as in No. 27. This is the length of the between-jump interval containing the given time t. For each ω, this is one of the random variables T_k described in §7.2. Does $\tau(t)$ have the same exponential distribution as all the T_k's? [This is a nice example where logic must take precedence over "intuition," and is often referred to as a paradox. The answer should be easy from No. 27. For further discussion at a level slightly more advanced than this book, see Chung, "The Poisson process as renewal process," *Periodica Mathematica Hungarica*, Vol. 2 (1972), pp. 41–48.

29. Use Chebyshev's inequality to show that if X and Y are two arbitrary random variables satisfying $E\{(X - Y)^2\} = 0$, then we have $P(X = Y) = 1$, namely X and Y are almost surely identical. [Hint: $P(|X - Y| > \epsilon) = 0$ for any $\epsilon > 0$.]

30. Recall the coefficient of correlation $\rho(X, Y)$ from p. 169. Show that if $\rho(X, Y) = 1$, then the two "normalized" random variables:

$$\tilde{X} = \frac{X - E(X)}{\sigma(X)}, \qquad \tilde{Y} = \frac{Y - E(Y)}{\sigma(Y)}$$

are almost surely identical. What if $\rho(X, Y) = -1$? [Hint: compute $E\{(\tilde{X} - \tilde{Y})^2\}$ and use No. 29.]

Appendix 2

Stirling's Formula and De Moivre-Laplace's Theorem

In this appendix we complete some details in the proof of Theorem 5, establish Stirling's formula (7.3.3) and relate it to the normal integral (7.4.1). We begin with an estimate.

Lemma. *If* $|x| \leq 2/3$, *then*

$$\log (1 + x) = x - \frac{x^2}{2} + \theta(x)$$

where $|\theta(x)| \leq |x|^3$.

Proof: We have by Taylor's series for $\log (1 + x)$:

$$\log (1 + x) = x - \frac{x^2}{2} + \sum_{n=3}^{\infty} (-1)^{n-1} \frac{x^n}{n}.$$

Hence $\theta(x)$ is equal to the series above and

$$\theta(x) \leq \sum_{n=3}^{\infty} \frac{|x|^n}{n} \leq \frac{1}{3} \sum_{n=3}^{\infty} |x|^n = \frac{|x|^3}{3(1 - |x|)}.$$

For $|x| \leq 2/3$, $3(1 - |x|) \geq 1$ and the lemma follows. The choice of the constant 2/3 is a matter of convenience; a similar estimate holds for any constant <1.

We will use the Lemma first to complete the proof of Theorem 5, by showing that the omitted terms in the two series expansions in (7.3.17) may indeed be ignored as $n \to \infty$. When n is sufficiently large the two quantities in (7.3.17′) will be $\leq 2/3$. Consequently the Lemma is applicable and the contribution from the "tails" of the two series, represented by dots there, is bounded by

$$k \left|\frac{\sqrt{npq}x_k}{k}\right|^3 + (n - k) \left|\frac{\sqrt{npq}x_k}{n - k}\right|^3.$$

Since $pq < 1$ and $|x_k| \leq A$, this does not exceed

$$\frac{n^{3/2}}{k^2} A^3 + \frac{n^{3/2}}{(n - k)^2} A^3,$$

which clearly tends to zero as $n \to \infty$, by (7.3.15). Therefore the tails vanish in the limit and we are led to (7.3.18) as shown there.

Next we shall prove, as a major step toward Stirling's formula, the relation below:

(A.2.1) $$\lim_{n\to\infty}\left\{\log n! - \left(n+\frac{1}{2}\right)\log n + n\right\} = C$$

where C is a constant to be determined later. Let d_n denote the quantity between the braces in (A.2.1), then a simple computation gives

$$d_n - d_{n+1} = \left(n+\frac{1}{2}\right)\log\left(1+\frac{1}{n}\right) - 1.$$

Using the notation in the Lemma, we write this as

$$\left(n+\frac{1}{2}\right)\left(\frac{1}{n} - \frac{1}{2n^2} + \theta\left(\frac{1}{n}\right)\right) - 1 = \left(n+\frac{1}{2}\right)\theta\left(\frac{1}{n}\right) - \frac{1}{4n^2},$$

and consequently by the Lemma with $x = 1/n$, $n \geq 2$:

$$|d_n - d_{n+1}| \leq \left(n+\frac{1}{2}\right)\frac{1}{n^3} + \frac{1}{4n^2} = \frac{2n+1}{2n^3} + \frac{1}{4n^2}.$$

Therefore the series $\sum_n |d_n - d_{n+1}|$ converges by the comparison test. Now recall that an absolutely convergent series is convergent, which means the partial sum tends to a finite limit, say C_1. Thus we have

$$\lim_{N\to\infty}\sum_{n=1}^{N}(d_n - d_{n+1}) = C_1;$$

but the sum above telescopes into $d_1 - d_{N+1}$, and so

$$\lim_{N\to\infty} d_{N+1} = d_1 - C_1,$$

and we have proved the assertion in (A.2.1) with $C = d_1 - C_1$. It follows that

$$\lim_{n\to\infty}\frac{n!e^n}{n^{n+(1/2)}} = e^C,$$

or if $K = e^C$:

(A.2.2) $$n! \sim K n^{n+(1/2)} e^{-n}.$$

If we compare this with (7.3.3) we see that it remains to prove that $K = \sqrt{2\pi}$ to obtain Stirling's formula. But observe that even without this evaluation of the constant K, the calculations in Theorems 5 and 6 of §7.3 are valid provided we replace $\sqrt{2\pi}$ by K everywhere. In particular, formula (7.3.19) with $a = -b$ becomes

(A.2.3) $$\lim_{n\to\infty} P\left(\left|\frac{S_n - np}{\sqrt{npq}}\right| \leq b\right) = \frac{1}{K}\int_{-b}^{b} e^{-x^2/2}\,dx.$$

On the other hand, we may apply Theorem 11 (Chebyshev's inequality) with $X = (S_n - np)/\sqrt{npq}$, $E(X) = 0$, $E(X^2) = 1$, to obtain the inequality:

(A.2.4) $$P\left(\left|\frac{S_n - np}{\sqrt{npq}}\right| \leq b\right) \geq 1 - \frac{1}{b^2}.$$

Combining the last two relations and remembering that a probability cannot exceed one, we obtain,

$$1 - \frac{1}{b^2} \leq \frac{1}{K}\int_{-b}^{b} e^{-x^2/2}\,dx \leq 1.$$

Letting $b \to \infty$ we conclude that

(A.2.5) $$K = \int_{-\infty}^{\infty} e^{-x^2/2}\,dx.$$

Since the integral above has the value $\sqrt{2\pi}$ by (7.4.1), we have proved that $K = \sqrt{2\pi}$.

Another way of evaluating K is via the Wallis's product formula given in many calculus texts (see e.g., Courant-John, *Introduction to calculus and analysis*, Vol. 1, New York: Interscience Publishers, 1965). If this is done then the argument above gives (A.2.5) with $K = \sqrt{2\pi}$, so that the formula for the normal integral (7.4.1) follows. This justifies the heuristic argument mentioned under (7.4.1), and shows the intimate relation between the two results named in the title of this appendix.

Chapter 8

From Random Walks to Markov Chains

8.1. Problems of the wanderer or gambler

The simplest *random walk* may be described as follows. A particle moves along a line by steps; each step takes it one unit to the right or to the left with probabilities p and $q = 1 - p$ respectively where $0 < p < 1$. For verbal convenience we suppose that each step is taken in a unit of time so that the nth step is made instantaneously at time n; furthermore we suppose that the possible positions of the particle are the set of all integers on the coordinate axis. This set is often referred to as the "integer lattice" on $R^1 = (-\infty, \infty)$ and will be denoted by I. Thus the particle executes a walk on the lattice, back and forth, and continues *ad infinitum*. If we plot its position X_n as a function of the time n, its *path* is a zigzag line of which some samples are shown below in Figure 30.

A more picturesque language turns the particle into a wanderer or drunkard and the line into an endless street divided into blocks. In each unit of time, say 5 minutes, he walks one block from street corner to corner, and at each corner he may choose to go ahead or turn back with probabilities p or q. He is then taking a random walk and his track may be traced on the street with a lot of doubling and re-doubling. This language suggests an immediate extension to a more realistic model where there are vertical as well as horizontal streets, regularly spaced as in parts of New York City. In this case each step may take one of the four possible directions as in Figure 31. This scheme corresponds to a random walk on the integer lattice of the plane R^2. We shall occasionally return to this below, but for the most part confine our discussion to the simplest situation of one dimension.

A mathematical formulation is near at hand. Let ξ_n be the nth step taken or *displacement*, so that

$$\xi_n = \begin{cases} +1 \text{ with probability } p, \\ -1 \text{ with probability } q; \end{cases} \tag{8.1.1}$$

and the ξ_n's are independent random variables. If we denote the initial position by X_0, then the position at time n (or after n steps) is just

$$X_n = X_0 + \xi_1 + \cdots + \xi_n. \tag{8.1.2}$$

Thus the random walk is represented by the sequence of random variables $\{X_n, n \geq 0\}$ which is a stochastic process in discrete time. In fact, $X_n - X_0$ is a sum of independent Bernoullian random variables much studied in

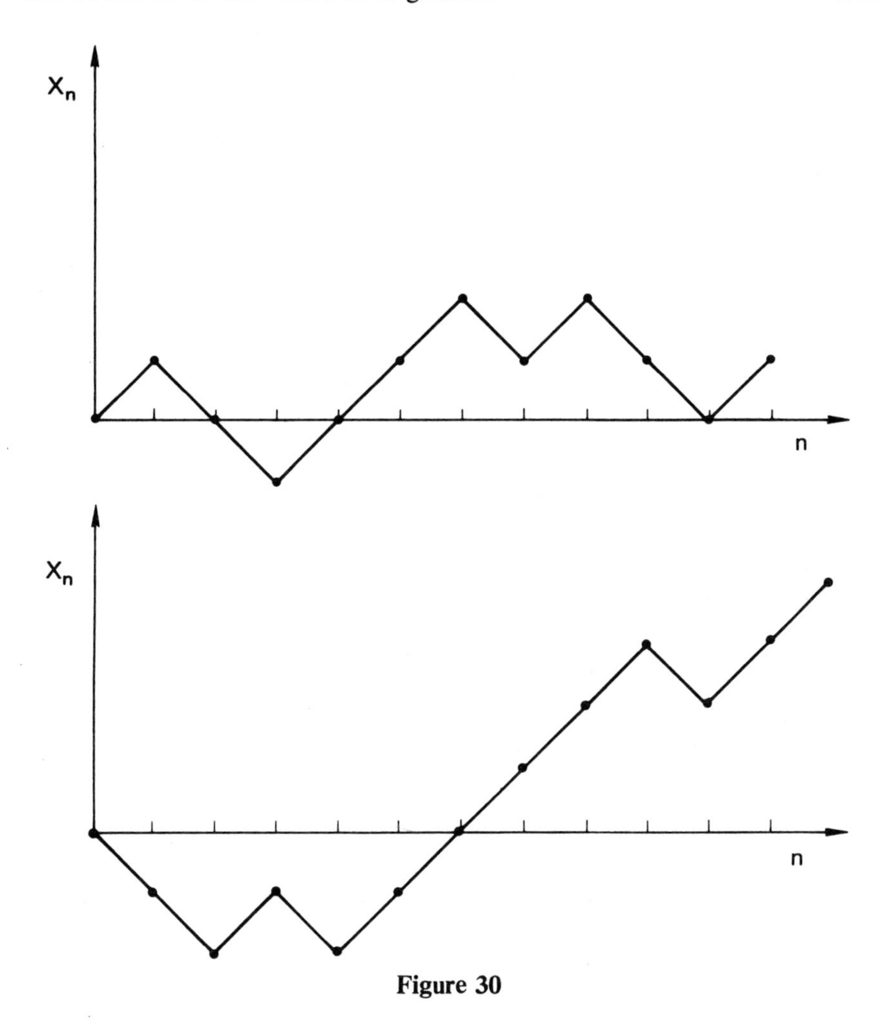

Figure 30

Chapters 5, 6 and 7. We have changed our previous notation (see e.g. (6.3.11)) to the present one in (8.1.2) to conform with later usage in §8.3. But apart from this what is new here?

The answer is that our point of view will be new. We are going to study the entire walk, or process, as it proceeds, or develops, in the course of time. In other words, each path of the particle or wanderer will be envisioned as a possible development of the process subject to the probability laws imposed on the motion. Previously we have been interested mostly in certain quantitative characteristics of X_n (formerly S_n) such as its mean, variance and distribution. Although the subscript n there is arbitrary and varies when $n \to \infty$, a probability like $P(a \leq X_n \leq b)$ concerns only the variable X_n taken one at a time, so to speak. Now we are going to probe deeper into the structure of the *sequence* $\{X_n, n \geq 0\}$ by asking questions which involve

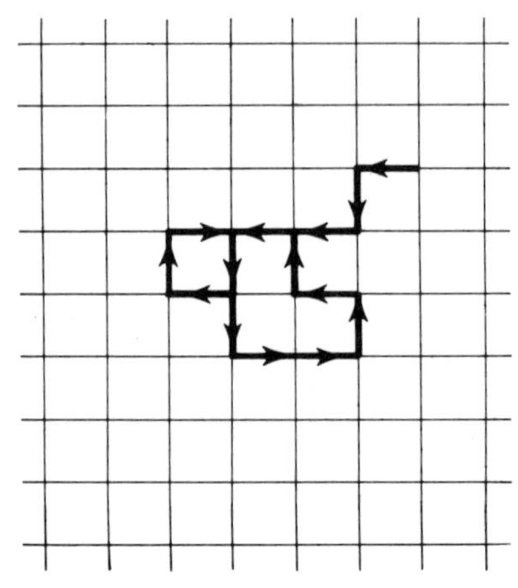

Figure 31

many of them all at once. Here are some examples. Will the moving particle ever "hit" a given point? If so, how long will this take, and will it happen before or after the particle hits some other point? One may also ask how frequently the particle hits a point or a set; how long it stays within a set, etc. Some of these questions will be made precise below and answered. In the meantime you should let your fancy go free and think up a few more such questions, and perhaps relate them to concrete models of practical significance.

Let us begin with the following problem.

Problem 1. Consider the interval $[0, c]$ where $c = a + b$ and $a \geq 1$, $b \geq 1$. If the particle starts at the point "a" what is the probability that it will hit one endpoint of the interval before the other?

This is a famous problem in another setting, discussed by Fermat and Pascal and solved in general by Montmart. Two gamblers Peter and Paul play a series of games in which Peter wins with probability p and Paul wins with probability q, and the outcomes of the successive games are assumed to be independent. For instance they may toss a coin repeatedly or play ping-pong or chess in which their skills are rated as p to q. The loser pays a dollar each time to the winner. Now if Peter has \$ a and Paul has \$ b at the outset and they continue to play until one of them is ruined (bankrupt), what is the probability that Peter will be ruined?

In this formulation the position of the particle at any time n becomes the number of dollars Peter has after n games. Each step to the right is \$1 won by him, each step to the left is \$1 lost. If the particle reaches 0 before c, then Peter has lost all his initial capital and is ruined; on the other hand if the particle reaches c before 0, then Paul has lost all his capital and is ruined. The game terminates when one of these eventualities occurs. Hence the historical name of "gambler's ruin problem."

We are now going to solve Problem 1. The solution depends on the following smart "put," for $1 \leq j \leq c - 1$:

(8.1.3) $u_j =$ the probability that the particle will reach 0 before c, when it starts from j.

The problem is to find u_a, but since "a" is arbitrary we really need all the u_j's. Indeed the idea is to exploit the relations between them and trap them together. These relations are given by the following set of *difference equations:*

$$u_j = pu_{j+1} + qu_{j-1}, \quad 1 \leq j \leq c - 1 \tag{8.1.4}$$

together with the *boundary conditions:*

$$u_0 = 1, \quad u_c = 0. \tag{8.1.5}$$

To argue (8.1.4), think of the particle as being at j and consider what will happen after taking one step. With probability p it will then be at $j + 1$, under which hypothesis the (conditional) probability of reaching 0 before c will be u_{j+1}; similarly with probability q it will be at $j - 1$, under which hypothesis the said probability will be u_{j-1}. Hence the *total* probability u_j is equal to the sum of the two terms on the right side of (8.1.4), by an application of Proposition 2 of §5.2. This argument spelled out in the extremal cases $j = 1$ and $j = c - 1$ entails the values of u_0 and u_c given in (8.1.5). These are not included in (8.1.3) and strictly speaking are *not* well-defined by the verbal description given there, although it makes sense by a kind of extrapolation.

The rest of our work is purely algebraic. Since $p + q = 1$ we may write the left member of (8.1.4) as $pu_j + qu_j$; after a transposition the equation becomes

$$q(u_j - u_{j-1}) = p(u_{j+1} - u_j).$$

Using the abbreviations

$$r = \frac{q}{p}, \quad d_j = u_j - u_{j+1};$$

we obtain the basic recursion between successive *differences* below:

$$d_j = rd_{j-1}. \tag{8.1.6}$$

Iterating we get $d_j = r^j d_0$; then summing by telescoping:

$$\begin{aligned} 1 = u_0 - u_c &= \sum_{j=0}^{c-1} (u_j - u_{j+1}) \\ &= \sum_{j=0}^{c-1} d_j = \sum_{j=0}^{c-1} r^j d_0 = \frac{1 - r^c}{1 - r} d_0 \end{aligned} \tag{8.1.7}$$

provided that $r \neq 1$. Next we have similarly

$$(8.1.8)\qquad \begin{aligned} u_j &= u_j - u_c = \sum_{i=j}^{c-1} (u_i - u_{i+1}) \\ &= \sum_{i=j}^{c-1} d_i = \sum_{i=j}^{c-1} r^i d_0 = \frac{r^j - r^c}{1 - r} d_0. \end{aligned}$$

It follows that

$$(8.1.9)\qquad u_j = \frac{r^j - r^c}{1 - r^c}, \quad 0 \leq j \leq c.$$

In case $r = 1$ we get from the penultimate terms in (8.1.7) and (8.1.8) that

$$(8.1.10)\qquad \begin{aligned} 1 &= cd_0, \\ u_j &= (c - j)d_0 = \frac{c - j}{c}; \\ u_a &= \frac{b}{c}. \end{aligned}$$

One half of Problem 1 has been completely solved; it remains to find

$v_j =$ the probability that the particle will reach c before 0, when it starts from j.

Exactly the same argument shows that the set of equations in (8.1.4) will be valid when the u's are replaced by v's, while the boundary conditions in (8.1.5) are merely interchanged: $v_0 = 0$, $v_c = 1$. Hence we can find all v_j by a similar method, which you may wish to carry out as an excellent exercise. However, there are quicker ways without this effort.

One way is perhaps easier to understand by thinking in terms of the gamblers. If we change p into q (namely r into $1/r$), and at the same time j into $c - j$ (because when Peter has \$ j, Paul has \$ $(c - j)$, and vice versa), then their roles are interchanged and so u_j will go over into v_j (not v_{c-j}, why?). Making these changes in (8.1.9) and (8.1.10), we obtain

$$v_j = \frac{1 - r^j}{1 - r^c} \quad \text{if} \quad p \neq q;$$

$$v_j = \frac{j}{c} \quad \text{if} \quad p = q.$$

Now it is a real pleasure to see that in both cases we have

$$(8.1.11)\qquad u_j + v_j = 1, \quad 0 \leq j \leq c.$$

Thus as a by-product, we have solved the next problem which may have occurred to you in the course of the preceding discussion.

Problem 2. If the particle starts inside the interval $[0, c]$, what is the probability that it will ever reach the boundary?

Since the boundary consists of the two endpoints 0 and c, the answer is given by (8.1.11) and is equal to one. In terms of the gamblers, this means that one of them is bound to be ruined sooner or later if the game is continued without a time limit; in other words it cannot go on forever. Now you can object that surely it is conceivable for Peter and Paul to seesaw endlessly as e.g. indicated by the sequence $+1-1+1-1+1-1\ldots$. The explanation is that while this eventuality is a *logical possibility* its probability is equal to zero as just shown. Namely, it will *almost never* happen in the sense discussed at the end of §7.6, and this is all we can assert.

Next, let us mention that Problem 2 can be solved without the intervention of Problem 1. Indeed, it is clear the question raised in Problem 2 is a more broad "qualitative" one which should not depend on the specific numerical answers demanded by Problem 1. It is not hard to show that even if the ξ's in (8.1.2) are replaced by independent random variables with an arbitrary common distribution, which are not identically zero, so that we have a generalized random walk with all kinds of possible steps, the answer to Problem 2 is still the same in the broader sense that the particle will sooner or later get out of any finite interval (see e.g. [Chung 1; Theorem 9.2.3]). Specializing to the present case where the steps are ± 1, we see that the particle must go through one of the endpoints before it can leave the interval $[0, c]$. If this conclusion tantamount to (8.1.11) is accepted with or without a proof, then of course we get $v_j = 1 - u_j$ without further calculation.

Let us state the answer to Problem 2 as follows.

Theorem 1. *For any random walk (with arbitrary p), the particle will almost surely*† *not remain in any finite interval forever.*

As a consequence, we can define a random variable which denotes the waiting time until the particle reaches the boundary. This is sometimes referred to as "absorption time" if the boundary points are regarded as "absorbing barriers," namely the particle is supposed to be stuck there as soon as it hits them. In terms of the gamblers, it is also known as the "duration of play." Let us put for $1 \leq j \leq c - 1$:

(8.1.12) $\quad S_j =$ the first time when the particle reaches 0 or c starting from j;

and denote its expectation $E(S_j)$ by e_j. The answer to Problem 2 asserts that S_j is almost surely finite, hence it is a random variable taking positive integer values. [Were it possible for S_j to be infinite it would not be a random variable as defined in §4.2, since "$+\infty$" is not a number. However, we shall not

† In general, "almost surely" means "with probability one".

elaborate on the sample space on which S_j is defined; it is not countable!] The various e_j's satisfy a set of relations like the u_j's, as follows:

$$e_j = pe_{j+1} + qe_{j-1} + 1, \quad 1 \leq j \leq c - 1, \qquad (8.1.13)$$
$$e_0 = 0, \; e_c = 0.$$

The argument is similar to that for (8.1.4) and (8.1.5), provided we explain the additional constant "1" on the right side of the first equation above. This is the unit of time spent in taking the one step involved in the argument from j to $j \pm 1$.

The complete solution of (8.1.13) may be carried out directly as before, or more expeditiously by falling back on a standard method in solving difference equations detailed in Exercise 13 below. Since the general solution is not enlightening we will indicate the direct solution only in the case $p = q = 1/2$, which is needed in later discussion. Let $f_j = e_j - e_{j+1}$, then

$$f_j = f_{j-1} + 2, \quad f_j = f_0 + 2j,$$
$$0 = \sum_{j=0}^{c-1} f_j = c(f_0 + c - 1).$$

Hence $f_0 = 1 - c$, and after a little computation,

$$e_j = \sum_{i=j}^{c-1} f_i = \sum_{i=j}^{c-1} (1 - c + 2i) = j(c - j). \qquad (8.1.14)$$

Since the random walk is symmetric, the expected absorption time should be the same when the particle is at distance j from 0, or from c (thus at distance $c - j$ from 0), hence it is *a priori* clear that $e_j = e_{c-j}$ which checks out with (8.1.14).

8.2. Limiting schemes

We are now ready to draw important conclusions from the preceding formulas. First of all, we will convert the interval $[0, c]$ into the half-line $[0, \infty)$ by letting $c \to +\infty$. It follows from (8.1.9) and (8.1.10) that

$$\lim_{c \to \infty} u_j = \begin{cases} r^j & \text{if } r < 1; \\ 1 & \text{if } r \geq 1. \end{cases} \qquad (8.2.1)$$

Intuitively, this limit should mean the probability that the particle will reach 0 before "it reaches $+\infty$", starting from j; or else the probability that Peter will be ruined when he plays against an "infinitely rich" Paul, who cannot be ruined. Thus it simply represents the probability that the particle will ever reach 0 from j, or that of Peter's eventual ruin when his capital is \$ j. This interpretation is correct and furnishes the answer to the following problem which is a sharpening of Problem 2.

Problem 3. If the particle starts from a (≥ 1), what is the probability that it will ever hit 0?

The answer is 1 if $p \leq q$; and $(q/p)^a$ if $p > q$. Observe that when $p \leq q$ the particle is at least as likely to go left as to go right, so the first conclusion is most plausible. Indeed, in case $p < q$ we can say more by invoking the law of large numbers in its strong form given in §7.6. Remembering our new notation in (8.1.2) and that $E(\xi_n) = p - q$, we see that in the present context (7.6.17) becomes the assertion that almost surely we have

$$\lim_{n\to\infty} \frac{(X_n - X_0)}{n} = p - q < 0.$$

This is a much stronger assertion than that $\lim_{n\to\infty} X_n = -\infty$. Now our particle moves only one unit at a time, hence it can go to $-\infty$ only by passing through *all* the points to the left of the starting point. In particular it will almost surely hit 0 from a.

In case $p > q$ the implication for gambling is curious. If Peter has a definite advantage, then even if he has only \$1 and is playing against an unruinable Paul, he still has a chance $1 - q/p$ to escape ruin forever. Indeed, it can be shown that in this happy event Peter will win big in the following precise sense, where X_n denotes his fortune after n games:

$$P\{X_n \to +\infty \mid X_n \neq 0 \quad \text{for all} \quad n\} = 1.$$

[This is a conditional probability given the event $\{X_n \neq 0$ for all $n\}$.] Is this intuitively obvious? Theorem 1 helps the argument here but does not clinch it.

When $p = q = 1/2$ the argument above does not apply, and since in this case there is symmetry between left and right, our conclusion may be stated more forcefully as follows.

Theorem 2. *Starting from any point in a symmetric random walk, the particle will almost surely hit any point any number of times.*

Proof: Let us write $i \Rightarrow j$ to mean that starting from i the particle will almost surely hit j, where $i \in I$, $j \in I$. We have already proved that if $i \neq j$, then $i \Rightarrow j$. Hence also $j \Rightarrow i$. But this implies $j \Rightarrow j$ by the obvious diagram $j \Rightarrow i \Rightarrow j$. Hence also $i \Rightarrow j \Rightarrow j \Rightarrow j \Rightarrow j \ldots$, which means that starting from i the particle will hit j as many times as we desire, and note that $j = i$ is permitted here.

We shall say briefly that the particle will hit any point in its range I *infinitely often;* and that the random walk is *recurrent* (or *persistent*). These notions will be extended to Markov chains in §8.4.

In terms of gambling, Theorem 2 has the following implication. If the game is fair, then Peter is almost sure to win any amount set in advance as his goal, provided he can afford to go into debt for an arbitrarily large amount. For Theorem 2 only guarantees that he will eventually win say \$1000000 without any assurance as to how much he may have lost before he gains this goal. Not a very useful piece of information this—but strictly fair from the point of view of Paul! More realistic prediction is given in (8.1.10), which may be rewritten as

(8.2.2) $$u_a = \frac{b}{a+b}, \quad v_a = \frac{a}{a+b};$$

which says that the chance of Peter winning his goal b before he loses his entire capital a is in the exact inverse proportion of a to b. Thus if he has \$100, his chance of winning \$1000000 is equal to 100/1000100 or about one in ten thousand. This is about the state of affairs when he plays in a casino, even if the house does not reserve an advantage over him.

Another wrinkle is added when we let $c \to +\infty$ in the definition of e_j. The limit then represents the expected time that the particle starting at $j\,(\geq 1)$ will first reach 0 (without any constraint as to how far it can go to the right of j). Now this limit is infinite according to (8.1.14). This means, even if Peter has exactly \$1 and is playing against an infinitely rich casino, he can "expect" to play a long, long time provided the game is fair. This assertion sounds fantastic as stated in terms of a single gambler, whereas the notion of mathematical expectation takes on practical meaning only through the law of large numbers applied to "ensembles." It is common knowledge that on any given day many small gamblers walk away from the casino with pocketed gains—they have happily escaped ruin because the casino did not have sufficient time to ruin them in spite of its substantial profit margin!

Let us mention another method to derive (8.2.2) which is stunning. In the case $p = q$ we have $E(\xi_n) = 0$ for every n, and consequently we have from (8.1.2) that

(8.2.3) $$E(X_n) = E(X_0) + E(\xi_1) + \cdots + E(\xi_n) = a.$$

In terms of the gamblers this means that Peter's expected capital remains constant throughout the play since the game is fair. Now consider the duration of play S_a in (8.1.12). It is a random variable which takes positive integer values. Since (8.2.3) is true for every such value might it not remain so when we substitute S_a for n there? This is in general risky business but it happens to be valid here by the special nature of S_a as well as that of the process $\{X_n\}$. We cannot justify it here (see Appendix 3) but will draw the conclusion. Clearly X_{S_a} takes only the two value 0 and c by its definition; let

(8.2.4) $$P(X_{S_a} = 0) = \rho, \quad P(X_{S_a} = c) = 1 - \rho.$$

Then

$$E(X_{S_a}) = \rho \cdot 0 + (1 - \rho) \cdot c = (1 - \rho)c.$$

Hence $E(X_{S_a}) = a$ means

$$\rho = 1 - \frac{a}{c} = \frac{b}{a + b}.$$

in agreement with (8.2.2). Briefly stated, the argument above says that the game remains fair up to and including the time of its termination. Is this intuitively obvious?

We now proceed to describe a limiting procedure which will lead from the symmetric random walk to *Brownian motion*. The English botanist Brown observed (1826) that microscopic particles suspended in a liquid are subject to continual molecular impacts and execute zigzag movements. Einstein and Smoluchovski found that in spite of their apparent irregularity these movements can be analyzed by laws of probability, in fact the displacement over a period of time follows a normal distribution. Einstein's result (1906) amounted to a derivation of the central limit theorem (see §7.4) by the method of differential equations. The study of Brownian motion as a stochastic process was undertaken by Wiener† in 1923, preceded by Bachelier's heuristic work, and soon was developed into its modern edifice by Paul Lévy and his followers. Together with the Poisson process (§7.2) it constitutes one of the two fundamental *species* of stochastic processes, in both theory and application. Although the mathematical equipment allowed in this book is not adequate to treat the subject properly, it is possible to give an idea how the Brownian motion process can be arrived at through random walk and to describe some of its basic properties.

The particle in motion observed by Brown moved of course in three dimensional space, but we can think of its projection on a coordinate axis. Since numerous impacts are received per second, we will shorten the unit of time; but we must also shorten the unit of length in such a way as to lead to the correct model. Let δ be the new time-unit, in other words the time between two successive impacts. Thus in our previous language t/δ steps are taken by the particle in old time t. Each step is still a symmetrical Bernoullian random variable but we now suppose that the step is of magnitude $\sqrt{\delta}$, namely for all k:

$$P(\xi_k = \sqrt{\delta}) = P(\xi_k = -\sqrt{\delta}) = \frac{1}{2}.$$

We have then

$$E(\xi_k) = 0, \quad \sigma^2(\xi_k) = \frac{1}{2}(\sqrt{\delta})^2 + \frac{1}{2}(-\sqrt{\delta})^2 = \delta.$$

Let $X_0 = 0$ so that by (8.1.2)

† Norbert Wiener (1894-1964), renowned U.S. mathematician, father of cybernetics.

(8.2.5)
$$X_t = \sum_{k=1}^{t/\delta} \xi_k.$$

If δ is much smaller than t, t/δ is large and may be thought of as an integer. Hence we have by Theorem 4 of §6.3:

(8.2.6)
$$E(X_t) = 0, \quad \sigma^2(X_t) = \frac{t}{\delta} \cdot \delta = t.$$

Furthermore if t is fixed and $\delta \to 0$, then by the DeMoivre-Laplace central limit theorem (Theorem 6 of §7.3), X_t will have the normal distribution $N(0, t)$. This means we are letting our approximate scheme, in which the particle moves a distance of $\pm\sqrt{\delta}$ with equal probability in old time δ, go to the limit as $\delta \to 0$. This limiting scheme is the Brownian motion, also called *Wiener process*, and here is its formal definition.

Definition of Brownian Motion. A family of random variables $\{X(t)\}$, indexed by the continuous variable t ranging over $[0, \infty)$ is called the *Brownian Motion* iff it satisfies the following conditions:

(i) $X(0) = 0$;
(ii) the increments $X(s_i + t_i) - X(s_i)$, over an arbitrary finite set of disjoint intervals $(s_i, s_i + t_i)$, are independent random variables;
(iii) for each $s \geq 0$, $t \geq 0$, $X(s + t) - X(s)$ has the normal distribution $N(0, t)$.

For each constant a, the process $\{X(t) + a\}$, where $X(t)$ is just defined, is called the *Brownian motion starting at a.*

We have seen that the process constructed above by a limiting passage from symmetric random walks has the property (iii). Property (ii) comes from the fact that increments over disjoint intervals are obtained by summing the displacements ξ_k in disjoint blocks; hence the sums are independent by a remark made after Proposition 6 of §5.5.

The definition above should be compared with that of a Poisson process given in §7.2, the only difference being in (iii). However, by the manner in which a Poisson process is constructed there, we know the general appearance of its paths as described under Figure 29. The situation is far from obvious for Brownian motion. It is one of Wiener's major discoveries that almost all its paths are continuous; namely, for almost all ω, the function $t \to X(t, \omega)$ is a continuous function of t in $[0, \infty)$. In practice, we can discard the null set of ω's which yield discontinuous functions from the sample space Ω, and simply stipulate that all Brownian paths are continuous. This is a tremendously useful property which may well be added to the definition above. On the other hand, Wiener also proved that almost every path is nowhere differentiable, i.e. the curve does not have a tangent anywhere—

which only goes to show that one cannot rely on intuition any more in these matters.

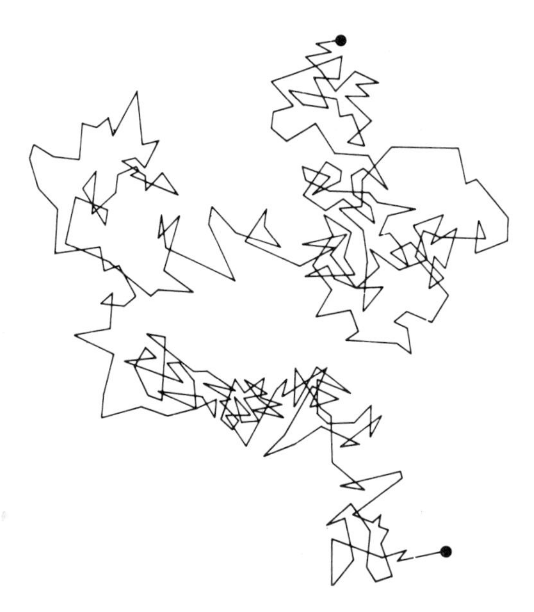

Fig. 32. Brownian movement. Observations made at equal time intervals. The real path is even more complicated.

However, it is not hard to guess the answers to our previous questions restated for Brownian motion. In fact, the analogue in Theorem 1 holds: starting at any point, the path will go through any other point infinitely many times. Note that because of the continuity of the path this will follow from the "intermediate value theorem" in calculus once we show that it will reach out as far as we wish. Since each approximating random walk has this property, it is obvious that the Brownian motion does too. Finally, let us show that the formula (8.2.2) holds also for Brownian motion, where u_a and v_a retain the same meanings as before but now a and c are arbitrary numbers such that $0 < a < c$. Consider the Brownian motion starting at a; then it follows from property (i) that $E(X_t) = a$ for all $t \geq 0$, which is just the continuous analogue of (8.2.3). Now we substitute T_a for t to get $E(X_{T_a}) = a$ as before. This time, the continuity of paths assures us that at the instant T_a, the position of the particle must be exactly at 0 or at c. In fact, the word "reach" used in the definition of u_a, v_a and T_a would have to be explained more carefully if the path could jump over the boundary. Thus we can again write (8.2.4) and get the same answer as for the symmetric random walk.

8.3. Transition probabilities

The model of random walks can be greatly generalized to that of *Markov chains*, named after A. A. Markov (see §7.6). As the saying goes, one may fail to see the forest on account of the trees. By doing away with some cumbersome and incidental features of special cases, a general theory emerges which is clearer and simpler and covers a wider range of applications. The remainder of this chapter is devoted to the elements of such a theory.

We continue to use the language of a moving particle as in the random walk scheme, and denote its range by I. This may now be a finite or infinite set of integers, and it will soon be apparent that in general no geometric or algebraic structure (such as right or left, addition and subtraction) is required of I. Thus it may be an arbitrary countable set of elements, provided that we extend our definition of random variables to take values in such a set. [In §4.2 we have defined a random variable to be numerically valued.] We shall call I the *state space* and an element of it a *state*. For example, in physical chemistry a state may be a certain level of energy for an atom; in public opinion polls it may be one of the voter's possible states of mind, etc. The particle moves from state to state and the probability law governing its change of states or transition will be prescribed, as follows. There is a set of *transition probabilities* p_{ij}, where $i \in I$, $j \in I$, such that: if the particle is in the state i at any time, *regardless of what state it has been in before then*, the probability that it will be in the state j after one step is given by p_{ij}. In symbols, if X_n denotes the state of the particle at time n, then we have

$$(8.3.1) \qquad P\{X_{n+1} = j \mid X_n = i; A\} = P\{X_{n+1} = j \mid X_n = i\} = p_{ij},$$

for an arbitrary event A determined by $\{X_0, \ldots, X_{n-1}\}$ alone. For instance A may be a completely specified "past" of the form "$X_0 = i_0, X_1 = i_1, \ldots, X_{n-1} = i_{n-1}$," or a more general past event where the states $i_0, \ldots, i_{n-1}$ are replaced by sets of states: "$X_0 \in J_0, X_1 \in J_1, \ldots, X_{n-1} \in J_{n-1}$." In the latter case some of these sets may be taken to be the whole space I, so that the corresponding random variables are in effect omitted from the conditioning; thus "$X_0 \in J_0, X_1 \in I, X_2 \in J_2$" is really just "$X_0 \in J_0, X_2 \in J_2$." The first equation in (8.3.1) renders the precise meaning of the phrase "regardless of prior history," and is known as the *Markov property*. The second equation says that the conditional probability there does not depend on the value of n; this is referred to as the *stationarity* (or *temporal homogeneity*) of the transition probabilities. Together they yield the following definition.

Definition of Markov chain. A stochastic process $\{X_n, n \in N^0\}$† taking

† It may be more convenient in some verbal descriptions to begin with $n = 1$ rather than $n = 0$.

values in a countable set I is called a *homogeneous Markov chain*, or *Markov chain with stationary transition probabilities*, iff (8.3.1) holds.

If the first equation in (8.3.1) holds without the second, then the Markov chain is referred to as being "non-homogeneous," in which case the probability there depends also on n and must be denoted by $p_{ij}(n)$, say. Since we shall treat only a homogeneous chain we mean this case when we say "Markov chain" or "chain" below without qualification.

As a consequence of the definition, we can write down the probabilities of successive transitions. Whenever the particle is in the state i_0, and regardless of its prior history, the *conditional* probability that it will be in the states $i_1, i_2, \ldots, i_n$, in the order given, during the next n steps may be suggestively denoted by the left member below and evaluated by the right member:

$$(8.3.2) \qquad P\{.\,.\,.\,.\,.\, i_0 \to i_1 \to i_2 \to \cdots \to i_n\} = p_{i_0 i_1} p_{i_1 i_2} \cdots p_{i_{n-1} i_n},$$

where the five dots at the beginning serve to indicate the irrelevant and forgotten past. This follows by using (8.3.1) in the general formula (5.2.2) for joint probabilities; for instance,

$$\begin{aligned} P\{X_4 = j, X_5 = k, X_6 = l \mid X_3 = i\} &= P\{X_4 = j \mid X_3 = i\} \\ &\cdot P\{X_5 = k \mid X_3 = i, X_4 = j\}\, P\{X_6 = l \mid X_3 = i, X_4 = j, X_5 = k\} \\ &= P\{X_4 = j \mid X_3 = i\}\, P\{X_5 = k \mid X_4 = j\}\, P\{X_6 = l \mid X_5 = k\} \\ &= p_{ij} p_{jk} p_{kl}. \end{aligned}$$

Moreover, we may adjoin any event A determined by $\{X_0, X_1, X_2\}$ alone behind the bars in the first two members above without affecting the result. This kind of calculation shows that: given the state of the particle *at any time*, its prior history is not only irrelevant to the next transition as postulated in (8.3.1), but equally so to any future transitions. Symbolically, for any event B determined by $\{X_{n+1}, X_{n+2}, \ldots\}$, we have

$$(8.3.3) \qquad P\{B \mid X_n = i; A\} = P\{B \mid X_n = i\}$$

as an extension of the Markov property. But that is not yet the whole story; there is a further and more sophisticated extension revolving around the three little words "at any time" italicized above, which will be needed and explained later.

It is clear from (8.3.2) that all probabilities concerning the chain are determined by the transition probabilities, provided that it starts from a fixed state, e.g., $X_0 = i$. More generally we may randomize the initial state by putting

$$P\{X_0 = i\} = p_i, \quad i \in I.$$

Then $\{p_i, i \in I\}$ is called the *initial distribution* of the chain and we have for arbitrary states $i_0, i_1, \ldots, i_n$:

(8.3.4) $$P\{X_0 = i_0, X_1 = i_1, \ldots, X_n = i_n\} = p_{i_0} p_{i_0 i_1} \cdots p_{i_{n-1} i_n}$$

as the joint distribution of random variables of the process. Let us pause to take note of the special case where for every $i \in I$ and $j \in I$ we have

$$p_{ij} = p_j.$$

The right member of (8.3.4) then reduces to $p_{i_0} p_{i_1} \ldots p_{i_n}$, and we see that the random variables $X_0, X_1, \ldots X_n$ are independent with the common distribution given by $\{p_j, j \in I\}$. Thus, a sequence of independent, identically distributed and countably valued random variables is a special case of Markov chain, which has a much wider scope. The basic concept of such a scheme is due to Markov who introduced it around 1907.

It is clear from the definition of p_{ij} that we have:

(8.3.5) $$\begin{aligned} &\text{(a)}\ \ p_{ij} \geq 0 && \text{for every } i \text{ and } j; \\ &\text{(b)}\ \ \sum_{j \in I} p_{ij} = 1 && \text{for every } i. \end{aligned}$$

Indeed it can be shown that these are the only conditions that must be satisfied by the p_{ij}'s in order that they be the transition probabilities of a homogeneous Markov chain. In other words, such a chain can be constructed to have a given matrix satisfying those conditions as its transition matrix. Examples are collected at the end of the section.

Let us denote by $p_{ij}^{(n)}$ the probability of transition from i to j in exactly n steps, namely:

(8.3.6) $$p_{ij}^{(n)} = P\{X_n = j \mid X_0 = i\}.$$

Thus $p_{ij}^{(1)}$ is our previous p_{ij} and we may add

$$p_{ij}^{(0)} = \delta_{ij} = \begin{cases} 0 & \text{if } i \neq j, \\ 1 & \text{if } i = j; \end{cases}$$

for convenience. The δ_{ij} above is known as *Kronecker's symbol* which you may have seen in linear algebra. We proceed to show that for $n \geq 1$, $i \in I$, $k \in I$, we have

(8.3.7) $$p_{ik}^{(n)} = \sum_j p_{ij} p_{jk}^{(n-1)} = \sum_j p_{ij}^{(n-1)} p_{jk},$$

where the sum is over I, an abbreviation which will be frequently used below. To argue this let the particle start from i, and consider the outcome after taking one step. It will then be in the state j with probability p_{ij}; and conditioned on this hypothesis, it will go to the state k in $n - 1$ more steps with probability $p_{jk}^{(n-1)}$, regardless of what i is. Hence the first equation in (8.3.7) is obtained by summing over all j according to the general formula for total probabilities; see (5.2.3) or (5.2.4). The second equation in (8.3.7) is proved in a similar way by considering first the transition in $n - 1$ steps, followed by one more step.

For $n = 2$, (8.3.7) becomes

$$p_{ik}^{(2)} = \sum_j p_{ij} p_{jk}, \tag{8.3.8}$$

which suggests the use of matrices. Let us arrange the p_{ij}'s in the form of a matrix

$$\Pi = [p_{ij}], \tag{8.3.9}$$

so that p_{ij} is the element at the ith row and jth column. Recall that the elements of Π satisfy the conditions in (8.3.5). Such a matrix is called *stochastic*. Now the product of two square matrices $\Pi_1 \times \Pi_2$ is another such matrix whose element at the ith row and jth column is obtained by multiplying the corresponding elements of the ith row of Π_1 with those of the jth column of Π_2, and then adding all such products. In case both Π_1 and Π_2 are the same Π, this yields precisely the right member of (8.3.8). Therefore we have

$$\Pi^2 = \Pi \times \Pi = [p_{ij}^{(2)}],$$

and it follows by induction on n and (8.3.7) that

$$\Pi^n = \Pi \times \Pi^{n-1} = \Pi^{n-1} \times \Pi = [p_{ij}^{(n)}].$$

In other words, the *n-step transition probabilities* $p_{ij}^{(n)}$ are just the elements in the nth power of Π. If I is the finite set $\{1, 2, \ldots, r\}$, then the rule of multiplication described above is of course the same as the usual one for square matrices (or determinants) of order r. When I is an infinite set the same rule applies but we must make sure that the resulting infinite series such as the one in (8.3.8) are all convergent. This is indeed so, by virtue of (8.3.7). We can now extend the latter as follows. For $n \in N^0$, $m \in N^0$ and $i \in I$, $k \in I$, we have

$$p_{ik}^{(n+m)} = \sum_j p_{ij}^{(n)} p_{jk}^{(m)}. \tag{8.3.10}$$

This set of equations is known as the *Chapman-Kolmogorov equations*. [Sydney Chapman, 1888–1970, English applied mathematician.] It is simply an expression of the law of exponentiation for powers of Π:

$$\Pi^{n+m} = \Pi^n \times \Pi^m,$$

and can be proved, either by induction on m from (8.3.7), purely algebraically, or by a probabilistic argument along the same line as that for (8.3.7). Finally, let us record the trivial equation, valid for each $n \in N^0$ and $i \in I$:

(8.3.11)
$$\sum_j p_{ij}^{(n)} = 1.$$

The matrix Π^n may be called the *n-step transition matrix*. Using $p_{ij}^{(n)}$ we can express joint probabilities when some intermediate states are not specified. An example will make this clear:

$$P\{X_4 = j, X_6 = k, X_9 = l \mid X_2 = i\} = p_{ij}^{(2)} p_{jk}^{(2)} p_{kl}^{(3)}.$$

We are now going to give some illustrative examples of homogeneous Markov chains, and one which is non-homogeneous.

Example 1. $I = \{\ldots, -2, -1, 0, 1, 2, \ldots\}$ is the set of all integers.

(8.3.12)
$$p_{ij} = \begin{cases} p & \text{if } j = i + 1 \\ q & \text{if } j = i - 1 \\ 0 & \text{otherwise;} \end{cases}$$

$$\Pi = \begin{bmatrix} \cdots\cdots \\ \ldots q\,0\,p\,0\,0 \ldots \\ \ldots 0\,q\,0\,p\,0 \ldots \\ \ldots 0\,0\,q\,0\,p \ldots \\ \cdots\cdots \end{bmatrix}$$

where $p + q = 1$, $p \geq 0$, $q \geq 0$. This is the *free random walk* discussed in §8.1. In the extreme cases $p = 0$ or $q = 0$, it is of course deterministic (almost surely).

Example 2. $I = \{0, 1, 2, \ldots\}$ is the set of nonnegative integers; p_{ij} is the same as in Example 1 for $i \neq 0$, but $p_{00} = 1$ which entails $p_{0j} = 0$ for all $j \neq 0$. This is the random walk with one *absorbing state* 0. It is the model appropriate for Problem 3 in §8.2. The absorbing state corresponds to the ruin (state of bankruptcy) of Peter, whereas Paul is infinitely rich so that I is unlimited to the right.

Example 3. $I = \{0, 1, \ldots, c\}$, $c \geq 2$.

$$\Pi = \begin{bmatrix} 1\,0\,0 \ldots\ldots \\ q\,0\,p\,0 \ldots\ldots \\ 0\,q\,0\,p\,0 \ldots\ldots \\ \ldots\ldots 0\,q\,0\,p \\ \ldots\ldots 0\,0\,0\,1 \end{bmatrix}$$

For $1 \leq i \leq c - 1$, the p_{ij}'s are the same as in Example 1, but

(8.3.13) $$p_{00} = 1, \quad p_{cc} = 1.$$

This is the random walk with two *absorbing barriers* 0 and c, and is appropriate for Problem 1 of §8.1. Π is a square matrix of order $c + 1$.

Example 4. In Example 3 replace (8.3.13) by

$$p_{01} = 1, \quad p_{c,c-1} = 1.$$

$$\Pi = \begin{bmatrix} 0\,1\,0\,0\,\ldots \\ q\,0\,p\,0\,\ldots \\ \ldots\ldots\ldots \\ \ldots\,0\,q\,0\,p \\ \ldots\,0\,0\,1\,0 \end{bmatrix}$$

This represents a random walk with two *reflecting barriers* such that after the particle reaches either endpoint of the interval $[0, c]$, it is bound to turn back at the next step. In other words, either gambler will be given a \$1 reprieve whenever he becomes bankrupt, so that the game can go on forever—for fun! We may also eliminate the two states 0 and c, and let $I = \{1, 2, \ldots, c - 1\}$,

$$p_{11} = q, \quad p_{12} = p; \quad p_{c-1,c-2} = q, \quad p_{c-1,c-1} = p$$

$$\Pi = \begin{bmatrix} q\,p\,0\,0\,\ldots \\ q\,0\,p\,0\,\ldots \\ \ldots\ldots\ldots \\ \ldots\,0\,q\,0\,p \\ \ldots\,0\,0\,q\,p \end{bmatrix}$$

Example 5. Let $p \geq 0$, $q \geq 0$, $r \geq 0$ and $p + q + r = 1$. In Examples 1 to 4 replace each row of the form $(\ldots q0p \ldots)$ by $(\ldots qrp \ldots)$. This means that at each step the particle may stay put, or that the game may be a draw, with probability r. When $r = 0$ this reduces to the preceding examples.

Example 6. Let $\{\xi_n, n \geq 0\}$ be a sequence of independent integer-valued random variables such that all except possibly ξ_0 have the same distribution given by $\{a_k, k \in I\}$, where I is the set of all integers. Define X_n as in (8.1.2): $X_n = \sum_{k=0}^{n} \xi_k$, $n \geq 0$. Since $X_{n+1} = X_n + \xi_{n+1}$, and ξ_{n+1} is independent of $X_0, X_1, \ldots, X_n$, we have for any event A determined by $X_0, \ldots, X_{n-1}$ alone:

$$\begin{aligned} P\{X_{n+1} = j \mid A; X_n = i\} &= P\{\xi_{n+1} = j - i \mid A; X_n = i\} \\ &= P\{\xi_{n+1} = j - i\} = a_{j-i}. \end{aligned}$$

Hence $\{X_n, n \geq 0\}$ constitutes a homogeneous Markov chain with the transition matrix $[p_{ij}]$, where

(8.3.14) $$p_{ij} = a_{j-i}.$$

The initial distribution is the distribution of ξ_0 which need not be the same as $\{a_k\}$. Such a chain is said to be *spatially homogeneous* as p_{ij} depends only on the difference $j - i$. Conversely, suppose $\{X_n, n \geq 0\}$ is a chain with the transition matrix given in (8.3.14), then we have

$$P\{X_{n+1} - X_n = k \mid X_n = i\} = P\{X_{n+1} = i + k \mid X_n = i\} = p_{i,i+k} = a_k.$$

It follows that if we put $\xi_{n+1} = X_{n+1} - X_n$, then the random variables $\{\xi_n, n \geq 1\}$ are independent (why?) and have the common distribution $\{a_k\}$. Thus a spatially as well as temporally homogeneous Markov chain is identical with the successive partial sums of independent and identically distributed integer-valued random variables. The study of the latter has been one of our main concerns in previous chapters.

In particular, Example 1 is the particular case of Example 6 with $a_1 = p$, $a_{-1} = q$; we may add $a_0 = r$ as in Example 5.

Example 7. For each $i \in I$ let p_i and q_i be two nonnegative numbers satisfying $p_i + q_i = 1$. Take I to be the set of all integers and put

(8.3.15) $$p_{ij} = \begin{cases} p_i & \text{if } j = i + 1, \\ q_i & \text{if } j = i - 1, \\ 0 & \text{otherwise.} \end{cases}$$

In this model the particle can move only to neighboring states as in Example 1, but the probabilities may now vary with the position. The model can be generalized as in Example 5 by allowing also the particle to stay put at each position i with probability r_i, with $p_i + q_i + r_i = 1$. Observe that this example contains also Examples 2, 3 and 4 above. The resulting chain is no longer representable as sums of independent steps as in Example 6. For a full discussion of the example see [Chung; 2].

Example 8. (Ehrenfest model). This may be regarded as a particular case of Example 7 in which we have $I = \{0, 1, \ldots, c\}$ and

(8.3.16) $$p_{i,i+1} = \frac{c - i}{c}, \quad p_{i,i-1} = \frac{i}{c}.$$

It can be realized by an urn scheme as follows. An urn contains c balls, each of which may be red or black; a ball is drawn at random from it and replaced by one of the other color. The state of the urn is the number of black balls in it. It is easy to see that the transition probabilities are as given above and the interchange can go on forever. P. and T. Ehrenfest used the model to study the transfer of heat between gas molecules. Their original urn scheme is slightly more complicated (see Exercise 14).

Example 9. Let $I = \{0, 1, 2, \ldots\}$ and

$$p_{i,0} = p_i, \ p_{i,i+1} = 1 - p_i \quad \text{for } i \in I.$$

The p_i's are arbitrary numbers satisfying $0 < p_i < 1$. This model is used to study a recurring phenomenon which is represented by the state 0. Each transition may signal an occurrence of the phenomenon, or else prolong the waiting time by one time unit. It is easy to see that the event "$X_n = k$" means that the last time $\leq n$ when the phenomenon occurred is at time $n - k$, where $0 \leq k \leq n$; in other words there has been a waiting period equal to k units since that occurrence. In the particular case where all p_i are equal to p we have

$$P\{X_v \neq 0 \quad \text{for } 1 \leq v \leq n - 1; \ X_n = 0 \mid X_0 = 0\} = (1 - p)^{n-1}p.$$

This gives the geometric waiting time discussed in Example 8 of §4.4.

Example 10. Let I be the integer lattice in R^d, the Euclidean space of d dimensions. This is a countable set. We assume that: starting at any lattice point, the particle can go only to one of the $2d$ neighboring points in one step, with various (not necessarily equal) probabilities. For $d = 1$ this is just Example 1, for $d = 2$ this is the street wanderer mentioned in §8.1. In the latter case we may represent the states by (i, i') where i and i' are integers; then we have

$$p_{(i,i')(j,j')} = \begin{cases} p_1 & \text{if } j = i + 1, j' = i'; \\ p_2 & \text{if } j = i - 1, j' = i'; \\ p_3 & \text{if } j = i, j' = i' + 1; \\ p_4 & \text{if } j = i, j' = i' - 1; \end{cases}$$

where $p_1 + p_2 + p_3 + p_4 = 1$. If all these four probabilities are equal to $1/4$ the chain is a symmetric two-dimensional random walk. Will the particle still hit every lattice point with probability one? Will it do the same in three dimensions? These questions will be answered in the next section.

Example 11. (non-homogeneous Markov chain). Consider the Pólya urn scheme described in §5.4 with $c \geq 1$. The number of black balls in the urn is called its state so that "$X_n = i$" means that after n drawings and insertions there are i black balls in the urn. Clearly each transition either increases this number by c or leaves it unchanged, and we have

$$(8.3.17) \quad P\{X_{n+1} = j \mid X_n = i; A\} = \begin{cases} \dfrac{i}{b + r + nc} & \text{if } j = i + c, \\ 1 - \dfrac{i}{b + r + nc} & \text{if } j = i, \\ 0 & \text{otherwise;} \end{cases}$$

where A is any event determined by the outcomes of the first $n - 1$ drawings. The probability above depends on n as well as i and j, hence the process is a non-homogeneous Markov chain. We may also allow $c = -1$, which is the case of sampling without replacement and yields a finite sequence of $\{X_n: 0 \leq n \leq b + r\}$.

Example 12. It is trivial to define a process which is not Markovian. For instance in Example 8 or 11, let $X_n = 0$ or 1 according as the nth ball drawn is red or black. Then it is clear that the probability of "$X_{n+1} = 1$" given the values of $X_1, \ldots, X_n$ will not in general be the same as given the value of X_n alone. Indeed the latter probability is not very useful.

8.4. Basic structure of Markov chains

We begin a general study of the structure of homogeneous Markov chains by defining a binary relation between the states. We say "i leads to j" and write "$i \rightsquigarrow j$" iff there exists $n \geq 1$ such that $p_{ij}^{(n)} > 0$; we say "i communicates with j" and write "$i \leftrightsquigarrow j$" iff we have both $i \rightsquigarrow j$ and $j \rightsquigarrow i$. The relation "$\rightsquigarrow$" is transitive, namely if $i \rightsquigarrow j$ and $j \rightsquigarrow k$ then $i \rightsquigarrow k$. This follows from the inequality

$$p_{ik}^{(n+m)} \geq p_{ij}^{(n)} p_{jk}^{(m)} \tag{8.4.1}$$

which is an algebraic consequence of (8.3.10), but perhaps even more obvious from its probabilistic meaning. For if it is possible to go from i to j in n steps, and also possible to go from j to k in m steps, then it is possible by combining these steps to go from i to k in $n + m$ steps. Here and henceforth we shall use such expressions as "it is possible" or "one can" to mean *with positive probability;* but observe that even in the trivial argument just given the Markov property has been used and cannot be done without. The relation "$\leftrightsquigarrow$" is clearly both symmetric and transitive and may be used to divide the states into disjoint classes as follows.

Definition of Class. A *class of states* is a subset of the state space such that any two states (distinct or not) in the class communicate with each other.

This kind of classification may be familiar to you under the name of "equivalence classes." But here the relation "$\leftrightsquigarrow$" is not necessarily reflexive; in other words there may be a state which does not lead to itself, hence it does not communicate with any state. Such states are simply unclassified! On the other hand, a class may consist of a single state i: this is the case when $p_{ii} = 1$. Such a state is called an *absorbing state.* Two classes which are not identical must be disjoint, because if they have a common element they must merge into one class via that element.

For instance in Examples 1, 4, 5, 8 and 9 all states form a single class provided $p > 0$ and $q > 0$; as also in Example 7 provided $p_i > 0$ and $q_i > 0$

for all i. In Example 2 there are two classes: the absorbing state 0 as a singleton and all the rest as another class. Similarly in Example 3 there are three classes. In Example 6 the situation is more complicated. Suppose for instance the a_k's are such that $a_k > 0$ if k is divisible by 5, and $a_k = 0$ otherwise. Then the state space I can be decomposed into five classes. Two states i and j belong to the same class if and only if $i - j$ is divisible by 5. In other words, these classes coincide with the *residue classes modulo* 5. It is clear that in such a situation it would be more natural to take one of these classes as the *reduced* state space, because if the particle starts from any class it will (almostly surely) remain in that class forever, so why bother dragging in those other states it will never get to?

In probability theory, particularly in Markov chains, the first instance of occurrence of a sequence of events is an important notion. Let j be an arbitrary state and consider the first time that the particle enters it, namely:

$$T_j(\omega) = \min \{n \geq 1 \mid X_n(\omega) = j\}, \tag{8.4.2}$$

where the right member reads as follows: the minimum positive value of n such that $X_n = j$. For some sample point ω, $X_n(\omega)$ may never be j, so that no value of n exists in the above and T_j is not really defined for that ω. In such a case we shall define it by the decree: $T_j(\omega) = \infty$. In common language, "it will never happen" may be rendered into "one can wait until eternity (or 'hell freezes over')." With this convention T_j is a random variable which may take the value ∞. Let us denote the set $\{1, 2, \ldots, \infty\}$ by N_∞. Then T_j takes values in N_∞; this is a slight extension of our general definition in §4.2.

We proceed to write down the probability distribution of T_j. For simplicity of notation we shall write $P_i\{\cdots\}$ for probability relations associated with a Markov chain starting from the state i. We then put, for $n \in N_\infty$:

$$f_{ij}^{(n)} = P_i\{T_j = n\}, \tag{8.4.3}$$

and

$$f_{ij}^* = \sum_{n=1}^{\infty} f_{ij}^{(n)} = P_i\{T_j < \infty\}. \tag{8.4.4}$$

Remember that $\sum_{n=1}^{\infty}$ really means $\sum_{1 \leq n < \infty}$; since we wish to stress the fact that the value ∞ for the superscript is not included in the summation. It follows that

$$P_i\{T_j = \infty\} = f_{ij}^{(\infty)} = 1 - f_{ij}^*. \tag{8.4.5}$$

Thus $\{f_{ij}^{(n)}, n \in N_\infty\}$ is the probability distribution of T_j for the chain starting from i.

We can give another more explicit expression for $f_{ij}^{(n)}$ etc., as follows:

$$
(8.4.6)\quad
\begin{aligned}
f_{ij}^{(1)} &= p_{ij} = P_i\{X_1 = j\},\\
f_{ij}^{(n)} &= P_i\{X_v \neq j \text{ for } 1 \leq v \leq n-1;\ X_n = j\},\ n \geq 2;\\
f_{ij}^{(\infty)} &= P_i\{X_v \neq j \text{ for all } v \geq 1\},\\
f_{ij}^{*} &= P_i\{X_v = j \text{ for some } v \geq 1\}.
\end{aligned}
$$

Note that we may have $i = j$ in the above, and "for some v" means "for at least one value of v."

The random variable T_j is called the *first entrance time into the state j;* the terms "first passage time" and "first hitting time" are also used. It is noteworthy that by virtue of homogeneity we have

$$
(8.4.7)\quad f_{ij}^{(n)} = P\{X_{m+v} \neq j \text{ for } 1 \leq v \leq n-1;\ X_{m+n} = j \mid X_m = i\}
$$

for any m for which the conditional probability is defined. This kind of interpretation will be constantly used without specific mention.

The key formula connecting the $f_{ij}^{(n)}$ and $p_{ij}^{(n)}$ will now be given.

Theorem 3. *For any i and j, and* $1 \leq n < \infty$, *we have*

$$
(8.4.8)\quad p_{ij}^{(n)} = \sum_{v=1}^{n} f_{ij}^{(v)} p_{jj}^{(n-v)}.
$$

Proof: This result is worthy of a formal treatment in order to bring out the basic structure of a homogeneous Markov chain. Everything can be set down in a string of symbols:

$$
\begin{aligned}
p_{ij}^{(n)} &= P_i\{X_n = j\} = P_i\{T_j \leq n;\ X_n = j\} = \sum_{v=1}^{n} P_i\{T_j = v;\ X_n = j\}\\
&= \sum_{v=1}^{n} P_i\{T_j = v\} P_i\{X_n = j \mid T_j = v\}\\
&= \sum_{v=1}^{n} P_i\{T_j = v\} P_i\{X_n = j \mid X_1 \neq j, \ldots, X_{v-1} \neq j, X_v = j\}\\
&= \sum_{v=1}^{n} P_i\{T_j = v\} P\{X_n = j \mid X_v = j\}\\
&= \sum_{v=1}^{n} P_i\{T_j = v\} P_j\{X_{n-v} = j\}\\
&= \sum_{v=1}^{n} f_{ij}^{(v)} p_{jj}^{(n-v)}.
\end{aligned}
$$

Let us explain each equation above. The first is the definition of $p_{ij}^{(n)}$; the second because $\{X_n = j\}$ implies $\{T_j \leq n\}$; the third because the events $\{T_j = v\}$ for $1 \leq v \leq n$ are disjoint; the fourth is by definition of conditional probability; the fifth is by the meaning of $\{T_j = v\}$ as given in (8.4.6); the sixth by the Markov property in (8.3.1) since $\{X_1 \neq j, \ldots, X_{v-1} \neq j\}$, as well as $\{X_0 = i\}$ implicit in the notation P_i, constitutes an event prior to the time v; the seventh is by the temporal homogeneity of a transition from j to j in $n - v$ steps; the eighth is just notation. The proof of Theorem 3 is therefore completed.

True, a quicker verbal account can be and is usually given for (8.4.8), but if you spell out the details and pause to ask "why" at each stage, it will come essentially to a rough translation of the derivation above. This is a pattern of argument much used in a general context in the advanced theory of Markov processes, so a thorough understanding of the simplest case as this one is well worth the pains.

For $i \neq j$, the formula (8.4.8) relates the transition matrix elements at (i, j) to the diagonal element at (j, j). There is a dual formula which relates them to the diagonal element at (i, i). This is obtained by an argument involving the *last exit from i* as the dual of *first entrance into j*. It is slightly more tricky in its conception and apparently known only to a few specialists. We will present it here for the sake of symmetry—and mathematical beauty. Actually the formula is a powerful tool in the theory of Markov chains, although it is not necessary for our discussions here.

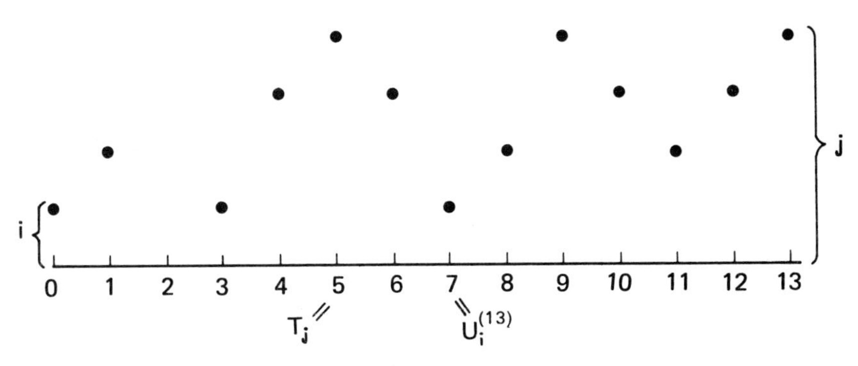

Figure 33

Define for $n \geq 1$:

$$U_i^{(n)}(\omega) = \max \{0 \leq v \leq n \mid X_v(\omega) = i\}; \tag{8.4.9}$$

namely $U_i^{(n)}$ is the last exit time from the state i before or at the given time n. This is the dual of T_j but complicated by its dependence on n. Next we introduce the counterpart of $f_{ij}^{(n)}$, as follows:

$$\begin{aligned} g_{ij}^{(1)} &= p_{ij}, \\ g_{ij}^{(n)} &= P_i\{X_v \neq i \text{ for } 1 \leq v \leq n-1; X_n = j\}, \quad 2 \leq n < \infty. \end{aligned} \tag{8.4.10}$$

Thus $g_{ij}^{(n)}$ is the probability of going from i to j in n steps without going through i *again* (for $n = 1$ the restriction is automatically satisfied since $i \neq j$). Contrast this with $f_{ij}^{(n)}$ which may now be restated as the probability of going from i to j in n steps without going through j *before*. Both kinds of probability impose a taboo on certain passages and are known as *taboo probabilities* (see [Chung 2; §1.9] for a fuller discussion). We can now state the following result.

Theorem 4. *For $i \neq j$, and $n \geq 1$, we have*

$$p_{ij}^{(n)} = \sum_{v=0}^{n-1} p_{ii}^{(v)} g_{ij}^{(n-v)}. \tag{8.4.11}$$

Proof: We shall imitate the steps in the proof of Theorem 3 as far as possible, thus:

$$p_{ij}^{(n)} = P_i\{X_n = j\} = P_i\{0 \leq U_i^{(n)} \leq n-1, X_n = j\} = \sum_{v=0}^{n-1} P_i\{U_i^{(n)} = v; X_n = j\}$$

$$= \sum_{v=0}^{n-1} P_i\{X_v = i, X_u \neq i \text{ for } v+1 \leq u \leq n-1; X_n = j\}$$

$$= \sum_{v=0}^{n-1} P_i\{X_v = i\} P_i\{X_u \neq i \text{ for } 1 \leq u \leq n-v-1; X_{n-v} = j\}$$

$$= \sum_{v=0}^{n-1} p_{ii}^{(v)} g_{ij}^{(n-v)}.$$

The major difference lies in the fourth equation above, but this is obvious from the meaning of $U_i^{(n)}$. We leave the rest to the reader.

We put also

$$g_{ij}^* = \sum_{n=1}^{\infty} g_{ij}^{(n)}. \tag{8.4.12}$$

However, while each term in the series above is a probability, it is not clear whether the series converges (it does, provided $i \rightsquigarrow j$; see Exercise 33 below). In fact, g_{ij}^* may be seen to represent the expected number of entrances in j between two successive entrances in i.

Theorems 3 and 4 may be called respectively the *first entrance* and *last exit decomposition formulas*. Used together they work like the two hands of a human being, though one can do many things with one hand tied behind one's back, as we shall see later. Here as a preliminary ambidextrous application let us state the following little proposition as a lemma.

Lemma. *$i \rightsquigarrow j$ is equivalent to* $f_{ij}^* > 0$ and to $g_{ij}^* > 0$.

Proof: If $f_{ij}^* = 0$, then $f_{ij}^{(n)} = 0$ for every n and it follows from (8.4.8) that $p_{ij}^{(n)} = 0$ for every n. Hence it is false that $i \rightsquigarrow j$. Conversely if $f_{ij}^* > 0$, then $f_{ij}^{(n)} > 0$ for some n; since $p_{ij}^{(n)} \geq f_{ij}^{(n)}$ from the meaning of these two probabilities, we get $p_{ij}^{(n)} > 0$ and so $i \rightsquigarrow j$.

Now the argument for g_{ij}^* is *exactly* the same when we use (8.4.11) in lieu of (8.4.8), demonstrating the beauty of dual thinking.

Let us admit that the preceding proof is unduly hard in the case of f_{ij}^*, since a little reflection should convince us that "$i \rightsquigarrow j$" and "$f_{ij}^* > 0$" both mean: "it is possible to go from i to j in some steps," (see also Exercise 31). However, is it equally obvious that "$g_{ij}^* > 0$" means the same thing? The latter says that it is possible to go from i to j in some steps without going through i again. Hence the asserted equivalence will imply this: if it is possible to go from i to j then it is also possible to do so without first returning to i. For example, since one can drive from New York to San Francisco, does it follow that one can do that without coming back for repairs, forgotten items, or a temporary postponement? Is this so obvious that no proof is needed?

An efficient way to exploit the decomposition formulas is to introduce generating functions associated with the sequences $\{p_{ij}^{(n)}, n \geq 0\}$ (see §6.5):

$$P_{ij}(z) = \sum_{n=0}^{\infty} p_{ij}^{(n)} z^n,$$

$$F_{ij}(z) = \sum_{n=1}^{\infty} f_{ij}^{(n)} z^n,$$

$$G_{ij}(z) = \sum_{n=1}^{\infty} g_{ij}^{(n)} z^n,$$

where $|z| < 1$. We have then by substitution from (8.4.8) and inverting the order of summation:

$$\begin{aligned} P_{ij}(z) &= \delta_{ij} + \sum_{n=1}^{\infty} \left(\sum_{v=1}^{n} f_{ij}^{(v)} p_{jj}^{(n-v)} \right) z^v z^{n-v} \\ &= \delta_{ij} + \sum_{v=1}^{\infty} f_{ij}^{(v)} z^v \sum_{n=0}^{\infty} p_{jj}^{(n-v)} z^{n-v} \\ &= \delta_{ij} + F_{ij}(z) P_{jj}(z). \end{aligned} \tag{8.4.13}$$

The inversion is justified because both series are absolutely convergent for $|z| < 1$. In exactly the same way we obtain for $i \neq j$:

$$P_{ij}(z) = P_{ii}(z) G_{ij}(z). \tag{8.4.14}$$

The first application is to the case $i = j$.

Theorem 5. *For any state i we have $f_{ii}^* = 1$ if and only if*

$$\sum_{n=0}^{\infty} p_{ii}^{(n)} = \infty; \tag{8.4.15}$$

if $f_{ii}^* < 1$, then we have

$$\sum_{n=0}^{\infty} p_{ii}^{(n)} = \frac{1}{1 - f_{ii}^*}. \tag{8.4.16}$$

Proof: From (8.4.13) with $i = j$ and solving for $P_{ii}(z)$ we obtain

(8.4.17) $$P_{ii}(z) = \frac{1}{1 - F_{ii}(z)}.$$

If we put $z = 1$ above and observe that

$$P_{ii}(1) = \sum_{n=0}^{\infty} p_{ii}^{(n)}, \quad F_{ii}(1) = f_{ii}^*;$$

both assertions of the theorem follow. Let us point out that strictly speaking we must let $z \uparrow 1$ in (8.4.17) (why?), and use the following theorem from calculus. If $c_n \geq 0$ and the power series $C(z) = \sum_{n=0}^{\infty} c_n z^n$ converges for $|z| < 1$, then $\lim_{z \uparrow 1} C(z) = \sum_{n=0}^{\infty} c_n$, finite or infinite. This important result is called an *Abelian theorem* (after the great Norwegian mathematician Abel) and will be used again later.

The dichotomy in Theorem 5 yields a fundamental property of a state.

Definition of recurrent and nonrecurrent state. A state i is called *recurrent* iff $f_{ii}^* = 1$, and *nonrecurrent* iff $f_{ii}^* < 1$.

The adjectives "persistent" and "transient" are used by some authors for "recurrent" and "nonrecurrent." For later use let us insert a corollary to Theorem 5 here.

Corollary to Theorem 5. *If j is nonrecurrent, then $\sum_{n=0}^{\infty} p_{ij}^{(n)} < \infty$ for every i. In particular, $\lim_{n\to\infty} p_{ij}^{(n)} = 0$ for every i.*

Proof: For $i = j$, this is just (8.4.16). If $i \neq j$, this follows from (8.4.13) since

$$P_{ij}(1) = F_{ij}(1)P_{jj}(1) \leq P_{jj}(1) < \infty.$$

It is easy to show that two communicating states are either both recurrent or both nonrecurrent. Thus either property pertains to a class and may be called a *class property*. To see this let $i \backsim j$, then there exists $m \geq 1$ and $m' \geq 1$ such that $p_{ij}^{(m)} > 0$ and $p_{ji}^{(m')} > 0$. Now the same argument for (8.4.1) leads to the inequality:

$$p_{jj}^{(m'+n+m)} \geq p_{ji}^{(m')} p_{ii}^{(n)} p_{ij}^{(m)}.$$

Summing this over $n \geq 0$, we have

(8.4.18) $$\sum_{n=0}^{\infty} p_{jj}^{(n)} \geq \sum_{n=0}^{\infty} p_{jj}^{(m'+n+m)} \geq p_{ji}^{(m')} \left(\sum_{n=0}^{\infty} p_{ii}^{(n)} \right) p_{ij}^{(m)}.$$

If i is recurrent, then by (8.4.15) the last term above is infinite, hence so is the first term, and this means j is recurrent by Theorem 5. Since i and j are interchangeable we have proved our assertion regarding recurrent states. The assertion regarding nonrecurrent states then follows because nonrecurrence is just the negation of recurrence.

The preceding result is nice and useful, but we need a companion which says that it is impossible to go from a recurrent to a nonrecurrent state. [The reverse passage is possible as shown by Example 3 of §8.3.] This result lies deeper and will be proved twice below by different methods. The first relies on the dual Theorems 3 and 4.

Theorem 6. *If i is recurrent and $i \rightsquigarrow j$, then j is also recurrent.*

Proof: There is nothing to prove if $i = j$, hence we may suppose $i \neq j$. We have by (8.4.13) and (8.4.14):

$$P_{ij}(z) = F_{ij}(z)P_{jj}(z), \quad P_{ij}(z) = P_{ii}(z)G_{ij}(z),$$

from which we infer

$$F_{ij}(z)P_{jj}(z) = P_{ii}(z)G_{ij}(z). \tag{8.4.19}$$

If we let $z \uparrow 1$ as at the end of proof of Theorem 5, we obtain:

$$F_{ij}(1)P_{jj}(1) = P_{ii}(1)G_{ij}(1) = \infty$$

since $G_{ij}(1) > 0$ by the Lemma and $P_{ii}(1) = \infty$ by Theorem 5. Since $F_{ij}(1) > 0$ by the Lemma we conclude that $P_{jj}(1) = \infty$, hence j is recurrent by Theorem 5. This completes the proof of Theorem 6 but let us note that the formula (8.4.19) written in the form

$$\frac{P_{ii}(z)}{P_{jj}(z)} = \frac{F_{ij}(z)}{G_{ij}(z)}$$

leads to other interesting results when $z \uparrow 1$, called "ratio limit theorems" (see [Chung 2; §I.9]).

8.5. Further developments

To probe the depth of the notion of recurrence we now introduce a new "transfinite" probability, that of entering a given state *infinitely often:*

$$q_{ij} = P_i\{X_n = j \text{ for an infinite number of values of } n\}. \tag{8.5.1}$$

We have already encountered this notion in Theorem 2 of §8.2; in fact the latter asserts in our new notation that $q_{ij} = 1$ for every i and j in a symmetric random walk. Now what exactly does "infinitely often" mean? It means "again and again, without end," or more precisely: "given any large number, say m, it will happen more than m times." This need not strike you as anything hard to grasp, but it may surprise you that if we want to express q_{ij} in symbols, it looks like this (cf. the end of §1.3):

$$q_{ij} = P_i\left\{\bigcap_{m=1}^{\infty} \bigcup_{n=m}^{\infty} [X_n = j]\right\}.$$

For comparison let us write also

$$f_{ij}^* = P_i\left\{\bigcup_{n=1}^{\infty} [X_n = j]\right\}.$$

However, we will circumvent such formidable formalities in our discussion below.

To begin with, it is trivial from the meaning of the probabilities that

$$q_{ij} \leq f_{ij}^* \tag{8.5.2}$$

because "infinitely often" certainly entails "at least once." The next result is crucial.

Theorem 7. *For any state i, we have*

$$q_{ii} = \begin{cases} 1 & \text{if } i \text{ is recurrent,} \\ 0 & \text{if } i \text{ is nonrecurrent.} \end{cases}$$

Proof: Put $X_0 = i$, and $\alpha = f_{ii}^*$. Then α is the probability of at least one return to i. At the moment of the first return, the particle is in i and its prior history is irrelevant; hence from that moment on it will move as if making a fresh start from i ("like a newborn baby"). If we denote by R_m the event of "at least m returns," then this implies that the conditional probability $P(R_2 \mid R_1)$ is the same as $P(R_1)$ and consequently

$$P(R_2) = P(R_1R_2) = P(R_1)P(R_2 \mid R_1) = \alpha\cdot\alpha = \alpha^2.$$

Repeating this argument, we have by induction for $m \geq 1$:

$$P(R_{m+1}) = P(R_mR_{m+1}) = P(R_m)P(R_{m+1} \mid R_m) = \alpha^m\cdot\alpha = \alpha^{m+1}.$$

Therefore the probability of infinitely many returns is equal to

$$\lim_{m\to\infty} P(R_m) = \lim_{m\to\infty} \alpha^m = \begin{cases} 1 & \text{if } \alpha = 1, \\ 0 & \text{if } \alpha < 1, \end{cases} \tag{8.5.3}$$

proving the theorem.

Now is a good stopping time to examine the key point in the preceding proof:

$$P(R_{m+1} \mid R_m) = \alpha$$

which is explained by considering the moment of the mth return to the initial state i and starting anew from that moment on. The argument works because whatever has happened prior to the moment is irrelevant to future happenings. [Otherwise one can easily imagine a situation in which previous returns tend to inhibit a new one, such as visiting the same old tourist attraction.] This seems to be justified by the Markov property except for one essential caveat. Take $m = 1$ for definiteness; then the moment of the first return is precisely the T_i defined in (8.4.2), and the argument above is based on applying the Markovian assumption (8.3.3) at the moment T_i. But T_i is a random variable, its value depends on the sample point ω; can we substitute it for the constant time n in those formulas? You might think that since the latter holds true for *any n*, and $T_i(\omega)$ is equal to *some n* whatever ω may be, such a substitution must be "OK". (Indeed we have made a similar substitution in §8.2 without justification.) The fallacy in this thinking is easily exposed,† but here we will describe the type of random variables for which the substitution is legitimate.

Given the homogeneous Markov chain $\{X_n, n \in N^0\}$, a random variable T is said to be *optional* [or a *stopping time*] iff for each n, the event $\{T = n\}$ is determined by $\{X_0, X_1, \ldots, X_n\}$ alone. An event is *prior to T* iff it is determined by $\{X_0, X_1, \ldots, X_{T-1}\}$, and *posterior to T* iff it is determined by $\{X_{T+1}, X_{T+2}, \ldots\}$. (When $T = 0$ there is no prior event to speak of.) The state of the particle at the moment T is of course given by X_T [note: this is the random variable $\omega \rightarrow X_{T(\omega)}(\omega)$]. In case T is a constant n, these notions agree with our usual interpretation of "past" and "future" relative to the "present" moment n. In the general case they may depend on the sample point. There is nothing far-fetched in this; for instance phrases such as "pre-natal care," "post-war construction" or "the day after the locusts" contain an uncertain and therefore random date. When a gambler decides that he will bet on red "after black has appeared three times in a row," he is dealing with X_{T+1} where the value of T is a matter of chance. However, it is essential to observe that these relative notions make sense by virtue of the way an optional T is defined. Otherwise if the determination of T involves the future as well as the past and present, then "pre-T" and "post-T" will be mixed up and serve no useful purpose. If the gambler can foresee the future, he would not need probability theory! In this sense an optional time has also been described as being "independent of the future"; it must have been decided upon as an "option" without the advantage of clairvoyance.

We can now formulate the following extension of (8.3.3). For any optional T, any event A prior to T and any event B posterior to T, we have

$$P\{B \mid X_T = i; A\} = P\{B \mid X_T = i\}; \tag{8.5.4}$$

† E.g., suppose $X_0 = i_0 \neq k$, and take $T = T_k - 1$ in (8.5.5) below. Since $X_{T+1} = k$ the equation cannot hold in general.

and in particular for any states i and j:

(8.5.5) $$P\{X_{T+1} = j \mid X_T = i;\ A\} = p_{ij}.$$

This is known as the *strong Markov property*. It is actually implied by the apparently weaker form given in (8.3.3), hence also in the original definition (8.3.1). Probabilists used to announce the weak form and use the strong one without mentioning the difference. Having flushed the latter out in the open *we will accept it as the definition for a homogeneous Markov chain*. For a formal proof see [Chung 2; §I.13]. Let us observe that the strong Markov property was needed as early as in the proof of Theorem 2 of §8.2, where it was deliberately concealed in order not to sound a premature alarm. Now it is time to look back with understanding.

To return to Theorem 7 we must now verify that the T_i used in the proof there is indeed optional. This has been effectively shown in (8.4.6), for the event

$$\{T_i = n\} = \{X_v \neq i \text{ for } 1 \leq v \leq n-1;\ X_n = i\}$$

is clearly determined by $\{X_1, \ldots, X_n\}$ only. This completes the rigorous proof of Theorem 7, to which we add a corollary.

Corollary to Theorem 7. *For any i and j,*

$$q_{ij} = \begin{cases} f_{ij}^* & \text{if } j \text{ is recurrent,} \\ 0 & \text{if } j \text{ is nonrecurrent.} \end{cases}$$

Proof: This follows at once from the theorem and the relation:

(8.5.6) $$q_{ij} = f_{ij}^* q_{jj}.$$

For, to enter j infinitely many times means to enter it at least once and then return to it infinitely many times. As in the proof of Theorem 8, the reasoning involved here is based on the strong Markov property.

The next result shows the power of "thinking infinite."

Theorem 8. *If i is recurrent and $i \rightsquigarrow j$, then*

(8.5.7) $$q_{ij} = q_{ji} = 1.$$

Proof: The conclusion implies $i \leftrightsquigarrow j$, and that j is recurrent by the corollary above. Thus the following proof contains a new proof of Theorem 6.

Let us note that for any two events A and B, we have $A \subset AB \cup B^c$ and consequently

(8.5.8) $$P(A) \leq P(B^c) + P(AB).$$

Now consider

$$A = \{\text{enter } i \text{ infinitely often}\},$$
$$B = \{\text{enter } j \text{ at least once}\}.$$

Then $P_i(A) = q_{ii} = 1$ by Theorem 7 and $P_i(B^c) = 1 - f^*_{ij}$. As for $P_i(AB)$ this means the probability that the particle will enter j at some finite time and *thereafter* enter i infinitely many times, because "infinite minus finite is still infinite." Hence if we apply the strong Markov property at the first entrance time into j, we have

$$P(AB) = f^*_{ij} q_{ji}.$$

Substituting into the inequality (8.5.8), we obtain

$$1 = q_{ii} \leq 1 - f^*_{ij} + f^*_{ij} q_{ji}$$

and so

$$f^*_{ij} \leq f^*_{ij} q_{ji}.$$

Since $f^*_{ij} > 0$ this implies $1 \leq q_{ji}$, hence $q_{ji} = 1$. Since $q_{ji} \leq f^*_{ji}$ it follows that $f^*_{ji} = 1$, and so $j \rightsquigarrow i$. Thus i and j communicate and therefore j is recurrent by (8.4.18). Knowing this we may interchange the roles of i and j in the preceding argument to infer $q_{ij} = 1$.

Corollary. *In a recurrent class* (8.5.7) *holds for any two states i and j.*

When the state space of a chain forms a single recurrent class, we shall call the chain recurrent; similarly for "nonrecurrent". The state of affairs for a recurrent chain described in the preceding Corollary is precisely that for a symmetric random walk in Theorem 2 of §8.2. In fact, the latter is a particular case as we now proceed to show.

We shall apply the general methods developed above to the case of random walk discussed in §8.1, namely Example 1 of §8.3. We begin by evaluating $p_{ii}^{(n)}$. This is the probability that the particle returns to its initial position i in exactly n steps. Hence $p_{ii}^{(2n-1)} = 0$ for $n \geq 1$; and in the notation of (8.1.2)

$$p_{ii}^{(2n)} = P\{\xi_1 + \cdots + \xi_{2n} = 0\} = \binom{2n}{n} p^n q^n \tag{8.5.9}$$

by Bernoulli's formula (7.3.1), since there must be n steps to the right and n steps to the left, in some order. Thus we obtain the generating function

$$P_{ii}(z) = \sum_{n=0}^{\infty} \binom{2n}{n} (pqz^2)^n. \tag{8.5.10}$$

Recalling the general binomial coefficients from (5.4.4), we record the pretty identity:

$$\binom{-\frac{1}{2}}{n} = \frac{(-1)^n 1 \cdot 3 \cdots (2n-1)}{2^n \cdot n!} = \frac{(-1)^n}{2^{2n}} \binom{2n}{n} \tag{8.5.11}$$

where the second equation is obtained by multiplying both the denominator and numerator of its left member by $2^n \cdot n! = 2 \cdot 4 \cdots (2n)$. Substituting into (8.5.10), we arrive at the explicit analytic formula:

$$P_{ii}(z) = \sum_{n=0}^{\infty} \binom{-\frac{1}{2}}{n} (-4pqz^2)^n = (1 - 4pqz^2)^{-1/2} \tag{8.5.12}$$

where the second member is the binomial (Taylor's) series of the third member.

It follows that

$$\sum_{n=0}^{\infty} p_{ii}^{(n)} = P_{ii}(1) = \lim_{z \uparrow 1} P_{ii}(z) = \lim_{z \uparrow 1} (1 - 4pqz^2)^{-1/2}. \tag{8.5.13}$$

Now $4pq = 4p(1 - p) \leq 1$ for $0 \leq p \leq 1$; and $= 1$ if and only if $p = 1/2$ (why?). Hence the series above diverges if $p = 1/2$, and converges if $p \neq 1/2$. By Theorem 5, i is recurrent if and only if $p = 1/2$. The calculations above do not depend on the integer i because of spatial homogeneity. Thus for $p = 1/2$ the chain is recurrent; otherwise it is nonrecurrent. In other words, the random walk is recurrent if and only if it is symmetric.

There is another method of showing this directly from (8.5.9), without the use of generating functions. For when $p = \dfrac{1}{2}$ we have

$$p_{ii}^{(2n)} = \binom{2n}{n} \frac{1}{2^{2n}} \sim \frac{1}{\sqrt{\pi n}}, \tag{8.5.14}$$

by (7.3.6) as an application of Stirling's formula. Hence by the comparison test for positive series, the series in (8.5.13) diverges because $\sum_n \dfrac{1}{\sqrt{n}}$ does. This method has the merit of being applicable to random walks in higher dimensions. Consider the symmetric random walk in R^2 (Example 10 of §8.3 with all four probabilities equal to 1/4). To return from any state (i, j) to (i, j) in $2n$ steps means that: for some k, $0 \leq k \leq n$, the particle takes, in some order, k steps each to the east and west, and $n - k$ steps each to the north and south. The probability for this, by the multinomial formula (6.4.6), is equal to:

$$\begin{aligned} p_{(i,j)(i,j)}^{(2n)} &= \frac{1}{4^{2n}} \sum_{k=0}^{n} \frac{(2n)!}{k!k!(n-k)!(n-k)!} \\ &= \frac{(2n)!}{4^{2n} n! n!} \sum_{k=0}^{n} \binom{n}{k}^2 = \frac{(2n)!}{4^{2n} n! n!} \binom{2n}{n} = \left[\frac{1}{2^{2n}} \binom{2n}{n}\right]^2, \end{aligned} \tag{8.5.15}$$

where in the penultimate equation we have used a formula given in Exercise 28 of Chapter 3. The fact that this probability turns out to be the exact square of the one in (8.5.14) is a pleasant coincidence. [It is not due to any apparent independence between the two components of the walk along the two coordinate axes.] It follows by comparison with (8.5.14) that

$$\sum_n p_{(i,j)(i,j)}^{(2n)} \sim \sum_n \frac{1}{\pi n} = \infty.$$

Hence another application of Theorem 5 shows that the symmetric random walk in the plane as well on the line is a recurrent Markov chain. A similar but more complicated argument shows that it is nonrecurrent in R^d for $d \geq 3$, because the probability analogous to that in (8.5.15) is bounded by $\frac{c}{n^{d/2}}$ (where c is a constant), and $\sum_n \frac{1}{n^{d/2}}$ converges for $d \geq 3$. These results were first discovered by Pólya in 1921. The non-symmetric case can be treated by using the normal approximation given in (7.3.13), but there the nonrecurrence is already implied by the strong law of large numbers as in R^1; see §8.2.

As another illustration, we will derive an explicit formula for $f_{ii}^{(n)}$ in case $p = \frac{1}{2}$. By (8.4.17) and (8.5.12), we have

$$F_{ii}(z) = 1 - \frac{1}{P_{ii}(z)} = 1 - (1 - z^2)^{1/2}.$$

Hence another expansion by means of binomial series gives

$$F_{ii}(z) = 1 - \sum_{n=0}^{\infty} \binom{\frac{1}{2}}{n}(-z^2)^n$$

$$= \frac{1}{2} z^2 + \sum_{n=2}^{\infty} \frac{1 \cdot 3 \cdots (2n-3)}{2^n \cdot n!} z^{2n}.$$

Thus $f_{ii}^{(2n-1)} = 0$; and

$$f_{ii}^{(2n)} = \frac{1}{2^{2n}} \binom{2n}{n} \frac{1}{2n-1}, \quad n \geq 1; \tag{8.5.16}$$

by a calculation similar to (8.5.11). In particular we have

n	1	2	3	4	5
$f_{ii}^{(2n)}$	$\frac{1}{2}$	$\frac{1}{8}$	$\frac{1}{16}$	$\frac{5}{128}$	$\frac{7}{256}$

Comparison with (8.5.14) shows that

$$f_{ii}^{(2n)} \sim \frac{1}{2\sqrt{\pi}\, n^{3/2}}$$

and so $\sum_{n=1}^{\infty} nf_{ii}^{(n)} = \infty$. This can also be gotten by calculating $F_{ii}'(1)$. Thus, although return is almost certain the expected time before return is infinite. This result will be seen in a moment to be equivalent to the remark made in §8.2 that $e_1 = \infty$.

We can calculate $f_{ij}^{(n)}$ for any i and j in a random walk by a similar method. However, sometimes a combinational argument is quicker and more revealing. For instance, we have

(8.5.17)
$$f_{00}^{(2n)} = \frac{1}{2}f_{10}^{(2n-1)} + \frac{1}{2}f_{-1,0}^{(2n-1)} = f_{10}^{(2n-1)} = f_{01}^{(2n-1)}.$$

To argue this let the particle start from 0 and consider the outcome of its first step as in the derivation of (8.1.4); then use the symmetry and spatial homogeneity to get the rest. The details are left to the reader.

8.6. Steady state

In this section we consider a recurrent Markov chain, namely we suppose that the state space forms a single recurrent class.

After the particle in such a chain has been in motion for a long time, it will be found in various states with various probabilities. Do these settle down to limiting values? This is what the physicists and engineers call a "steady state" (distribution).† They are accustomed to thinking in terms of an "ensemble" or large number of particles moving according to the same probability laws and independently of one another, such as in the study of gaseous molecules. In the present case the laws are those pertaining to a homogeneous Markov chain as discussed in the preceding sections. After a long time, the proportion (percentage) of particles to be found in each state gives approximately the steady-state probability of that state. [Note the double usage of the word "state" in the last sentence; we shall use "stationary" for the adjective "steady-state"]. In effect, this is the frequency interpretation of probability mentioned in Example 3 of §2.1, in which the limiting proportions are taken to determine the corresponding probabilities. In our language, if the particle starts from the state i, then the probability of the set of paths in which it moves to state j at time n, namely $\{\omega \mid X_n(\omega) = j\}$, is given by $P_i\{X_n = j\} = p_{ij}^{(n)}$. We are therefore interested in the asymptotic behavior of $p_{ij}^{(n)}$ as $n \to \infty$. It turns out that a somewhat more amenable quantity is its average value over a long period of time, namely:

(8.6.1)
$$\frac{1}{n+1}\sum_{v=0}^{n} p_{ij}^{(v)} \quad \text{or} \quad \frac{1}{n}\sum_{v=1}^{n} p_{ij}^{(v)}.$$

The difference between these two averages is negligible for large n but we shall use the former. This quantity has a convenient interpretation as follows. Fix

---

† Strange to relate, they call a "distribution" a "state"!

our attention on a particular state j and imagine that a counting device records the number of time units the particle spends in j. This is done by introducing the random variables below which count 1 for the state j but 0 for any other state:

$$\xi_v(j) = \begin{cases} 1 & \text{if } X_v = j, \\ 0 & \text{if } X_v \neq j. \end{cases}$$

We have used such indicators e.g. in (6.4.11). Next we put

$$N_n(j) = \sum_{v=0}^{n} \xi_v(j)$$

which represents the total *occupation time* of the state j in n steps. Now if E_i denotes the mathematical expectation associated with the chain starting from i [this is a conditional expectation; see end of §5.2], we have

$$E_i(\xi_v(j)) = p_{ij}^{(v)}$$

and so by Theorem 1 of §6.1:

$$E_i(N_n(j)) = \sum_{v=0}^{n} E_i(\xi_v(j)) = \sum_{v=0}^{n} p_{ij}^{(v)}. \tag{8.6.2}$$

Thus the quantity in (8.6.1) turns out to be the average expected occupation time.

In order to study this we consider first the case $i = j$ and introduce the *expected return time* from j to j as follows:

$$m_{jj} = E_j(T_j) = \sum_{v=1}^{\infty} v f_{jj}^{(v)} \tag{8.6.3}$$

where T_j is defined in (8.4.2). Since j is a recurrent state we know that T_j is almost surely finite, but its expectation may be finite or infinite. We shall see that the distinction between these two cases is essential.

Here is the heuristic argument linking (8.6.2) and (8.6.3). Since the time required for a return is m_{jj} units on the basis of expectation, there should be about n/m_{jj} such returns in a span of n time units on the same basis. In other words the particle spends about n/m_{jj} units of time in the state j during the first n steps, namely $E_j(N_n(j)) \approx n/m_{jj}$. The same argument shows that it makes no difference whether the particle starts from j or any other state i, because after the first entrance into j the initial i may be forgotten and we are concerned only with the subsequent returns from j to j. Thus we are led to the following limit theorem.

Theorem 9. *For any i and j we have*

$$\lim_{n\to\infty} \frac{1}{n+1} \sum_{v=0}^{n} p_{ij}^{(v)} = \frac{1}{m_{jj}}. \tag{8.6.4}$$

The argument indicated above can be made rigorous by invoking a general form of the strong law of large numbers (see §7.5), applied to the successive return times which form a sequence of independent and identically distributed random variables. Unfortunately the technical details are above the level of this book. There is another approach which relies on a powerful analytical result due to Hardy and Littlewood. [This is the same Hardy as in the Hardy-Weinberg theorem of §5.6.] It is known as a *Tauberian theorem* (after Tauber who first found a result of the kind) and may be stated as follows.

Theorem 10. *If* $A(z) = \sum_{n=0}^{\infty} a_n z^n$, *where* $a_n \geq 0$ *for all* n *and the series converges for* $0 \leq z < 1$, *then we have*

$$\lim_{n\to\infty} \frac{1}{n+1} \sum_{v=0}^{n} a_v = \lim_{z\to 1} (1 - z)A(z). \tag{8.6.5}$$

To get a feeling for this theorem, suppose all $a_n = c > 0$. Then

$$A(z) = c \sum_{n=0}^{\infty} z^n = \frac{c}{1 - z}$$

and the relation in (8.6.5) reduces to the trivial identity

$$\frac{1}{n+1} \sum_{v=0}^{n} c = c = (1 - z)\frac{c}{1 - z}.$$

Now take $A(z)$ to be

$$P_{ij}(z) = \sum_{n=0}^{\infty} p_{ij}^{(n)} z^n = F_{ij}(z)P_{jj}(z) = \frac{F_{ij}(z)}{1 - F_{jj}(z)},$$

where the last two equations come from (8.4.13) and (8.4.17). Then we have

$$\lim_{z\to 1} (1 - z)P_{ij}(z) = F_{ij}(1) \lim_{z\to 1} \frac{1 - z}{1 - F_{jj}(z)} = \lim_{z\to 1} \frac{1 - z}{1 - F_{jj}(z)}$$

since $F_{ij}(1) = f_{ij}^* = 1$ by Theorem 8 of §8.5. The last-written limit may be evaluated by l'Hospital rule:

$$\lim_{z\to 1} \frac{(1 - z)'}{(1 - F_{jj}(z))'} = \lim_{z\to 1} \frac{-1}{-F_{jj}'(z)} = \frac{1}{F_{jj}'(1)}$$

where "$'$" stands for differentiation with respect to z. Since $F_{jj}'(z) = \sum_{v=1}^{\infty} v f_{jj}^{(v)} z^{v-1}$ we have $F_{jj}'(1) = \sum_{v=1}^{\infty} f_{jj}^{(v)} = m_{jj}$, and so (8.6.4) is a special case of (8.6.5).

We now consider a finite state space I in order not to strain our mathematical equipment. The finiteness of I has an immediate consequence.

Theorem 11. *If I is finite and forms a single class (namely if there are only a finite number of states and they all communicate with each other), then the chain is necessarily recurrent.*

Proof: Suppose the contrary; then each state is transient and so almost surely the particle can spend only a finite number of time units in it by Theorem 7. Since the number of states is finite, the particle can spend altogether only a finite number of time units in the whole space I. But time increases *ad infinitum* and the particle has nowhere else to go. This absurdity proves that the chain must be recurrent. (What then can the particle do?)

Let $I = \{1, 2, \ldots, l\}$, and put

$$x = (x_1, \ldots, x_l)$$

which is a row vector of l components. Consider the "steady-state equation"

$$(8.6.6) \qquad x = x\Pi \quad \text{or} \quad x(\Delta - \Pi) = 0$$

where Δ is the identity matrix with l rows and l columns: $\Delta = (\delta_{ij})$, and Π is the transition matrix in (8.3.9). This is a system of l linear homogeneous equations in l unknowns. Now the determinant of the matrix $\Delta - \Pi$

$$\begin{vmatrix} 1 - p_{11} & -p_{12} & \cdots & -p_{1l} \\ -p_{21} & 1 - p_{22} & \cdots & -p_{2l} \\ \cdots & \cdots & \cdots & \cdots \\ -p_{l1} & -p_{l2} & \cdots & 1 - p_{ll} \end{vmatrix}$$

is equal to zero because the sum of all elements in each row is $1 - \sum_j p_{ij} = 0$. Hence we know from linear algebra that the system has a non-trivial solution, namely one which is not the zero vector. Clearly if x is a solution then so is $cx = (cx_1, cx_2, \ldots, cx_l)$ for any constant c. The following theorem identifies all solutions when I is a single finite class.

We shall write

$$(8.6.7) \qquad w_j = \frac{1}{m_{jj}}, \quad j \in I; \qquad w = (w_1, w_2, \ldots, w_l);$$

and $\sum_j$ for $\sum_{j \in I}$ below.

Theorem 12. *If I is finite and forms a single class, then*

(*i*) *w is a solution of (8.6.6);*
(*ii*) $\sum_j w_j = 1$;
(*iii*) *$w_j > 0$ for all j;*
(*iv*) *any solution of (8.6.6) is a constant multiple of w.*

Proof: We have from (8.3.7), for every $v \geq 0$:

$$p_{ik}^{(v+1)} = \sum_j p_{ij}^{(v)} p_{jk}.$$

Taking an average over v we get

$$\frac{1}{n+1} \sum_{v=0}^{n} p_{ik}^{(v+1)} = \sum_j \left(\frac{1}{n+1} \sum_{v=0}^{n} p_{ij}^{(v)} \right) p_{jk}.$$

The left member differs from $\frac{1}{n+1} \sum_{v=0}^{n} p_{ik}^{(v)}$ by $\frac{1}{n+1} (p_{ik}^{(n+1)} - p_{ik}^{(0)})$ which tends to zero as $n \to \infty$; hence its limit is equal to w_k by Theorem 9. Since I is finite we may let $n \to \infty$ term by term in the right member. This yields

$$w_k = \sum_j w_j p_{jk}$$

which is $w = w\Pi$; hence (i) is proved. We can now iterate:

$$w = w\Pi = (w\Pi)\Pi = w\Pi^2 = (w\Pi)\Pi^2 = w\Pi^3 = \ldots, \tag{8.6.8}$$

to obtain $w = w\Pi^n$, or explicitly for $n \geq 1$:

$$w_k = \sum_j w_j p_{jk}^{(n)}. \tag{8.6.9}$$

Next we have $\sum_j p_{ij}^{(v)} = 1$ for every i and $v \geq 1$. Taking an average over v we obtain

$$\frac{1}{n+1} \sum_{v=0}^{n} \sum_j p_{ij}^{(v)} = 1.$$

It follows that

$$\sum_j w_j = \sum_j \lim_{n \to \infty} \left(\frac{1}{n+1} \sum_{v=0}^{n} p_{ij}^{(v)} \right) = \lim_{n \to \infty} \frac{1}{n+1} \sum_{v=0}^{n} \sum_j p_{ij}^{(v)} = 1$$

where the second equation holds because I is finite. This establishes (ii) from which we deduce that at least one of the w_j's, say w_i, is positive. For any k we have $i \rightsquigarrow k$ and so there exists n such that $p_{ik}^{(n)} > 0$. Using this value of n in (8.6.9), we see that w_k is also positive. Hence (iii) is true. Finally suppose x

is any solution of (8.6.6). Then $x = x\Pi^v$ for every $v \geq 1$ by iteration as before, and

$$x = \frac{1}{n+1} \sum_{v=0}^{n} x\Pi^v$$

by averaging. In explicit notation this is

$$x_k = \sum_j x_j \left(\frac{1}{n+1} \sum_{v=0}^{n} p_{jk}^{(v)} \right).$$

Letting $n \to \infty$ and using Theorem 9 we obtain

$$x_k = \left(\sum_j x_j \right) w_k.$$

Hence (iv) is true with $c = \sum_j x_j$. Theorem 12 is completely proved.

We call $\{w_j, j \in I\}$ the *stationary* (*steady-state*) *distribution* of the Markov chain. It is indeed a probability distribution by (ii). The next result explains the meaning of the word "stationary."

Theorem 13. *Suppose that we have for every* j,

$$P\{X_0 = j\} = w_j, \tag{8.6.10}$$

then the same is true when X_0 *is replaced by any* X_n, $n \geq 1$. *Furthermore the joint probability*

$$P\{X_{n+v} = j_v, 0 \leq v \leq l\} \tag{8.6.11}$$

for arbitrary j_v, *is the same for all* $n \geq 0$.

Proof: We have by (8.6.9),

$$P\{X_n = j\} = \sum_i P\{X_0 = i\} P_i\{X_n = j\} = \sum_i w_i p_{ij}^{(n)} = w_j.$$

Similarly the probability in (8.6.11) is equal to

$$P\{X_n = j_0\} p_{j_0 j_1} \cdots p_{j_{l-1} j_l} = w_{j_0} p_{j_0 j_1} \cdots p_{j_{l-1} j_l}$$

which is the same for all n.

Thus, with the stationary distribution as its initial distribution, the chain becomes a stationary process as defined in §5.4. Intuitively, this means that if a system is in its steady state, it will hold steady indefinitely there, that is, so far as distributions are concerned. Of course changes go on in the system, but they tend to balance out to maintain an over-all equilibrium. For instance, many ecological systems have gone through millions of years of transitions and may be considered to have reached their stationary phase—until human intervention abruptly altered the course of evolution. However, if the new process is again a homogeneous Markov chain as supposed here, then it too will settle down to a steady state in due time according to our theorems.

The practical significance of Theorem 12 is that it guarantees a solution of (8.6.6) which satisfies the conditions (ii) and (iii). In order to obtain this solution, we may proceed as follows. Discard one of the l equations and solve the remaining equations for $w_2, \ldots, w_l$ in terms of w_1. These are of the form $w_j = c_j w_1$, $1 \leq j \leq l$, where $c_1 = 1$. The desired solution is then given by

$$w_j = \frac{c_j}{\sum_{j=1}^{l} c_j}, \quad 1 \leq j \leq l.$$

Example 13. A switch may be *on* or *off;* call these two positions states 1 and 2. After each unit of time the state may hold or change, but the respective probabilities depend only on the present position. Thus we have a homogeneous Markov chain with $I = \{1, 2\}$ and

$$\Pi = \begin{bmatrix} p_{11} & p_{12} \\ p_{21} & p_{22} \end{bmatrix}$$

where all elements are supposed to be positive. The steady-state equations are

$$\begin{aligned} (1 - p_{11})x_1 - p_{21}x_2 &= 0, \\ -p_{12}x_1 + (1 - p_{22})x_2 &= 0. \end{aligned}$$

Clearly the second equation is just the negative of the first and may be discarded. Solving the first equation we get

$$x_2 = \frac{1 - p_{11}}{p_{21}} x_1 = \frac{p_{12}}{p_{21}} x_1.$$

Thus

$$w_1 = \frac{p_{21}}{p_{12} + p_{21}}, \quad w_2 = \frac{p_{12}}{p_{12} + p_{21}}.$$

In view of Theorem 9, this means: in the long run the switch will be on or off for a total amount of time in the ratio of $p_{21}:p_{12}$.

Example 14. At a carnival Daniel won a prize for free rides on the merry-go-round. He therefore took "infinitely many" rides but each time when the bell rings he moves onto the next hobby-horse forward or backward, with probability p or $q = 1 - p$. What proportion of time was he on each of these horses?

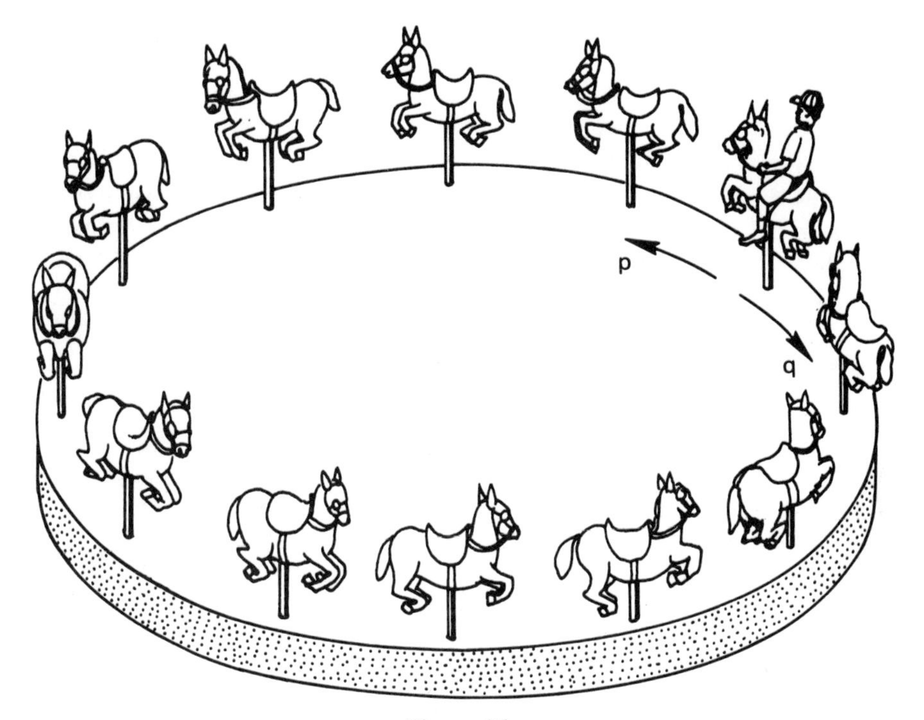

Figure 34

This may be described as "random walk on a circle." The transition matrix looks like this:

$$\begin{bmatrix} 0 & p & 0 & 0 & \cdots & 0 & 0 & q \\ q & 0 & p & 0 & & 0 & 0 & 0 \\ 0 & q & 0 & p & & 0 & 0 & 0 \\ \cdots & \cdots & \cdots & \cdots & \cdots & \cdots & \cdots & \cdots \\ 0 & 0 & 0 & 0 & \cdots & 0 & p & 0 \\ 0 & 0 & 0 & 0 & \cdots & q & 0 & p \\ p & 0 & 0 & 0 & \cdots & 0 & q & 0 \end{bmatrix}$$

The essential feature of this matrix is that the elements in each column (as well as in each row) add up to one. In general notation, this means that we have for every $j \in I$:

$$\sum_{i \in I} p_{ij} = 1. \tag{8.6.12}$$

Such a matrix is called *doubly stochastic*. Now it is trivial that under the condition (8.6.12), $x = (1, 1, \ldots, 1)$ where all components are equal to one, is a solution of the equation (8.6.6). Since the stationary distribution w must be a multiple of this by (iv) of Theorem 12, and also satisfy (iii), it must be

$$w = \left(\frac{1}{l}, \frac{1}{l}, \ldots, \frac{1}{l}\right)$$

where as before l is the number of states in I. This means if Daniel spent 4 hours on the merry-go-round and there are 12 horses, his occupation time of each horse is about 20 minutes, provided that he changed horses sufficiently many times to make the limiting relation in (8.6.4) operative.

For a recurrent Markov chain in an infinite state space, Theorem 12 must be replaced by a drastic dichotomy as follows:

(a) either all $w_j > 0$, then (ii) and (iii) hold as before, and Theorem 13 is also true;
(b) or all $w_j = 0$.

The chain is said to be *positive-recurrent* (or *strongly ergodic*) in case (a), and *null-recurrent* (or *weakly ergodic*) in case (b). The symmetric random walk discussed in §8.1-2 is an example of the latter (see Exercise 38 below). It can be shown (see [Chung 2; §I.7]) that if the equation (8.6.6) has a solution $x = (x_1, x_2, \ldots)$ satisfying the condition $0 < \sum_j |x_j| < \infty$, then in fact all $x_j > 0$ and the stationary distribution is given by

$$w_j = \frac{x_j}{\sum_j x_j}, \quad j \in I.$$

The following example illustrates this result.

Example 15. Let $I = \{0, 1, 2, \ldots\}$, and $p_{ij} = 0$ for $|i - j| > 1$, whereas the other p_{ij}'s are arbitrary positive numbers. These must then satisfy the equation

$$p_{j,j-1} + p_{jj} + p_{j,j+1} = 1 \tag{8.6.13}$$

for every j. This may be regarded as a special case of Example 7 in §8.3 with $p_{0,-1} = 0$ and a consequent reduction of state space. It may be called a *simple birth-and-death process* (in discrete time) in which j is the population size and $j \to j + 1$ or $j \to j - 1$ corresponds to a single birth or death. The equation (8.6.6) becomes:

$$\begin{aligned} x_0 &= x_0 p_{00} + x_1 p_{10}, \\ x_j &= x_{j-1} p_{j-1,j} + x_j p_{jj} + x_{j+1} p_{j+1,j}, \quad j \geq 1. \end{aligned} \tag{8.6.14}$$

This is an infinite system of linear homogeneous equations, but it is clear that all possible solutions can be obtained by assigning an arbitrary value to x_0, and then solve for $x_1, x_2, \ldots$ successively from the equations. Thus we get

$$x_1 = \frac{p_{01}}{p_{10}} x_0,$$

$$x_2 = \frac{1}{p_{21}} \{x_1(1 - p_{11}) - x_0 p_{01}\} = \frac{p_{01}(1 - p_{11} - p_{10})}{p_{21}p_{10}} x_0 = \frac{p_{01}p_{12}}{p_{10}p_{21}} x_0.$$

It is easy to guess (perhaps after a couple more steps) that we have in general

$$(8.6.15) \qquad x_j = c_j x_0 \quad \text{where} \quad c_0 = 1, \quad c_j = \frac{p_{01}p_{12} \cdots p_{j-1,j}}{p_{10}p_{21} \cdots p_{j,j-1}}, \quad j \geq 1.$$

To verify this by induction, let us assume that $p_{j,j-1}x_j = p_{j-1,j}x_{j-1}$; then we have by (8.6.14) and (8.6.13):

$$\begin{aligned} p_{j+1,j}x_{j+1} &= (1 - p_{jj})x_j - p_{j-1,j}x_{j-1} \\ &= (1 - p_{jj} - p_{j,j-1})x_j = p_{j,j+1}x_j. \end{aligned}$$

Hence this relation holds for all j and (8.6.15) follows. We have therefore

$$(8.6.16) \qquad \sum_{j=0}^{\infty} x_j = \left(\sum_{j=0}^{\infty} c_j\right) x_0,$$

and the dichotomy cited above is as follows, provided that the chain is recurrent. It is easy to see that this is true in case (a).

Case (a). If $\sum_{j=0}^{\infty} c_j < \infty$, then we may take $x_0 = 1$ to obtain a solution satisfying $\sum_j x_j < \infty$. Hence the chain is positive-recurrent and the stationary distribution is given by

$$w_j = \frac{c_j}{\sum_{j=0}^{\infty} c_j}, \quad j \geq 0.$$

Case (b). If $\sum_{j=0}^{\infty} c_j = \infty$, then for any choice of x_0, either $\sum_j |x_j| = \infty$ or $\sum_j |x_j| = 0$ by (8.6.16). Hence $w_j = 0$ for all $j \geq 0$, and the chain is null-recurrent or transient.

The preceding example may be modified to reduce the state space to a finite set by letting $p_{c,c+1} = 0$ for some $c \geq 1$. A specific case of this is Example 8 of §8.3 which will now be examined.

Example 16. Let us find the stationary distribution for the Ehrenfest model. We can proceed exactly as in Example 15, leading to the formula (8.6.15) but this time it stops at $j = c$. Substituting the numerical values from (8.3.16), we obtain

$$c_j = \frac{c(c-1)\cdots(c-j+1)}{1\cdot 2\cdots j} = \binom{c}{j}, \quad 0 \le j \le c.$$

We have $\sum_{j=0}^{c} c_j = 2^c$ from (3.3.7); hence

$$w_j = \frac{1}{2^c}\binom{c}{j}, \; 0 \le j \le c.$$

This is just the binomial distribution $B\left(c, \frac{1}{2}\right)$.

Thus the steady state in Ehrenfest's urn may be simulated by coloring the c balls red or black with probability 1/2 each, and independently of one another; or again by picking them at random from an infinite reservoir of red and black balls in equal proportions.

Next, recalling (8.6.3) and (8.6.7), we see that the mean recurrence times are given by

$$m_{jj} = 2^c \binom{c}{j}^{-1}, \quad 0 \le j \le c.$$

For the extreme cases $j = 0$ (no black ball) and $j = c$ (no red ball) this is equal to 2^c which is enormous even for $c = 100$. It follows (see Exercise 42) that the expected time for a complete reversal of the composition of the urn is very long indeed. On the other hand, the chain is recurrent; hence starting e.g. from an urn containing all black balls, it is almost certain that they will eventually be all replaced by red balls at some time in the Ehrenfest process, and vice versa. Since the number of black balls can change only one at a time the composition of the urn must go through all intermediate "phases" again and again. The model was originally conceived to demonstrate the reversibility of physical processes, but with enormously long cycles for reversal. "If we wait long enough, we shall grow younger again!"

Finally, let us describe without proof a further possible decomposition of a recurrent class. The simplest illustration is that of the classical random walk. In this case the state space of all integers may be divided into two subclasses: the even integers and the odd integers. At one step the particle must move from one subclass to the other, so that the alternation of the two subclasses is a deterministic part of the transition. In general, for each recurrent class C containing at least two states, there exists a unique positive integer d, called the *period* of the class, with the following properties:

(a) for every $i \in C$, $p_{ii}^{(n)} = 0$ if $d \nmid n$†; on the other hand, $p_{ii}^{(nd)} > 0$ for all sufficiently large n (how large depending on i);

† "$d \nmid n$" reads "d does not divide n".

(b) for every $i \in C$ and $j \in C$, there exists an integer r, $1 \leq r \leq d$, such that $p_{ij}^{(n)} = 0$ if $d \nmid n - r$; on the other hand, $p_{ij}^{(nd+r)} > 0$ for all sufficiently large n (how large depending on i and j).

Fixing the state i, we denote by C_r the set of all states j associated with the same number r in (b), for $1 \leq r \leq d$. These are disjoint sets whose union is C. Then we have the deterministic cyclic transition:

$$C_1 \to C_2 \to \cdots \to C_d \to C_1.$$

Here is the diagram of such an example with $d = 4$:

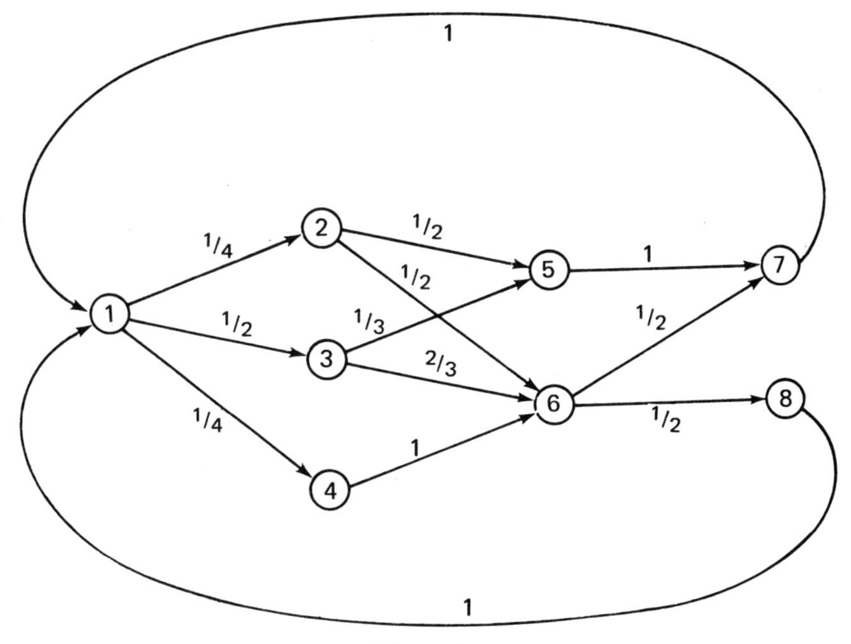

Figure 35

where the transition probabilities between the states are indicated by the numbers attached to the directed lines joining them.

The period d of C can be found as follows. Take any $i \in C$ and consider the set of all $n \geq 1$ such that $p_{ii}^{(n)} > 0$. Among the common divisors of this set there is a greatest one: this is equal to d. The fact that this number is the same for all choices of i is part of the property of the period. Incidentally, the decomposition described above holds for any class which is stochastically closed (see §8.7 for definition); thus the free random walk has period 2 whether it is recurrent or transient.

When $d = 1$ the class is said to be *aperiodic*. A sufficient condition for this is: there exists an integer m such that all elements in Π^m are positive. For then it follows from the Chapman-Kolmogorov equations that the same is true of Π^n for all $n \geq m$, and so property (a) above implies that $d = 1$. In this case the fundamental limit theorem given in (8.6.4) can be sharpened as follows:

(8.6.17)
$$\lim_{n\to\infty} p_{ij}^{(n)} = \frac{1}{m_{jj}};$$

namely the limit of averages may be replaced by a strict individual limit. In general if the period is d, and i and j are as in (b) above, then

(8.6.18)
$$\lim_{n\to\infty} p_{ij}^{(nd+r)} = \frac{d}{m_{jj}}.$$

We leave it to the reader to show: granted that the limit above exists, its value must be that shown there as a consequence of (8.6.4). Actually (8.6.18) follows easily from the particular case (8.6.17) if we consider d steps at a time in the transition of the chain, so that it stays in a fixed subclass. The sharp result above was first proved by Markov who considered only a finite state space, and was extended by Kolmogorov in 1936 to the infinite case. Several different proofs are now known; see [Chung 2; §I.6] for one of them.

8.7. Winding up (or down?)

In this section we shall give some idea of the general behavior of a homogeneous Markov chain when there are both recurrent and transient states. Let R denote the set of all recurrent states, T the set of all transient states, so that $I = R \cup T$. We begin with a useful definition: a set of states will be called [*stochastically*] *closed* iff starting from any state in the set the particle will remain forever in the set. Here and hereafter we shall omit the tedious repetition of the phrase "almost surely" when it is clearly indicated. The salient features of the global motion of the particle may be summarized as follows.

(i) A recurrent class is closed. Hence, once the particle enters such a class it will stay there forever.

(ii) A finite set of transient states is not closed. In fact, starting from such a set the particle will eventually move out and stay out of it.

(iii) If T is finite then the particle will eventually enter into one of the various recurrent classes.

(iv) In general the particle will be absorbed into the recurrent classes with total probability α, and remain forever in T with probability $1 - \alpha$, where $0 \leq \alpha \leq 1$.

Let us prove assertion (i). The particle cannot go from a recurrent state to any transient state by Theorem 6; and it cannot go to any recurrent state in a different class because two states from different classes do not communicate by definition, hence one does not lead to the other by Theorem 8 if these states are recurrent. Therefore from a recurrent class the particle can only move within the class. Next, the truth of assertion (ii) is contained in the proof of Theorem 11, according to which the particle can only spend a finite number of time units in a finite set of transient states. Hence from a certain instant on it will be out of the set. Assertion (iii) is a consequence of (ii) and is illustrated by Example 3 of §8.3 (gambler's ruin problem). Assertion (iv) states an obvious alternative on account of (i), and is illustrated by Example 1 of §8.3 with $p > 1/2$, in which case $\alpha = 0$; or by Example 9. In the latter case it is clear that starting from $i \geq 1$, either the particle will be absorbed in the state 0 with probability f^*_{i0} in the notation of (8.4.6); or it will move steadily through the infinite set of transient states $\{i + 1, i + 2, \ldots\}$ with probability $1 - f^*_{i0}$.

Let us further illustrate some of the possibilities by a simple numerical example.

Example 17. Let the transition matrix be as follows:

$$
\begin{array}{c|ccc:cc:ccc}
 & 1 & 2 & 3 & 4 & 5 & 6 & \cdot & \cdot\ \cdot \\
\hline
1 & \frac{1}{8} & \frac{3}{8} & \frac{1}{4} & \frac{1}{4} & 0 & 0 & 0 & \cdots \\
2 & 0 & \frac{1}{2} & 0 & 0 & \frac{1}{3} & \frac{1}{6} & 0 & \cdots \\
3 & \frac{1}{5} & \frac{3}{10} & 0 & \frac{1}{5} & \frac{1}{5} & 0 & \frac{1}{10} & \cdots \\
\hdashline
4 & & \textcircled{O} & & \frac{1}{2} & \frac{1}{2} & & \textcircled{O} & \\
5 & & & & 1 & 0 & & & \\
\hdashline
6 & & & & & & & & \\
\vdots & & \textcircled{O} & & \textcircled{O} & & & \textcircled{R_2} & \\
\end{array}
\tag{8.7.1}
$$

The state space may be finite or infinite according to the specification of R_2, which may be the transition matrix of any recurrent Markov chain such as Example 4 or 8 of §8.3, or Example 1 there with $p = 1/2$.

Here $T = \{1, 2, 3\}$, $R_1 = \{4, 5\}$ and R_2 are two distinct recurrent classes. The theory of communication between states implies that the four blocks of 0's in the matrix will be preserved when it is raised to any power. Try to confirm this fact by a few actual schematic multiplications. On the other hand, some of the single 0's will turn positive in the process of multiplication. There are actually two distinct transient classes: $\{1, 3\}$ and $\{2\}$; it is possible to go from the first to the second but not vice versa. [This is not important; in fact, a transient class which is not closed is not a very useful entity. It was defined to be a class in §8.4 only by the force of circumstance!] All three transient states lead to both R_1 and R_2, but it would be easy to add another which leads to only one of them. The problem of finding the various absorption probabilities can be solved by the general procedure below.

Let $i \in T$ and C be a recurrent class. Put for $n \geq 1$:

$$y_i^{(n)} = \sum_{j \in C} p_{ij}^{(n)} = P_i\{X_n \in C\}. \tag{8.7.2}$$

This is the probability that the particle will be in C at time n, given that it starts from i. Since C is closed it will then also be in C at time $n + 1$; thus $y_i^{(n)} \leq y_i^{(n+1)}$ and so by the monotone sequence theorem in calculus the limit exists as $n \to \infty$:

$$y_i = \lim_{n \to \infty} y_i^{(n)} = P_i\{X_n \in C \text{ for some } n \geq 1\},$$

(why the second equation?) and gives the probability of absorption.

Theorem 14. *The $\{y_i\}$ above satisfies the system of equations:*

$$x_i = \sum_{j \in T} p_{ij}x_j + \sum_{j \in C} p_{ij}, \quad i \in T. \tag{8.7.3}$$

If T is finite, it is the unique solution of this system. Hence it can be computed by standard method of linear algebra.

Proof: Let the particle start from i, and consider its state j after one step. If $j \in T$, then the Markov property shows that the conditional probability of absorption becomes y_j; if $j \in C$, then it is already absorbed; if $j \in (I - T) - C$, then it can never be absorbed in C. Taking into account these possibilities, we get

$$y_i = \sum_{j \in T} p_{ij}y_j + \sum_{j \in C} p_{ij} \cdot 1 + \sum_{j \in (I-T)-C} p_{ij} \cdot 0.$$

This proves the first assertion of the theorem. Suppose now T is the finite set $\{1, 2, \ldots, t\}$. The system (8.7.3) may be written in matrix form as follows:

$$(\Delta_T - \Pi_T)x = y^{(1)}, \tag{8.7.4}$$

where Δ_T is the identity matrix indexed by $T \times T$; Π_T is the restriction of Π on $T \times T$, and $y^{(1)}$ is given in (8.7.2). According to a standard result in linear algebra, the equation above has a unique solution if and only if the matrix $\Delta_T - \Pi_T$ is nonsingular, namely it has an inverse $(\Delta_T - \Pi_T)^{-1}$, and then the solution is given by

$$x = (\Delta_T - \Pi_T)^{-1}y^{(1)}. \tag{8.7.5}$$

Suppose the contrary, then the same result asserts that there is a nonzero solution to the associated homogeneous equation. Namely there is a column vector $v = (v_1, \ldots, v_t) \neq (0, \ldots, 0)$ satisfying

$$(\Delta_T - \Pi_T)v = 0, \quad \text{or} \quad v = \Pi_T v.$$

It follows by iteration that

$$v = \Pi_T(\Pi_T v) = \Pi_T^2 v = \Pi_T^2(\Pi_T v) = \Pi_T^3 v = \ldots,$$

and so for every $n \geq 1$:

$$v = \Pi_T^n v.$$

[cf. (8.6.8) but observe the difference between right-hand and left-hand multiplications.] This means

$$v_i = \sum_{j \in T} p_{ij}^{(n)} v_j, \quad i \in T.$$

Letting $n \to \infty$ and using the Corollary to Theorem 5 we see that every term in the sum converges to zero and so $v_i = 0$ for all $i \in T$, contrary to hypothesis. This contradiction establishes the nonsingularity of $\Delta_T - \Pi_T$ and consequently the existence of a unique solution given by (8.7.5). Since $\{y_i, i \in T\}$ is a solution the theorem is proved.

For Example 17 above, the equations in (8.7.3) for absorption probabilities into R_1 are:

$$\begin{aligned}
x_1 &= \frac{1}{8}x_1 + \frac{3}{8}x_2 + \frac{1}{4}x_3 + \frac{1}{4}\\
x_2 &= \frac{1}{2}x_2 + \frac{1}{3}\\
x_3 &= \frac{1}{5}x_1 + \frac{3}{10}x_2 + \frac{2}{5}.
\end{aligned}$$

We get x_2 at once from the second equation, and then x_1, x_3 from the others:

$$x_1 = \frac{26}{33}, \quad x_2 = \frac{2}{3}, \quad x_3 = \frac{25}{33}.$$

For each i, the absorption probabilities into R_1 and R_2 add up to one, hence those for R_2 are just $1 - x_1$, $1 - x_2$, $1 - x_3$. This is the unique solution to another system of equations in which the constant terms above are replaced by $0, \frac{1}{6}, \frac{1}{10}$. You may wish to verify this as it is a good habit to double-check these things, at least once in a while.

It is instructive to remark that the problem of absorption into recurrent classes can always be reduced to that of absorbing states. For each recurrent class may be merged into a single absorbing state since we are not interested in the transitions within the class; no state in the class leads outside, whereas the probability of entering the class at one step from any transient state i is precisely the $y_i^{(1)}$ used above. Thus, the matrix in (8.7.1) may be converted to the following one:

$$\begin{bmatrix} \frac{1}{8} & \frac{3}{8} & \frac{1}{4} & \frac{1}{4} & 0 \\ 0 & \frac{1}{2} & 0 & \frac{1}{3} & \frac{1}{6} \\ \frac{1}{5} & \frac{3}{10} & 0 & \frac{2}{5} & \frac{1}{10} \\ 0 & 0 & 0 & 1 & 0 \\ 0 & 0 & 0 & 0 & 1 \end{bmatrix}$$

in which the last two states $\{4\}$ and $\{5\}$ take the place of R_1 and R_2. The absorption probabilities become just f_{i4}^* and f_{i5}^* in the notation of (8.4.6). The two systems of equations remain of course the same.

When T is finite and there are exactly two absorbing states there is another interesting method. As before let $T = \{1, 2, \ldots, t\}$ and let the absorbing states be denoted by 0 and $t + 1$, so that $I = \{0, 1, \ldots, t + 1\}$. The method depends on the discovery of a positive nonconstant solution of the equation $(\Delta - \Pi)x = 0$, namely some such $v = (v_0, v_1, \ldots, v_{t+1})$ satisfying

$$v_i = \sum_{j=0}^{t+1} p_{ij} v_j, \quad i = 0, 1, \ldots, t + 1. \tag{8.7.6}$$

Observe that the two equations for $i = 0$ and $i = t + 1$ are automatically true for any v, because $p_{0j} = \delta_{0j}$ and $p_{t+1,j} = \delta_{t+1,j}$; also that $v_i = 1$ is always a solution of the system, but it is constant. Now iteration yields

$$v_i = \sum_{j=0}^{t+1} p_{ij}^{(n)} v_j$$

for all $n \geq 1$; letting $n \to \infty$ and observing that

$$\lim_{n\to\infty} p_{ij}^{(n)} = 0 \quad \text{for } 1 \leq j \leq t;$$

$$\lim_{n\to\infty} p_{ij}^{(n)} = f_{ij}^* \quad \text{for } j = 0 \quad \text{and} \quad j = t + 1;$$

we obtain

$$v_i = f_{i0}^* v_0 + f_{i,t+1}^* v_{t+1}. \tag{8.7.7}$$

Recall also that

$$1 = f_{i0}^{*} + f_{i,t+1}^{*}. \tag{8.7.8}$$

We claim that $v_0 \neq v_{t+1}$, otherwise it would follow from the last two equations that $v_i = v_0$ for all i, contrary to the hypothesis that v is nonconstant. Hence we can solve these equations as follows:

$$f_{i0}^{*} = \frac{v_i - v_{t+1}}{v_0 - v_{t+1}}, \quad f_{i,t+1}^{*} = \frac{v_0 - v_i}{v_0 - v_{t+1}}. \tag{8.7.9}$$

Example 18. Let us return to Problem 1 of §8.1, where $t = c - 1$. If $p \neq q$, then $v_i = (q/p)^i$ is a nonconstant solution of (8.7.6). This is trivial to verify but you may well demand to know how on earth did we discover such a solution? The answer in this case is easy (but motivated by knowledge of difference equations used in §8.1): try a solution of the form λ^i and see what λ must be. Now if we substitute this v_i into (8.7.9) we get f_{i0}^{*} equal to the u_i in (8.1.9).

If $p = q = \frac{1}{2}$, then $v_i = i$ is a nonconstant solution of (8.7.6) since

$$i = \frac{1}{2}(i + 1) + \frac{1}{2}(i - 1). \tag{8.7.10}$$

This leads to the same answer as given in (8.1.10). The new solution has to do with the idea of a martingale (see Appendix 3). Here is another similar example.

Example 19. The following model of random reproduction was introduced by S. Wright in his genetical studies (see e.g. [Karlin] for further details). In a *haploid* organism the genes occur singly rather than in pairs as in the diploid case considered in §5.6. Suppose $2N$ genes of types A and a (the alleles) are selected from each generation. The number of A-genes is the state of the Markov chain and the transition probabilities are given below: $I = \{0, 1, \ldots, 2N\}$, and

$$p_{ij} = \binom{2N}{j}\left(\frac{i}{2N}\right)^{j}\left(1 - \frac{i}{2N}\right)^{2N-j}. \tag{8.7.11}$$

Thus if the number of A-genes in any generation is equal to i, then we may suppose that there is an infinite pool of both types of genes in which the proportion of A to a is as i: $2N - i$, and $2N$ independent drawings are made from it to give the genes of the next generation. We are therefore dealing with $2N$ independent Bernoullian trials with success probability $i/2N$, which results in the binomial distribution $B(2N; i/2N)$ in (8.7.11). It follows that (see (4.4.16) or (6.3.6)) the expected number of A-genes is equal to

$$\sum_{j=0}^{2N} p_{ij}j = 2N\frac{i}{2N} = i. \tag{8.7.12}$$

This means that the expected number of A-genes in the next generation is equal to the actual (but random) number of these genes in the present generation. In particular, this expected number remains constant through the successive generations. The situation is the same as in the case of a fair game discussed in §8.2 after (8.2.3). The uncertified trick used there is again applicable and in fact leads to exactly the same conclusion except for notation. However, now we can also apply the proven formula (8.7.9) which gives at once

$$f_{i0}^* = \frac{2N - i}{2N}, \quad f_{i,2N}^* = \frac{i}{2N}.$$

These are the respective probabilities that the population will wind up being pure a-type or A-type.

Our final example deals with a special but important kind of homogeneous Markov chain. Another specific example, queuing process, is outlined with copious hints in Exercises 29–31 below.

Example 20. A subatomic particle may split into several particles after a nuclear reaction; a male child bearing the family name may have a number of male children or none. These processes may be repeated many times unless extinction occurs. These are examples of a *branching process* defined below.

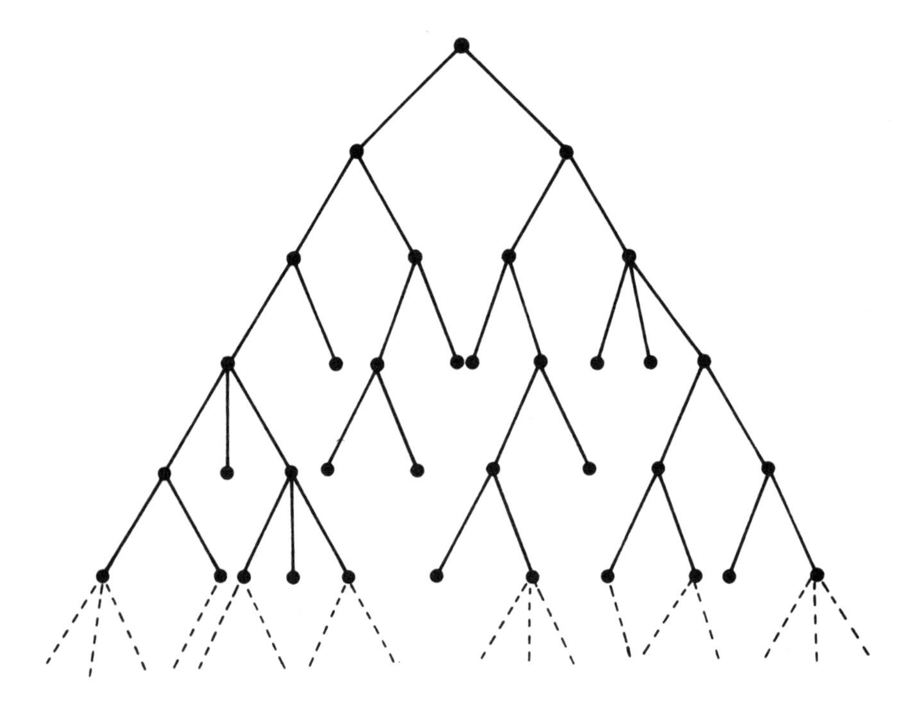

Figure 36

There is no loss of generality to assume that at the beginning there is exactly one particle: $X_0 = 1$. It gives rise to X_1 descendants of the first generation, where

$$P(X_1 = j) = a_j, \quad j = 0, 1, 2, \ldots. \tag{8.7.13}$$

Unless $X_1 = 0$, each of the particles in the first generation will give rise to descendants of the second generation, whose number follows the same probability distribution given in (8.7.13), and the actions of the various particles are assumed to be stochastically independent. What is the distribution of the number of particles of the second generation? Let the generating function of X_1 be g:

$$g(z) = \sum_{j=0}^{\infty} a_j z^j.$$

Suppose the number of particles in the first generation is equal to j, and we denote the numbers of their descendants by $Z_1, \ldots, Z_j$ respectively. Then by hypothesis these are independent random variables each having g as its generating function. The total number of particles in the second generation is $X_2 = Z_1 + \cdots + Z_j$ and this has the generating function g^j by Theorem 6 of §6.5. Recalling (6.5.12) and the definition of conditional expectation in (5.2.11), this may be written as follows:

$$E(z^{X_2} \mid X_1 = j) = g(z)^j, \tag{8.7.14}$$

and consequently by (5.2.12):

$$E(z^{X_2}) = \sum_{j=0}^{\infty} P(X_1 = j)E(z^{X_2} \mid X_1 = j) = \sum_{j=0}^{\infty} a_j g(z)^j = g(g(z)).$$

Let g_n be the generating function of X_n so that $g_1 = g$; then the above says $g_2 = g(g_1)$. Exactly the same argument gives $g_n = g(g_{n-1}) = g \circ g \circ \cdots \circ g$ (there are n appearances of g), where "$\circ$" denotes the composition of functions. In other words g_n is just the n-fold composition of g with itself. Using this new definition of g_n, we record this as follows:

$$g_n(z) = E(z^{X_n}) = \sum_{k=0}^{\infty} P(X_n = k)z^k. \tag{8.7.15}$$

Since the distribution of the number of descendants in each succeeding generation is determined solely by the number in the existing generation, regardless of past evolution, it is clear that the sequence $\{X_n, n \geq 0\}$ has the Markov property. It is a homogeneous Markov chain because the law of reproduction is the same from generation to generation. In fact, it follows from (8.7.14) that the transition probabilities are given below:

(8.7.16) $\quad p_{jk}$ = coefficient of z^k in the power series for $g(z)^j$.

To exclude trivial cases, let us now suppose that

$$0 < a_0 < a_0 + a_1 < 1. \tag{8.7.17}$$

The state space is then (why?) the set of all nonnegative integers. The preceding hypothesis implies that all states lead to 0 (why?) which is an absorbing state. Hence all states except 0 are transient but there are infinitely many of them. The general behavior under (iii) at the beginning of the section does not apply and only (iv) is our guide. [Observe that the term "particle" was used in a different context there.] Indeed, we will now determine the value of α which is called the *probability of extinction* in the present model.

Putting $z = 0$ in (8.7.15) we see that $g_n(0) = p_{10}^{(n)}$; on the other hand our general discussion about absorption tells us that

$$\alpha = \lim_{n\to\infty} p_{10}^{(n)} = \lim_{n\to\infty} g_n(0). \tag{8.7.18}$$

Since $g_n(0) = g(g_{n-1}(0))$, by letting $n \to \infty$ we obtain

$$\alpha = g(\alpha). \tag{8.7.19}$$

Thus the desired probability is a root of the equation $\varphi(z) = 0$ where $\varphi(z) = g(z) - z$; we shall call it simply a root of φ. Since $g(1) = 1$, one root is $z = 1$. Next we have

$$\varphi''(z) = g''(z) = \sum_{j=2}^{\infty} j(j-1)a_j z^{j-2} > 0$$

for $z > 0$, on account of (8.7.17). Hence the derivative φ' is an increasing function. Now recall *Rolle's theorem* from calculus: between two roots of a differentiable function there is at least one root of its derivative. It follows that φ cannot have more than two roots in $[0, 1]$, for then φ' would have more than one root which is impossible because φ' increases. Thus φ can have at most one root different from 1 in $[0, 1]$, and we have two cases to consider.†

Case 1. φ has no root in $[0, 1)$. Then since $\varphi(0) = a_0 > 0$, we must have $\varphi(z) > 0$ for all z in $[0, 1)$, for a continuous function cannot take both positive and negative values in an interval without vanishing somewhere. Thus we have

$$\varphi(1) - \varphi(z) < \varphi(1) = 0, \quad 0 \le z < 1;$$

and it follows that

† It is customary to draw two pictures to show the two cases below. The reader is invited to do this and see if he is more readily convinced than the author.

$$\varphi'(1) = \lim_{z \uparrow 1} \frac{\varphi(1) - \varphi(z)}{1 - z} \leq 0;$$

hence $g'(1) \leq 1$.

Case 2. φ has a unique root r in $[0, 1)$. Then by Rolle's theorem φ' must have a root s in $[r, 1)$, i.e., $\varphi'(s) = g'(s) - 1 = 0$, and since g' is an increasing function we have

$$g'(1) > g'(s) = 1.$$

To sum up: the equation $g(z) = z$ has a positive root less than 1 if and only if $g'(1) > 1$.

In Case 1, we must have $\alpha = 1$ since $0 \leq \alpha \leq 1$ and α is a root by (8.7.19). Thus the population is almost certain to become extinct.

In Case 2, we will show that α is the root $r < 1$. For $g(0) < g(r) = r$; and supposing for the sake of induction $g_{n-1}(0) < r$, then $g_n(0) = g(g_{n-1}(0)) < g(r) = r$ because g is an increasing function. Thus $g_n(0) < r$ for all n and so $\alpha \leq r$ by (8.7.18). But then α must be equal to r because both of them are roots of the equation in $[0, 1)$.

What will happen in Case 2 if the population escapes extinction? According to the general behavior under (iv) it must then remain forever in the transient states $\{1, 2, \ldots\}$ with probability $1 - \alpha$. Can its size sway back and forth from small to big and vice versa indefinitely? This question is answered by the general behavior under (ii), according to which it must stay out of every finite set $\{1, 2, \ldots, \ell\}$ eventually, no matter how large ℓ is. Therefore it must in fact become infinite (not necessarily monotonically, but as a limit), namely:

$$P\left\{\lim_{n\to\infty} X_n = +\infty \mid X_n \neq 0 \text{ for all } n\right\} = 1.$$

The conclusion is thus a "boom or bust" syndrome. The same is true of the gambler who has an advantage over an infinitely rich opponent (see §8.2): if he is not ruined he will also become infinitely rich. Probability theory contains a lot of such extreme results some of which are known as *zero-or-one* ("all or nothing") laws.

In the present case there is some easy evidence for the conclusions reached above. Let us compute the expectation of the population of the nth generation. Let $\mu = E(X_1)$ be the expected number of descendants of each particle. Observe that $\mu = g'(1)$ so that we have $\mu \leq 1$ in Case 1 and $\mu > 1$ in Case 2. Suppose $\mu < \infty$; then if the number of particles in the n-1st generation is j, the expected number of particles in the nth generation will be $j\mu$ (why?). Using conditional expectation, this may be written as,

$$E\{X_n \mid X_{n-1} = j\} = \mu j.$$

It follows from (5.2.12) that

$$E(X_n) = \sum_{j=0}^{\infty} \mu j P(X_{n-1} = j) = \mu E(X_{n-1}),$$

and consequently by iteration

$$E(X_n) = \mu^n E(X_0) = \mu^n.$$

Therefore we have

$$\lim_{n\to\infty} E(X_n) = \lim_{n\to\infty} \mu^n = \begin{cases} 0, & \text{if } \mu < 1, \\ 1, & \text{if } \mu = 1, \\ \infty, & \text{if } \mu > 1. \end{cases}$$

This tends to support our conclusion in Case 1 for certain extinction; in fact it is intuitively obvious that if $\mu < 1$ the population fails to be self-replacing on the basis of averages. The case $\mu = 1$ may be disposed of with a bit more insight, but let us observe that here we have the strange situation that $E(X_n) = 1$ for all n, but $P\left(\lim_{n\to\infty} X_n = 0\right) = 1$ by Case 1. In case $\mu > 1$ the crude interpretation would be that the population will certainly become infinite. But we have proved under Case 2 that there is a definite probability that it will die out as a dire contrast. This too is interesting in relating simple calculations to more sophisticated theory. These comments are offered at the closing of this book as an invitation to the reader for further wonderment about probability and its meaning.

Exercises

1. Let X_n be as in (8.1.2) with $X_0 = 0$. Find the following probabilities:
   (a) $P\{X_n \geq 0 \text{ for } n = 1, 2, 3, 4\}$,
   (b) $P\{X_n \neq 0 \text{ for } n = 1, 2, 3, 4\}$,
   (c) $P\{X_n \leq 2 \text{ for } n = 1, 2, 3, 4\}$,
   (d) $P\{|X_n| \leq 2 \text{ for } n = 1, 2, 3, 4\}$.
2. Let $Y_n = X_{2n}$ where X_n is as in No. 1. Show that $\{Y_n, n \geq 0\}$ is a Markov chain and find its transition matrix. Similarly for $\{Z_n, n \geq 0\}$ where $Z_n = X_{2n+1}$; what is its initial distribution?
3. Let a coin be tossed indefinitely; let H_n and T_n denote respectively the numbers of heads and tails obtained in the first n tosses. Put $X_n = H_n$, $Y_n = H_n - T_n$. Are these Markov chains? If so find the transition matrix.
4.* As in No. 3 let $Z_n = |H_n - T_n|$. Is this a Markov chain? [Hint: compute e.g., $P\{Y_{2n} = 2i \mid Z_{2n} = 2i\}$ by Bernoulli's formula, then $P\{Z_{2n+1} = 2i \pm 1 \mid Z_{2n} = 2i, Y_{2n} > 0\}$.]

5. Let the transition matrix be given below:

$$\text{(a)} \begin{pmatrix} \frac{1}{2} & \frac{1}{2} \\ \frac{1}{3} & \frac{2}{3} \end{pmatrix} \qquad \text{(b)} \begin{pmatrix} p_1 & q_1 & 0 \\ 0 & p_2 & q_2 \\ q_3 & 0 & p_3 \end{pmatrix}$$

Find $f_{11}^{(n)}, f_{12}^{(n)}, g_{12}^{(n)}$ for $n = 1, 2, 3$ (for notation see (8.4.6) and (8.4.10)).

6. In a model for the *learning process* of a rat devised by Estes, the rodent is said to be in state 1 if it has learned a certain trick (to get a peanut or avoid an electric shock), and to be in state 2 if it has not yet learned it. Suppose that once it becomes learned it will remain so, while if it is not yet learned it has a probability α of becoming so after each trial run. Write down the transition matrix and compute $p_{21}^{(n)}, f_{21}^{(n)}$ for all $n \geq 1$; and m_{21} (see (8.6.3) for notation).

7. Convince yourself that it is a trivial matter to construct a transition matrix in which there are any given number of transient and recurrent classes, each containing a given number of states, provided that either (a) I is infinite, or (b) I is finite but not all states are transient.

8. Given any transition matrix Π, show that it is trivial to enlarge it by adding new states which lead to old ones, but it is impossible to add any new state which communicates with any old one.

9. In the "double or nothing" game, you bet all you have and you have a fifty-fifty chance to double it or lose it. Suppose you begin with \$1 and decide to play this game up to n times (you may have to quit sooner because you are broke). Describe the Markov chain involved with its transition matrix.

10. Leo is talked into playing heads in a coin-tossing game in which the probability of heads is only 0.48. He decides that he will quit as soon as he is one ahead. What is the probability that he may never quit?

11. A man has two girl friends, one uptown and one downtown. When he wants to visit one of them for a weekend he chooses the uptown girl with probability p. Between two visits he stays home for a weekend. Describe the Markov chain with three states for his weekend whereabouts: "uptown," "home" and "downtown." Find the long-run frequencies of each. [This is the simplest case of Example 4 of §8.3, but here is a nice puzzle related to the scheme. Suppose that the man decides to let chance make his choice by going to the bus stop where buses go both uptown and downtown and jumping aboard the first bus that comes. Since he knows that buses run in both directions every fifteen minutes, he figures that these equal frequencies must imply $p = 1/2$ above. But after a while he realizes that he has been visiting uptown twice as frequently as downtown. How can this happen? This example carries an important lesson to the practicing statistician, namely that the relevant datum may not be what appears at first right.

Assume that the man arrives at the bus stop at random between 6 p.m. and 8 p.m. Figure out the precise bus schedules which will make him board the uptown buses with probability $p = 2/3$.]

12. Solve Problem 1 of §8.1 when there is a positive probability r of the particle remaining in its position at each step.

13.* Solve (8.1.13) when $p \neq q$ as follows. First determine the two values λ_1 and λ_2 such that $x_j = \lambda^j$ is a solution of $x_j = px_{j+1} + qx_{j-1}$. The general solution of this system is then given by $A\lambda_1^j + B\lambda_2^j$ where A and B are constants. Next find a particular solution of $x_j = px_{j+1} + qx_{j-1} + 1$ by trying $x_j = Cj$ and determine the constant C. The general solution of the latter system is then given by $A\lambda_1^j + B\lambda_2^j + Cj$. Finally determine A and B from the boundary conditions in (8.1.13).

14. The original Ehrenfest model is as follows. There are a total of N balls in two urns. A ball is chosen at random from the $2N$ balls from either urn and put into the other urn. Let X_n denote the number of balls in a fixed urn after n drawings. Show that this is a Markov chain having the transition probabilities given in (8.3.16) with $c = 2N$.

15. A scheme similar to that in No. 14 was used by Daniel Bernoulli [son of Johann, who was younger brother of Jakob] and Laplace to study the flow of incompressible liquids between two containers. There are N red and N black balls in two urns containing N balls each. A ball is chosen at random from each urn and put into the other. Find the transition probabilities for the number of red balls in a specified urn.

16. In certain ventures such as doing homework problems one success tends to reinforce the chance for another by imparting experience and confidence; in other ventures the opposite may be true. Anyway let us assume that the after effect is carried over only two consecutive trials so that the resulting sequence of successes and failures constitutes a Markov chain on two states $\{s,f\}$. Let

$$p_{ss} = \alpha, \quad p_{ff} = \beta,$$

where α and β are two arbitrary members between 0 and 1. Find the long-run frequency of successes.

17. The following model has been used for the study of *contagion*. Suppose that there are N persons some of whom are sick with influenza. The following assumptions are made:
    (a) when a sick person meets a healthy one, the chance is α that the latter will be infected;
    (b) all encounters are between two persons;
    (c) all possible encounters in pairs are equally likely;
    (d) one such encounter occurs in every (chosen) unit of time.

    Define a Markov chain for the spread of the disease and write down its transition matrix. [Are you overwhelmed by all these oversimplifying assumptions? Applied mathematics is built upon the shrewd selection and exploitation of such simplified models.]

18. The age of a light bulb is measured in days and fractions of a day do not count. If a bulb is burned out during the day then it is replaced by a new one at the beginning of the next day. Assume that a bulb which is alive at the beginning of the day, possibly one which has just been installed, has probability p of surviving at least one day so that its age will be increased by one. Assume also that the successive bulbs used lead independent lives. Let $X_0 = 0$ and X_n denote the age of the bulb which is being used at the beginning of the $n + 1$st day. (We begin with the first day, thus $X_1 = 1$ or 0 according as the initial bulb is still in place or not at the beginning of the second day.) The process $\{X_n, n \geq 0\}$ is an example of a *renewal process*. Show that it is a recurrent Markov chain, find its transition probabilities and stationary distribution. [Note: the life span of a bulb being essentially a continuous variable, a lot of words are needed to describe the scheme accurately in discrete time, and certain ambiguities must be resolved by common sense. It would be simpler and clearer to formulate the problem in terms of heads and tails in coin-tossing (how?), but then it would have lost the flavor of application!]
19. Find the stationary distribution for the random walk with two reflecting barriers (Example 4 of §8.3).
20. In a sociological study of "conformity" by B. Cohen, the following Markov chain model was used. There are four states: S_1 = consistently nonconforming, S_2 = indecisively nonconforming, S_3 = indecisively conforming, S_4 = consistently conforming. In a group experiment subjects were found to switch states after each session according to the following transition matrix:

	S_1	S_2	S_3	S_4
S_1	1	0	0	0
S_2	.06	.76	.18	0
S_3	0	.27	.69	.04
S_4	0	0	0	1

Find the probabilities of ultimate conversion from the "conflict" states S_2 and S_3 into the "resolution" states S_1 and S_4.

21. In a genetical model similar to Example 19 of §8.7, we have $I = \{0, 1, \ldots, 2N\}$ and

$$p_{ij} = \binom{2i}{j}\binom{2N - 2i}{N - j}\bigg/\binom{2N}{N}.$$

How would you describe the change of genotypes from one generation to another by some urn scheme? Find the absorption probabilities.

[Hint: compute $\sum_{j=0}^{2N} jp_{ij}$ by simplifying the binomial coefficients, or by Theorem 1 of §6.1.]

22. For the branching process in Example 20 of §8.7, if a_0, a_1 and a_2 are positive but the other a_j's are all zero, find the probability of extinction.
23. Suppose that the particles in the first generation of a branching process follow a probability law of splitting given by $\{b_j, j \geq 0\}$ which may be different from that of initial particle given by (8.7.13). What then is the distribution of the number of particles in the second generation?
24. A sequence of electric impulses is measured by a meter which records the highest voltage that has passed through it up to any given time. Suppose that the impulses are uniformly distributed over the range $\{1, 2, \ldots, \ell\}$. Define the associated Markov chain and find its transition matrix. What is the expected time until the meter records the maximum value ℓ? [Hint: argue as in (8.1.13) for the expected absorption time into the state ℓ; use induction after computing $e_{\ell-1}$ and $e_{\ell-2}$.]
25. In proof-reading a manuscript each reader finds at least one error. But if there are j errors when he begins, he will leave it with any number of errors between 0 and $j - 1$ with equal probabilities. Find the expected number of readers needed to discover all the errors. [Hint: $e_j = j^{-1}(e_1 + \cdots + e_{j-1}) + 1$, now simplify $e_j - e_{j-1}$.]
26. A deck of m cards may be shuffled in various ways. Let the state space be the $m!$ different orderings of the cards. Each particular mode of shuffling sends any state (ordering) into another. If the various modes are randomized this results in various transition probabilities between the states. Following my tip (a) in §3.4 for combinatorial problems, let us begin with $m = 3$ and the following two modes of shuffling:
(i) move the top card to the bottom, with probability p;
(ii) interchange the top and middle cards, with probability $1 - p$.
Write down the transition matrix. Show that it is doubly stochastic and all states communicate. Show that if either mode alone is used the states will not all communicate.
27. Change the point of view in No. 26 by fixing our attention on a particular card, say the queen of spades if the three cards are the king, queen and knight of spades. Let X_n denote its position after n shufflings. Show that this also constitutes a Markov chain with a doubly stochastic transition matrix.
28.* Now generalize Nos. 26 and 27: for any m and any randomized shuffling, the transition matrices in both formulations are doubly stochastic. [Hint: each mode of shuffling as a permutation on m cards has an inverse. Thus if it sends the ordering j into k then it sends some ordering i into j. For fixed j the correspondence $i = i(k)$ is one-to-one and $p_{ij} = p_{jk}$. This proves the result for the general case of No. 26. Next consider two orderings j_1 and j_2 with the fixed card in the topmost position, say. Each mode of shuffling which sends j_1 into an ordering

with the given card second from the top does the same to j_2. Hence the sum of probabilities of such modes is the same for j_1 or j_2, and gives the transition probability $1 \rightarrow 2$ for the displacement of the card in question.]

29.* Customers arrive singly at a counter and enter into a queue if it is occupied. As soon as one customer finishes the service for the next customer begins if there is anyone in the queue, or upon the arrival of the next customer if there is no queue. Assume that the service time is constant (e.g., a taped recording or automatic hand-dryer), then this constant may be taken as the unit of time. Assume that the arrivals follow a Poisson process with parameter α in this unit. For $n \geq 1$ let X_n denote the number of customers in the queue at the instant when the nth customer finishes his service. Let $\{Y_n, n \geq 1\}$ be independent random variables with the Poisson distribution $\pi(\alpha)$; see §7.1. Show that

$$X_{n+1} = (X_n - 1)^+ + Y_n, \quad n \geq 1;$$

where $x^+ = x$ if $x > 0$ and $x^+ = 0$ if $x \leq 0$. Hence conclude that $\{X_n, n \geq 1\}$ is a Markov chain on $\{0, 1, 2, \ldots\}$ with the following transition matrix:

$$\begin{bmatrix} c_0 & c_1 & c_2 & c_3 & \cdots \\ c_0 & c_1 & c_2 & c_3 & \cdots \\ 0 & c_0 & c_1 & c_2 & \cdots \\ 0 & 0 & c_0 & c_1 & \cdots \\ \cdots & \cdots & \cdots & \cdots & \end{bmatrix}$$

where $c_j = \pi_j(\alpha)$. [Hint: this is called a *queuing process* and $\{X_n, n \geq 1\}$ is an *imbedded Markov chain*. At the time when the nth customer finishes there are two possibilities. (i) The queue is not empty; then the $n + 1$st customer begins his service at once and during his (unit) service time Y_n customers arrive. Hence when he finishes the number in the queue is equal to $X_n - 1 + Y_n$. (ii) The queue is empty; then the counter is free and the queue remains empty until the arrival of the $n + 1$st customer. He begins service at once and during his service time Y_n customers arrive. Hence when he finishes the number in the queue is equal to Y_n. The Y_n's are independent and have $\pi(\alpha)$ as distribution, by Theorems 1 and 2 of §7.2.]

30.* Generalize the scheme in No. 29 as follows. The service time is a random variable S such that $P\{S = k\} = b_k$, $k \geq 1$. Successive service times are independent and identically distributed. Show that the conclusions of No. 29 hold with

$$c_j = \sum_{k=1}^{\infty} b_k \pi_j(k\alpha).$$

31.* In No. 29 or No. 30, let $\mu = \sum_{j=0}^{\infty} jc_j$. Prove that the Markov chain is transient, null-recurrent, or positive-recurrent according as $\mu < 1$, $\mu = 1$ or $\mu > 1$. [This result is due to Lindley; here are the steps for a proof within the scope of Chapter 8. In the notation of §8.4 let

$$F_{10}(z) = f(z), \quad g(z) = \sum_{j=0}^{\infty} c_j z^j.$$

(a) $F_{j,j-1}(z) = f(z)$ for all $j \geq 1$; because e.g. $f_{j,j-1}^{(4)} = P\{Y_n \geq 1,\ Y_n + Y_{n+1} \geq 2,\ Y_n + Y_{n+1} + Y_{n+2} \geq 3,\ Y_n + Y_{n+1} + Y_{n+2} + Y_{n+3} = 3 \mid X_n = j\}$

(b) $F_{j0}(z) = f(z)^j$ for $j \geq 1$, because the queue size can decrease only by one at a step;

(c) $f_{10}^{(1)} = c_0, f_{10}^{(v)} = \sum_{j=1}^{\infty} c_j f_{j0}^{(v-1)}$ for $v \geq 2$; hence

$$f(z) = c_0 z + \sum_{j=1}^{\infty} c_j z F_{j0}(z) = zg(f(z));$$

(d) $F_{00}(z) = zg(f(z))$ by the same token;

(e) if $f(1) = \rho$, then ρ is the smallest root of the equation $\rho = g(\rho)$ in $[0, 1]$; hence $F_{00}(1) = f(1) < 1$ or $= 1$ according as $g'(1) > 1$ or ≤ 1 by Example 4 of §8.7;

(f) $f'(1) = f'(1)g'(1) + g(1)$; hence if $g'(1) \leq 1$ then in the notation of (8.6.3), $m_{00} = F'_{00}(1) = f'(1) = \infty$ or $< \infty$ according as $g'(1) = 1$ or < 1. Q.E.D.

For more complicated queueing models see e.g., [Karlin].

32.* A company desires to operate s identical machines. These machines are subject to failure according to a given probability law. To replace these failed machines the company orders new machines at the beginning of each week to make up the total s. It takes one week for each new order to be delivered. Let X_n be the number of machines in working order at the beginning of the nth week and let Y_n denote the number of machines that fail during the nth week. Establish the recursive formula

$$X_{n+1} = s - Y_n$$

and show that $\{X_n, n \geq 1\}$ constitutes a Markov chain. Suppose that the failure law is uniform, i.e.:

$$P\{Y_n = j \mid X_n = i\} = \frac{1}{i+1}, \quad j = 0, 1, \ldots, i.$$

Find the transition matrix of the chain, its stationary distribution, and the expected number of machines in operation in the steady state.

33.* In No. 32 suppose the failure law is binomial:

$$P\{Y_n = j \mid X_n = i\} = \binom{i}{j} p^j(1-p)^{i-j}, \quad j = 0, 1, \ldots, i,$$

with some probability p. Answer the same questions as before. [These two problems about *machine replacement* are due to D. Iglehart.]

34. The matrix $[p_{ij}]$, $i \in I$, $j \in I$ is called *substochastic* iff for every i we have $\sum_{j \in I} p_{ij} \le 1$. Show that every power of such a matrix is also substochastic.

35. Show that the set of states C is stochastically closed if and only if for every $i \in C$ we have $\sum_{j \in C} p_{ij} = 1$.

36. Show that

$$\max_{0 \le n < \infty} P_i\{X_n = j\} \le P_i\left\{\bigcup_{n=0}^{\infty} [X_n = j]\right\} \le \sum_{n=0}^{\infty} P_i\{X_n = j\}.$$

Hence deduce that $i \rightsquigarrow j$ if and only if $f_{ij}^* > 0$.

37. Prove that if $q_{ij} > 0$, then $\sum_{n=0}^{\infty} p_{ij}^{(n)} = \infty$.

38.* Prove that if $j \rightsquigarrow i$, then $g_{ij}^* < \infty$. Give an example where $g_{ij}^* = \infty$. [Hint: show that $g_{ij}^{(n)} f_{ji}^{(v)} \le f_{ii}^{(n+v)}$ and choose v so that $f_{ji}^{(v)} > 0$.]

39. Prove that if there exists j such that $i \rightsquigarrow j$ but not $j \rightsquigarrow i$, then i is transient. [Hint: use Theorem 9; or argue as in the proof of Theorem 9 to get $q_{ii} \le p_{ij}^{(n)} \cdot 0 + (1 - p_{ij}^{(n)}) \cdot 1$ for every n.]

40. Define for arbitrary i, j and k in I and $n \ge 1$:

$${}_kp_{ij}^{(n)} = P_i\{X_v \ne k \text{ for } 1 \le v \le n-1;\ X_n = j\}.$$

Show that if $k = j$ this reduces to $f_{ij}^{(n)}$, while if $k = i$ it reduces to $g_{ij}^{(n)}$. In general, prove that

$$\sum_{\ell \ne k} {}_kp_{i\ell}^{(n)}\, {}_kp_{\ell j}^{(m)} = {}_kp_{ij}^{(n+m)}.$$

These are called *taboo probabilities* because the passage through k during the transition is taboo.

41. If the total number of states is r, and $i \rightsquigarrow j$, then there exists n such that $1 \le n \le r$ and $p_{ij}^{(n)} > 0$. [Hint: any sequence of states leading from i to j in which some k occurs twice can be shortened.]

42. Generalize the definition in (8.6.3) as follows:

$$m_{ij} = E_i(T_j) = \sum_{v=1}^{\infty} v f_{ij}^{(v)}.$$

Prove that $m_{ij} + m_{ji} \geq m_{ii}$ for any two states i and j. In particular, in Example 16 of §8.6, we have $m_{0c} \geq 2^{c-1}$.

43. Prove that the symmetric random walk is null-recurrent. [Hint: $p_{ij}^{(n)} = P\{\xi_1 + \cdots + \xi_n = j - i\}$; use (7.3.7) and the estimate following it.]
44. For any state i define the *holding time* in i as follows: $S = \max\{n \geq 1 \mid X_v = i, \text{for all } v = 1, 2, \ldots, n\}$. Find the distribution of S.
45.* Given the Markov chain $\{X_n, n \geq 1\}$ in which there is no absorbing state, define a new process as follows. Let n_1 be the smallest value of n such that $X_n \neq X_1$, n_2 the smallest value $> n_1$ such that $X_n \neq X_{n_1}$, n_3 the smallest value $> n_2$ such that $X_n \neq X_{n_2}$ and so on. Now put $Y_v = X_{n_v}$; show that $\{Y_v, v \geq 1\}$ is also a Markov chain and derive its transition matrix from that of $\{X_n, n \geq 1\}$. Prove that if a state is recurrent in one of them, then it is also recurrent in the other.
46. In the notation of No. 3, put $H_n^{(2)} = \sum_{v=1}^{n} H_v$. Show that $\{X_n\}$ does not form a Markov chain but if we define a process whose value Y_n at time n is given by the ordered pair of states (X_{n-1}, X_n), then $\{Y_n, n \geq 1\}$ is a Markov chain. What is its state space and transition matrix? The process $\{H_n^{(2)}, n \geq 0\}$ is sometimes called a *Markov chain of order* 2. How would you generalize this notion to a higher order?
47. There is a companion to the Markov property which shows it *in reverse time*. Let $\{X_n\}$ be a homogeneous Markov chain. For $n \geq 1$ let B be any event determined by $X_{n+1}, X_{n+2}, \ldots$. Show that we have for any two states i and j:

$$P\{X_{n-1} = j \mid X_n = i; B\} = P\{X_{n-1} = j \mid X_n = i\};$$

but this probability may depend on n. However, if $\{X_n\}$ is stationary as in Theorem 13, show that the probability above is equal to

$$\tilde{p}_{ij} = \frac{w_j p_{ji}}{w_i}$$

and so does not depend on n. Verify that $[\tilde{p}_{ij}]$ is a transition matrix. A homogeneous Markov chain with this transition matrix is said to be a *reverse chain* relative to the original one.

Appendix 3

Martingale

Let each X_n be a random variable having a finite expectation, and for simplicity we will suppose it to take integer values. Recall the definition of conditional expectation from the end of §5.2. Suppose that for every event A determined by $X_0, \ldots, X_{n-1}$ alone, and for each possible value i of X_n, we have

$$E\{X_{n+1} \mid A; X_n = i\} = i; \tag{A.3.1}$$

then the process $\{X_n, n \geq 0\}$ is called a *martingale*. This definition resembles that of a Markov chain given in (8.3.1) in the form of the conditioning, but the equation is a new kind of hypothesis. It is more suggestively exhibited in the symbolic form below:

$$E\{X_{n+1} \mid X_0, X_1, \ldots, X_n\} = X_n.$$

This means: for arbitrary given values of $X_0, X_1, \ldots, X_n$, the conditional expectation of X_{n+1} is equal to the value of X_n, regardless of the other values. The situation is illustrated by the symmetric random walk or the genetical model in Example 19 of §8.7. In the former case, if the present position of the particle is X_n, then its position after one step will be $X_n + 1$ or $X_n - 1$ with probability 1/2 each. Hence we have, whatever the value of X_n:

$$E\{X_{n+1} \mid X_n\} = \frac{1}{2}(X_n + 1) + \frac{1}{2}(X_n - 1) = X_n;$$

furthermore this relation remains true when we add to the conditioning the previous positions of the particle represented by $X_0, X_1, \ldots, X_{n-1}$. Thus the defining condition (A.3.1) for a martingale is satisfied. In terms of the gambler, it means that if the game is fair then at each stage his expected gain or loss cancel out so that his expected future worth is exactly equal to his present assets. A similar assertion holds true for the number of A-genes in the genetical model. More generally, when the condition (8.7.6) is satisfied, then the process $\{v(X_n)\ n \geq 0\}$ constitutes a martingale, where v is the function $i \to v(i)$, $i \in I$. Finally in Example 20 of §8.7, it is easy to verify that the normalized population size $\{X_n/\mu^n, n \geq 0\}$ is a martingale.

If we take A to be an event with probability one in (A.3.1), and use (5.2.12), we obtain

$$
\text{(A.3.2)} \qquad \begin{aligned} E(X_{n+1}) &= \sum_i P(X_n = i)E(X_{n+1} \mid X_n = i) \\ &= \sum_i P(X_n = i)i = E(X_n). \end{aligned}
$$

Hence in a martingale all the random variables have the same expectation. This is observed in (8.2.3), but the fact by itself is not significant. The following result from the theory of martingales covers the applications mentioned there and in §8.7. Recall the definition of an optional random variable from §8.5.

Theorem. If the martingale is bounded, namely if there exists a constant M such that $|X_n| \leq M$ for all n, then for any optional T we have

$$
\text{(A.3.3)} \qquad E(X_T) = E(X_0).
$$

In the case of Problem 1 of §8.1 with $p = 1/2$, we have $|X_n| \leq c$; in the case of Example 3 of §8.3 we have $|X_n| \leq 2N$. Hence the theorem is applicable and the absorption probabilities fall out from it as shown in §8.2.

The extension of (A.3.2) to (A.3.3) may be false for a martingale and an optional T, without some supplementary condition such as boundedness. In this respect, the theorem above differs from the strong Markov property discussed in §8.5. Here is a trivial but telling example for the failure of (A.3.3). Let the particle start from 0 and let T be the first extrance time into 1. Then T is finite by Theorem 2 of §8.2, hence X_T is well defined and must equal 1 by its definition. Thus $E(X_T) = 1$ but $E(X_0) = 0$.

Martingale theory was largely developed by J. L. Doob (1910–   ) and has become an important chapter of modern probability theory; for an introduction see [Chung 1; Chapter 9].

GENERAL REFERENCES

Chung, Kai Lai [1]. *A Course in Probability Theory*, Second Edition. Academic Press, New York, 1974.

——— [2]. *Markov Chains with Stationary Transition Probabilities*, Second Edition. Springer-Verlag, New York, 1967.

David, F. N. *Games, Gods and Gambling*. Hafner Publishing Co., New York, 1962.

Feller, William. [1] *An Introduction to Probability Theory and its Applications*, Vol. 1, Third Edition. John Wiley & Sons, New York, 1968.

——— [2]. *An Introduction to Probability Theory and its Applications*, Vol. 2, Second Edition. John Wiley & Sons, New York, 1971.

Keynes, John Maynard. *A Treatise on Probability*. Macmillan Co., London, 1921.

Karlin, Samuel. *A First Course in Stochastic Processes*, Academic Press, New York, 1966.

Råde, Lennart. et al. *The Teaching of Probability and Statistics*, Proceedings of the First CSMP International Conference, Edited by Lennart Råde, Almqvist & Wiksell Förlag AB, Stockholm, 1970.

Uspensky, J. V. *Introduction to Mathematical Probability*, McGraw-Hill Book Co., New York, 1937.

ANSWERS TO PROBLEMS

Chapter 1

7. $(A \cup B)(B \cup C) = ABC + ABC^c + A^cBC + A^cBC^c + AB^cC$; $A\backslash B = AB^cC + AB^cC^c$; {the set of ω which belongs to exactly one of the sets A, B, C} $= AB^cC^c + A^cBC^c + A^cB^cC$.
10. The dual is true.
14. Define $A \# B = A^c \cup B^c$, or $A^c \cap B^c$.
19. $I_{A\backslash B} = I_A - I_AI_B$; $I_{A-B} = I_A - I_B$.
20. $I_{A \cup B \cup C} = I_A + I_B + I_C - I_{AB} - I_{AC} - I_{BC} + I_{ABC}$.

Chapter 2

4. $P(A + B) \leq P(A) + P(B)$.
5. $P(S_1 + S_2 + S_3 + S_4) \geq P(S_1) + P(S_2) + P(S_3) + P(S_4)$.
11. Take $AB = \varnothing$, $P(A) > 0$, $P(B) > 0$.
13. 17.
14. 126.
15. $|A \cup B \cup C| = |A| + |B| + |C| - |AB| - |AC| - |BC| + |ABC|$.
16. $P(A \,\Delta\, B) = P(A) + P(B) - 2P(AB) = 2P(A \cup B) - P(A) - P(B)$.
17. Equality holds when m and n are relatively prime.
20. $p_n = \dfrac{1}{2^n}, n \geq 1$; $p_n = \dfrac{1}{n(n+1)}, n \geq 1$.
22. $\dfrac{14}{60}$.
24. If A is independent of itself, then $P(A) = 0$ or $P(A) = 1$; if A and B are disjoint and independent then $P(A)P(B) = 0$.
28. $p_1p_2q_3p_4q_5$ where p_k = probability that the kth coin falls heads, $q_k = 1 - p_k$. The probability of exactly 3 heads for 5 coins is equal to $\sum p_{k_1}p_{k_2}p_{k_3}q_{k_4}q_{k_5}$ where the sum ranges over the 10 unordered triples (k_1, k_2, k_3) of $(1, 2, 3, 4, 5)$ and (k_4, k_5) denotes the remaining unordered pair.

Chapter 3

1. $3 + 2$; $3 + 2 + (3 \times 2)$.
2. 3^2, $\binom{3+2-1}{2}$.

3. Three shirts are delivered in two different packages each of which may contain 0 to 3. If the shirts are distinguishable: 2^3; if not: $\binom{2+3-1}{3}$.
4. $3 \times 4 \times 3 \times 5 \times 3$; $3 \times 4 \times (3+1) \times (2+1) \times 3$.
5. $26^2 + 26^3$; 100.
6. 9^7.
7. $\binom{12}{6}$
8. $4!$; $2 \times 4!4!$.
9. $\binom{20}{3}$; $(20)_3$.
10. 35 (0 sum being excluded); 23.
11. $\frac{1}{2}$ if the missing ones are as likely to be of the same size as of different sizes; $\frac{2}{3}$ if each missing one is equally likely to be of any of the sizes.
12. $\frac{2}{3}$; $\frac{4!}{6!}$ or $\frac{2 \times 4!}{6!}$ according as the two keys are tried in one or both orders (how is the lost key counted?).
13. 20/216 (by enumeration); some interpreted "steadily increasing" to mean "forming an arithmetical progression," if you know what that means.
14. (a) $1/6^3$; (b) $\{6 \times 1 + 90 \times 3 + 120 \times 6\}/6^6$.
15. $\binom{6}{4}4!$; $\binom{6}{3}\binom{4}{3}3!$.
16. $1 - \left\{\binom{5}{0}\binom{5}{4} + \binom{5}{1}\binom{4}{3} + \binom{5}{2}\binom{3}{2} + \binom{5}{3}\binom{2}{1} + \binom{5}{4}\binom{1}{0}\right\} \Big/ \binom{10}{4}$.
17. From an outside lane: 3/8; from an inside lane: 11/16.
18. $\left(\frac{m-1}{m}\right)^n$; $\frac{(m-1)_n}{(m)_n}$.
19. $\frac{1}{6}, \frac{4}{6}, \frac{1}{6}$.
20. (a) $4 \Big/ \binom{18}{15}$; (b) $\binom{14}{11} \Big/ \binom{18}{15}$.
21. Assuming that neither pile is empty: (a) both distinguishable: $2^{10} - 2$; (b) books distinguishable but piles not: $(2^{10} - 2)/2$; (c) piles distinguishable but books not: 9; (d) both indistinguishable: 5.
22. $\frac{10!}{3!3!2!2!}$; $\frac{10!}{3!3!2!2!} \times \frac{4!}{2!2!}$; $\frac{10!}{3!3!2!2!} \times \frac{6!}{2!2!2!}$.
23. (a) $\binom{31}{15}^7 \binom{30}{15}^4 \binom{29}{15}(180)!/(366)_{180}$, (b) $(305)_{30}/(366)_{30}$.
24. $\binom{29}{10} \Big/ \binom{49}{30}$.
25. $\binom{n-100}{93}\binom{100}{7} \Big/ \binom{n}{100}$.

27. Divide the following numbers by $\binom{52}{5}$:

    (a) $4 \times \binom{13}{5}$; (b) 9×4^5; (c) 4×9;

    (d) 13×48; (e) $13 \times 12 \times 4 \times 6$.

29. Divide the following numbers by 6^6:

    $6; 6 \times 5 \times \frac{6!}{5!1!}; 6 \times 5 \times \frac{6!}{4!2!}; 6 \times \binom{5}{2} \times \frac{6!}{4!}$

    $\binom{6}{2} \times \frac{6!}{3!3!}; (6)_3 \times \frac{6!}{3!2!}; 6 \times \binom{5}{3} \times \frac{6!}{3!};$

    $\binom{6}{3} \times \frac{6!}{2!2!2!}; \binom{6}{2}\binom{4}{2} \times \frac{6!}{2!2!}; \binom{6}{1}\binom{5}{4} \times \frac{6!}{2!}; 6!$

    Add these up for a check; use your calculator if you have one.

30. Do the problem first for $n = 2$ by enumeration to see the situation. In general, suppose that the right pocket is found empty and there are k matches remaining in the left pocket. For $0 \leq k \leq n$ the probability of this event is equal to $\frac{1}{2^{2n-k}}\binom{2n-k}{n}\frac{1}{2}$. This must be multiplied by 2 because right and left may be interchanged. A cute corollary to the solution is the formula below:

$$\sum_{k=0}^{n} \frac{1}{2^{2n-k}}\binom{2n-k}{n} = 1.$$

Chapter 4

2. $P\{X + Y = k\} = 1/3$ for $k = 3, 4, 5$; same for $Y + Z$ and $Z + X$.
3. $P\{X + Y - Z = k\} = 1/3$ for $k = 0, 2, 4$;
   $P\{\sqrt{(X^2 + Y^2)Z} = x\} = 1/3$ for $x = \sqrt{13}, \sqrt{15}, \sqrt{20}$;
   $P\{Z/|X - Y| = 3\} = 1/3, P\{Z/|X - Y| = 1\} = 2/3$.
4. Let $P(\omega_j) = 1/10$ for $j = 1, 2$; $=1/5$ for $j = 3, 4$; $=2/5$ for $j = 5$; $X(\omega_j) = j$ for $1 \leq j \leq 5$; $Y(\omega_j) = \sqrt{3}$ for $j = 1, 4$; $=\pi$ for $j = 2, 5$; $=\sqrt{2}$ for $j = 3$.
5. Let $P(\omega_j) = p_j$, $X(\omega_j) = v_j$, $1 \leq j \leq n$.
6. $\{X + Y = 7\} = \{(1, 6), (2, 5), (3, 4), (4, 3), (5, 2), (6, 1)\}$.
8. $P\{Y = 14000 + 4n\} = 1/5000$ for $1 \leq n \leq 5000$; $E(Y) = 24002$.
9. $P\{Y = 11000 + 3n\} = 1/10000$ for $1 \leq n \leq 1000$;
   $P\{Y = 10000 + 4n\} = 1/10000$ for $1001 \leq n \leq 10000$;
   $E(Y) = 41052.05$.
10. $E(Y) = 29000 + 7000.e^{-2/7}$.
11. $\lambda e^{-\lambda x}$, $x > 0$.
12. $2xf(x^2)$, $x > 0$; $\frac{2x}{b-a}$ for $\sqrt{a} \leq x \leq \sqrt{b}$.
13. (i) $f\left(\frac{x-b}{a}\right)\frac{1}{|a|}$ if $a \neq 0$. (ii) $\frac{1}{2\sqrt{x}}\{f(\sqrt{x}) + f(-\sqrt{x})\}$, $x > 0$.

15. $c = 1/(1 - q^m)$.

16. $P(Y = j) = \binom{n}{\frac{n+j}{2}}\frac{1}{2^n}$, for $-n \le j \le n$ such that $n + j$ is even.

    $E(Y) = 0$.

17. $P(X = j) = \binom{11}{j}\binom{539}{25 - j}\bigg/\binom{550}{25}$, $0 \le j \le 25$.

18. If there are r rotten apples in a bushel of n apples and k are picked at random, the expected number of rotten ones among those picked is equal to kr/n.

19. $P(X \ge m) = \frac{1}{m}$, $E(X) = +\infty$.

20. 1.

21. Choose $v_n = (-1)^n 2^n/n$, $p_n = 1/2^n$.

23. According to the three hypotheses on p. 97: (1) $\sqrt{3}/2$; (2) 3/4; (3) 2/3.

24. 2.

26. $F_R(r) = \frac{r^2}{100}$, $f_R(r) = \frac{r}{50}$ for $0 \le r \le 100$; $E(R) = \frac{20}{3}$.

27. $Y = d \tan \theta$, where d is the distance from the muzzle to the wall and θ is the angle the pistol makes with the horizontal direction.

    $P(Y \le y) = \arctan \frac{y}{d}$; $E(Y) = +\infty$.

28. $E(2^X) = +\infty$.

29. If at most m tosses are allowed, then his expectation is m cents.

31. $P((X, Y) = (m, m')) = \binom{n}{2}^{-1}$ for $1 \le m < m' \le n$; $P(X = m) = (n - m)\binom{n}{2}^{-1}$; $P(Y = m') = (m' - 1)\binom{n}{2}^{-1}$; $P(Y - X = k) = (n - k)\binom{n}{2}^{-1}$, $1 \le k \le n - 1$.

32. Joint density of (X, Y) is $f(u, v) = \begin{cases} 2, & \text{if } 0 \le u < v \le 1; \\ 0, & \text{otherwise.} \end{cases}$

Chapter 5

1. $\frac{1050}{6145}$, $\frac{95}{1095}$.
2. $\frac{18826}{19400}$.
3. 5/9.
4. (a) 1/2; (b) 1/10.
5. 1/4; 1/4.
6. 1/4.

7. 1/2.
8. $2\beta(1 - \alpha + \beta)^{-1}$.
9. 6/11, 3/11, 2/11.
10. 400/568.
17. 1/2.
18. $p^3 + \dfrac{3}{2}p^3(1 - p)$.
19. 379/400.
20. $P(\text{no unbrella} \mid \text{rain}) = 2/9$; $P(\text{no rain} \mid \text{umbrella}) = 5/9$.
21. 27/43.
22. $[p^2 + (1 - p)^2]/[3p^2 + (1 - p)^2]$.
23. (a) 3/8; (b) 3/4; (c) 1/3.
25. (a) $\dfrac{1}{6}\sum_{n=1}^{6}\binom{n}{k}\dfrac{1}{2^n}$; (b) $\binom{n}{3}\dfrac{1}{2^n}\left\{\sum_{n=3}^{6}\binom{n}{3}\dfrac{1}{2^n}\right\}^{-1}$ for $3 \le n \le 6$.
26. The probabilities that the number is equal to 1, 2, 3, 4, 5, 6 are equal respectively to:
    (1) p_1^2; (2) $p_1p_2 + p_1^2p_2$; (3) $p_1p_3 + 2p_1p_2^2 + p_1^3p_3$;
    (4) $2p_1p_2p_3 + p_2^3 + 3p_1^2p_2p_3$; (5) $2p_2^2p_3 + 3p_1p_2^2p_3 + 3p_1^2p_3^2$;
    (6) $p_2p_3^2 + p_2^3p_3 + 6p_1p_2p_3^2$; (7) $3p_1p_3^3 + 3p_2^2p_3^2$; (8) $3p_2p_3^3$; (9) p_3^4. Tedious work? See Example 20 and Exercise No. 23 of Chapter 8 for general method.
27. $\left(\dfrac{4}{6}\right)^n - 2\left(\dfrac{3}{6}\right)^n + \left(\dfrac{2}{6}\right)^n$.
28. $\sum_{n=0}^{\infty} p_n\left(\sum_{k=0}^{n} p_k\right)$; $\sum_{n=0}^{\infty} p_n^2$.
29. 2/7.
30. $P(\text{maximum} < y \mid \text{minimum} < x) = y^2/(2x - x^2)$ if $y \le x$;
    $= (2xy - x^2)/(2x - x^2)$ if $y > x$.
31. 1/4.
33. (a) $(r + c)/(b + r + 2c)$; (b) $(r + 2c)/(b + r + 2c)$; (c), (e), (f): $(r + c)/(b + r + c)$; (d): same as (b).
34. $\{b_1(b_2 + 1)r_1 + b_1r_2(r_1 + 1) + r_1b_2(r_1 - 1) + r_1(r_2 + 1)r_1\}/(b_1 + r_1)^2(b_2 + r_2 + 1)$.
35. $\left(\sum_{k=1}^{N} k^{n+1}\right) \Big/ N\left(\sum_{k=1}^{N} k^n\right)$.
39. $(1 + p)^2/4$; $(1 + pq)/2$.
40.

	0	1	2
0	q	p	0
1	$q/2$	1/2	$p/2$
2	0	q	p

Chapter 6

1. \$.1175; \$.5875.
2. \$94000; \$306000.

3. $2\left(\frac{3}{13}+\frac{2}{12}+\frac{4}{13}+\frac{3}{14}+\frac{4}{14}\right)$.

4. 21; 35/2.

5. .5; 2.5.

6. 13/4; $4\left\{1-\binom{39}{13}\Big/\binom{52}{13}\right\}$.

7. $(6/7)^{25}$; $7\left\{1-\left(\frac{6}{7}\right)^{25}\right\}$.

8. (a) $1-(364/365)^{500}-500(364)^{499}/(365)^{500}$.
   (b) 500/365.
   (c) $365\left\{1-\left(\frac{364}{365}\right)^{500}\right\}$.
   (d) 365 p where p is the number in (a).

9. Expected number of boxes getting k tokens is equal to $m\binom{n}{k}\frac{(m-1)^{n-k}}{m^n}$;

   expected number of tokens alone in a box is equal to $n\left(\frac{m-1}{m}\right)^{n-1}$.

10. $P(n_j \text{ tokens in } j\text{th box for } 1 \le j \le m) = \frac{n!}{n_1! \cdots n_m!}\frac{1}{m^n}$

    where $n_1 + \cdots + n_m = n$.

11. 49.
12. 7/2.
13. $100p$; $10\sqrt{p(1-p)}$.
14. 46/5.
15. (a) $N+1$; (b) $\sum_{n=1}^{N+1}\frac{(N)_{n-1}}{N^{n-1}}$.
16. Let M denote the maximum. With replacement:

    $P(M=k)=\frac{k^n-(k-1)^n}{N^n}$, $1 \le k \le N$;

    $E(M)=\sum_{k=1}^{N}\left\{1-\left(\frac{k-1}{N}\right)^n\right\}$;

    without replacement:

    $P(M=k)=\binom{k-1}{n-1}\Big/\binom{N}{n}$, $n \le k \le N$;

    $E(M)=\frac{n(N+1)}{n+1}$.

17. (a) $nr/(b+r)$; (b) $(r^2+br+cnr)/(b+r)$.
19. $1/p$.
22. $E(X)=1/\lambda$.
23. $E(T)=\frac{a}{\lambda}+\frac{1-a}{\mu}$; $\sigma^2(T)=\frac{2a}{\lambda^2}+\frac{2(1-a)}{\mu^2}-\left(\frac{a}{\lambda}+\frac{1-a}{\mu}\right)^2$.
24. $E(T \mid T>n)=1/\lambda$.
25. (a) $1/5\lambda$; (b) $137/60\lambda$.
26. .4%.

27. $E(aX + b) = aE(X) + b$, $\sigma^2(aX + b) = a^2\sigma^2(X)$.
28. Probability that he quits winning is 127/128, having won \$1; probability that he quits because he does not have enough to double his last bet is 1/128, having lost \$127. Expectation is zero. So is it worth it? In the second case he has probability 1/256 of losing \$150, same probability of losing \$104, and probability 127/128 of winning \$1. Expectation is still zero.
29. E (maximum) $= n/(n + 1)$; E (minimum) $= 1/(n + 1)$; E (range) $= (n - 1)/(n + 1)$.
30. $g(z) = \prod_{j=1}^{n} (q_j + p_j z)$; $g'(1) = \sum_{j=1}^{n} p_j$.
31. $u_k = P\{S_n \leq k\}$; $g(z) = (q + pz)^n/(1 - z)$.
32. $g(z) = (1 - z^{2N+1})/(2N + 1)z^N(1 - z)$; $g'(1) = 0$.
33. $\binom{2n}{n} \frac{1}{4^n}$.
34. $g(z) = z^N \prod_{j=0}^{N-1} \frac{N - j}{N - jz}$; $g'(1) = N \sum_{j=0}^{N-1} \frac{1}{N - j}$.
35. $m_1 = g'(1)$; $m_2 = g''(1) + g'(1)$; $m_3 = g'''(1) + 3g''(1) + g'(1)$; $m_4 = g^{(iv)}(1) + 6g'''(1) + 7g''(1) + g'(1)$.
36. $(-1)^n L^{(n)}(0)$.
37. (a) $(1 - e^{-c\lambda})/c$, $\lambda > 0$; (b) $2(1 - e^{-c\lambda} - c\lambda e^{-c\lambda})/c^2\lambda^2$, $\lambda > 0$; (c) $(\lambda + 1)^{-n}$.
38. Laplace transform of S_n is equal to $\mu^n/(\lambda + \mu)^n$;

$$P(a < S_n < b) = \frac{1}{(n-1)!} \int_a^b u^{n-1} e^{-u} \, du.$$

Chapter 7

1. $1 - \frac{5}{3} e^{-2/3}$.
2. $\left(1 - \frac{4}{100}\right)^{25} \approx e^{-1}$.
3. $e^{-\alpha}\alpha^k/k!$ where $\alpha = 1000/324$.
4. $e^{-20} \sum_{k=20}^{30} \frac{(20)^k}{k!}$.
5. Let $\alpha_1 = 4/3$, $\alpha_2 = 2$. $P\{X_1 = j \mid X_1 + X_2 = 2\} = \frac{2!}{j!(2-j)!} \frac{\alpha_1^j \alpha_2^{2-j}}{(\alpha_1 + \alpha_2)^2}$ for $j = 0, 1, 2$.
6. If $(n + 1)p$ is not an integer, the maximum term of $B_k(n; p)$ occurs at $k = [(n + 1)p]$ where $[x]$ denotes the greatest integer not exceeding x; if $(n + 1)p$ is an integer, there are two equal maximum terms for $k = (n + 1)p - 1$ and $(n + 1)p$.

7. If α is not an integer, the maximum term of $\pi_k(\alpha)$ occurs at $k = [\alpha]$; if α is an integer, at $k = \alpha - 1$ and $k = \alpha$.
8. $\exp[-\lambda c + \alpha(e^{-\lambda h} - 1)]$.
9. $\pi_k(\alpha + \beta)$.
11. $e^{-50} \sum_{k=50}^{60} \frac{(50)^k}{k!}$.
12. $\frac{1}{(n-1)!2^n} \int_N^\infty u^{n-1} e^{-u/2}\, du$.
13. $\Phi\left(3\sqrt{\frac{12}{35}}\right) - \Phi\left(-2\sqrt{\frac{12}{35}}\right)$.
14. Find n such that $2\Phi\left(\frac{\sqrt{n}}{10}\right) - 1 \geq .95$. We may suppose $p > 1/2$ (for the tack I used.)
15. 537.
16. 475.
24. $\sqrt{\frac{1}{2\pi x}}\, e^{-x/2}$.
27. $P\{\delta'(t) > u\} = e^{-\alpha u}$; $P\{\delta(t) > u\} = e^{-\alpha u}$ for $u < t$; $= 0$ for $u \geq t$.
28. No!

Chapter 8

1. (a) $p^4 + 3p^3q + 2p^2q^2$; (b) $p^4 + 2p^3q + 2pq^3 + q^4$; (c) $1 - (p^4 + p^3q)$; (d) $1 - (p^4 + p^3q + pq^3 + q^4)$.
2. For Y_n: $I =$ the set of even integers; $p_{2i,2i+2} = p^2$, $p_{2i,2i} = 2pq$, $p_{2i,2i-2} = q^2$;
   For Z_n: $I =$ the set of odd integers; $p_{2i-1,2i+1} = p^2$, $p_{2i-1,2i-1} = 2pq$, $p_{2i-1,2i-3} = q^2$; $P\{Z_0 = 1\} = p$, $P\{Z_0 = -1\} = q$.
3. For X_n: $I =$ the set of nonnegative integers; $p_{i,i} = q$, $p_{i,i+1} = p$.
   For Y_n: $I =$ the set of all integers, $p_{i,i-1} = q$, $p_{i,i+1} = p$.
4. $P\{|Y_{2n+1}| = 2i + 1 \mid |Y_{2n}| = 2i\} = (p^{2i+1} + q^{2i+1})/(p^{2i} + q^{2i})$,
   $P\{|Y_{2n+1}| = 2i - 1 \mid |Y_{2n}| = 2i\} = (p^{2i}q + pq^{2i})/(p^{2i} + q^{2i})$.
5. (a)

n	1	2	3
$f_{11}^{(n)}$	$\frac{1}{2}$	$\frac{1}{6}$	$\frac{1}{9}$
$f_{12}^{(n)}$	$\frac{1}{2}$	$\frac{1}{4}$	$\frac{1}{8}$
$g_{12}^{(n)}$	$\frac{1}{2}$	$\frac{1}{3}$	$\frac{2}{9}$

(b)

n	1	2	3
$f_{11}^{(n)}$	p_1	0	$q_1q_2q_3$
$f_{12}^{(n)}$	q_1	p_1q_1	$p_1^2q_1$
$g_{12}^{(n)}$	q_1	q_1p_2	$q_1p_2^2$

6. $\begin{bmatrix} 1 & 0 \\ \alpha & 1-\alpha \end{bmatrix}$; $f_{21}^{(n)} = (1-\alpha)^{n-1}\alpha$; $p_{21}^{(n)} = 1 - (1-\alpha)^n$; $m_{21} = \frac{1}{\alpha}$.

9. $I = \{0; 2^i, 0 \le i \le n-1\}$

$p_{2^i, 2^{i+1}} = \frac{1}{2}$, $p_{2^i, 0} = \frac{1}{2}$ for $0 \le i \le n-1$.

10. 1/13.

11.

	U	H	D
U	0	1	0
H	p	0	q
D	0	1	0

$w_U = \frac{p}{2}$, $w_H = \frac{1}{2}$, $w_D = \frac{q}{2}$.

12. Same as given in (8.1.9) and (8.1.10).

13. $e_j = \frac{l(1-r^j)}{(p-q)(1-r^l)} - \frac{j}{p-q}$ where $r = \frac{q}{p}$.

15. $p_{i,i-1} = (i/N)^2$, $p_{i,i} = 2i(N-i)/N^2$, $p_{i,i+1} = ((N-i)/N)^2$;

$w_i = \binom{N}{i}^2 \Big/ \binom{2N}{N}$; $0 \le i \le N$.

16. $w_s = (1-\beta)/(2-\alpha-\beta)$; $w_f = (1-\alpha)/(2-\alpha-\beta)$.

18. $p_{j,j+1} = p$, $p_{j,0} = 1-p$; $w_j = p^j q$, $0 \le j \le \infty$.

19. Let $r = p/q$, $A^{-1} = 1 + p^{-1}\sum_{k=1}^{c-1} r^k + r^{c-1}$; then $w_0 = A$; $w_k = p^{-1}r^k A$, $1 \le k \le c-1$; $w_c = r^{c-1}A$.

20. $f_{21}^* = .721$; $f_{31}^* = .628$.

21. $f_{i,2N}^* = \frac{i}{2N}$, $f_{i,0}^* = 1 - \frac{i}{2N}$, $0 \le i \le 2N$.

22. $(1 - a_2 + \sqrt{a_1^2 - 2a_1 + 1 - 4a_0a_2})/2a_2$.

23. Coefficient of z^i in $g(h(z))$ where $h(z) = \sum_{j=0}^{\infty} b_j z^j$.

24. Let e_j denote the expected number of further impulses received until the meter registers the value l, when the meter reads j; then $e_j = l$ for $1 \le j \le l-1$, $e_l = 0$.

25. $e_m = \sum_{j=1}^{m} \frac{1}{j}$.

26.

	(123)	(132)	(213)	(231)	(312)	(321)
(123)	0	0	q	p	0	0
(132)	0	0	0	0	q	p
(213)	q	p	0	0	0	0
(231)	0	0	0	0	p	q
(312)	p	q	0	0	0	0
(321)	0	0	p	q	0	0

27.

	1	2	3
1	0	q	p
2	1	0	0
3	0	p	q

32. $p_{ij} = 1/(i+1)$ for $s - i \le j \le s$; $= 0$ otherwise;

$w_j = 2(j+1)/(s+1)(s+2)$ for $0 \le j \le s$; $\sum_{j=0}^{s} jw_j = 2s/3$.

33. $p_{ij} = \binom{i}{s-j} p^{s-j}(1-p)^{i-s+j}$ for $s - i \le j \le s$; $= 0$ otherwise;

$w_j = \binom{s}{j}\left(\frac{1}{1+p}\right)^j\left(\frac{p}{1+p}\right)^{s-j}$ for $0 \le j \le s$; $\sum_{j=0}^{s} jw_j = s/(1+p)$.

44. $P\{S = k \mid X_0 = i\} = p_{ii}^k(1 - p_{ii}^k)$, $k \ge 1$.

45. $\tilde{p}_{ij} = p_{ij}/(1 - p_{ii})$ for $i \ne j$; $\tilde{p}_{ii} = 0$.

46. $P\{(X_n, X_{n+1}) = (k, 2k - j + 1) \mid (X_{n-1}, X_n) = (j, k)\} = p$,

$P\{(X_n, X_{n+1}) = (k, 2k - j) \mid (X_{n-1}, X_n) = (j, k)\} = q$.

Let $H_n^{(3)} = \sum_{v=1}^{n} H_v^{(2)}$, then $\{H_n^{(3)}\}$ is a Markov chain of order 3, etc.

Table 1

Values of the standard normal distribution function

Table I **Values of the standard normal distribution function**

$$\Phi(x) = \int_{-\infty}^{x} \frac{1}{\sqrt{2\pi}} e^{-u^2/2}\, du = P(X \leq x)$$

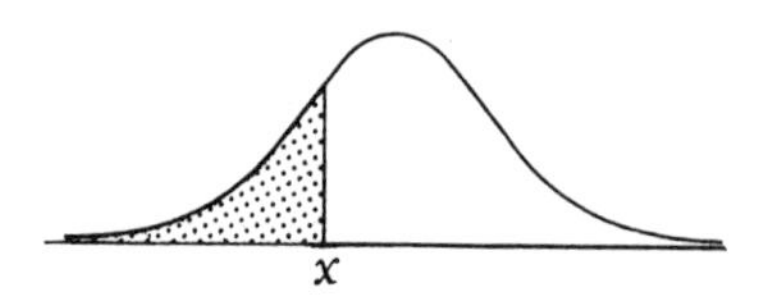

x	0	1	2	3	4	5	6	7	8	9
−3.	.0013	.0010	.0007	.0005	.0003	.0002	.0002	.0001	.0001	.0000
−2.9	.0019	.0018	.0017	.0017	.0016	.0016	.0015	.0015	.0014	.0014
−2.8	.0026	.0025	.0024	.0023	.0023	.0022	.0021	.0020	.0020	.0019
−2.7	.0035	.0034	.0033	.0032	.0031	.0030	.0029	.0028	.0027	.0026
−2.6	.0047	.0045	.0044	.0043	.0041	.0040	.0039	.0038	.0037	.0036
−2.5	.0062	.0060	.0059	.0057	.0055	.0054	.0052	.0051	.0049	.0048
−2.4	.0082	.0080	.0078	.0075	.0073	.0071	.0069	.0068	.0066	.0064
−2.3	.0107	.0104	.0102	.0099	.0096	.0094	.0091	.0089	.0087	.0084
−2.2	.0139	.0136	.0132	.0129	.0126	.0122	.0119	.0116	.0113	.0110
−2.1	.0179	.0174	.0170	.0166	.0162	.0158	.0154	.0150	.0146	.0143
−2.0	.0228	.0222	.0217	.0212	.0207	.0202	.0197	.0192	.0188	.0183
−1.9	.0287	.0281	.0274	.0268	.0262	.0256	.0250	.0244	.0238	.0233
−1.8	.0359	.0352	.0344	.0336	.0329	.0322	.0314	.0307	.0300	.0294
−1.7	.0446	.0436	.0427	.0418	.0409	.0401	.0392	.0384	.0375	.0367
−1.6	.0548	.0537	.0526	.0516	.0505	.0495	.0485	.0475	.0465	.0455
−1.5	.0668	.0655	.0643	.0630	.0618	.0606	.0594	.0582	.0570	.0559
−1.4	.0808	.0793	.0778	.0764	.0749	.0735	.0722	.0708	.0694	.0681
−1.3	.0968	.0951	.0934	.0918	.0901	.0885	.0869	.0853	.0838	.0823
−1.2	.1151	.1131	.1112	.1093	.1075	.1056	.1038	.1020	.1003	.0985
−1.1	.1357	.1335	.1314	.1292	.1271	.1251	.1230	.1210	.1190	.1170
−1.0	.1587	.1562	.1539	.1515	.1492	.1469	.1446	.1423	.1401	.1379
− .9	.1841	.1814	.1788	.1762	.1736	.1711	.1685	.1660	.1635	.1611
− .8	.2119	.2090	.2061	.2033	.2005	.1977	.1949	.1922	.1894	.1867
− .7	.2420	.2389	.2358	.2327	.2297	.2266	.2236	.2206	.2177	.2148
− .6	.2743	.2709	.2676	.2643	.2611	.2578	.2546	.2514	.2483	.2451
− .5	.3085	.3050	.3015	.2981	.2946	.2912	.2877	.2843	.2810	.2776
− .4	.3446	.3409	.3372	.3336	.3300	.3264	.3228	.3192	.3156	.3121
− .3	.3821	.3783	.3745	.3707	.3669	.3632	.3594	.3557	.3520	.3483
− .2	.4207	.4168	.4129	.4090	.4052	.4013	.3974	.3936	.3897	.3859
− .1	.4602	.4562	.4522	.4483	.4443	.4404	.4364	.4325	.4286	.4247
− .0	.5000	.4960	.4920	.4880	.4840	.4801	.4761	.4721	.4681	.4641

Table I Values of the standard normal distribution function

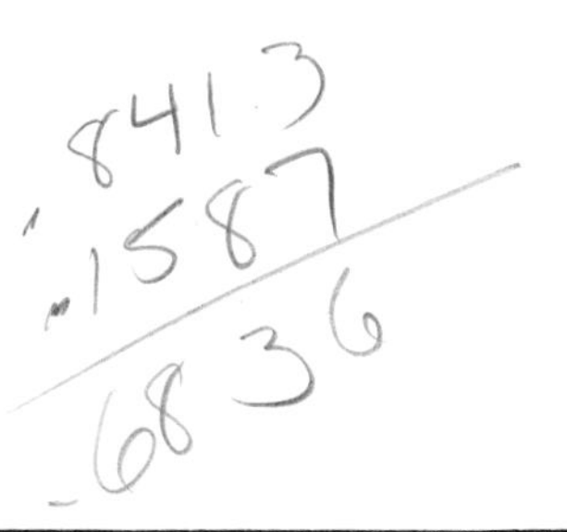

x	0	1	2	3	4	5	6	7	8	9
.0	.5000	.5040	.5080	.5120	.5160	.5199	.5239	.5279	.5319	.5359
.1	.5398	.5438	.5478	.5517	.5557	.5596	.5363	.5675	.5714	.5753
.2	.5793	.5832	.5871	.5910	.5948	.5987	.6026	.6064	.6103	.6141
.3	.6179	.6217	.6255	.6293	.6331	.6368	.6406	.6443	.6480	.6517
.4	.6554	.6591	.6628	.6664	.6700	.6736	.6772	.6808	.6844	.6879
.5	.6915	.6950	.6985	.7019	.7054	.7088	.7123	.7157	.7190	.7224
.6	.7257	.7291	.7324	.7357	.7389	.7422	.7454	.7486	.7517	.7549
.7	.7580	.7611	.7642	.7673	.7703	.7734	.7764	.7794	.7823	.7852
.8	.7881	.7910	.7939	.7967	.7995	.8023	.8051	.8078	.8106	.8133
.9	.8159	.8186	.8212	.8238	.8264	.8289	.8315	.8340	.8365	.8389
1.0	.8413	.8438	.8461	.8485	.8508	.8531	.8554	.8577	.8599	.8621
1.1	.8643	.8665	.8686	.8708	.8729	.8749	.8770	.8790	.8810	.8830
1.2	.8849	.8869	.8888	.8907	.8925	.8944	.8962	.8980	.8997	.9015
1.3	.9032	.9049	.9066	.9082	.9099	.9115	.9131	.9147	.9162	.9177
1.4	.9192	.9207	.9222	.9236	.9251	.9265	.9278	.9292	.9306	.9319
1.5	.9332	.9345	.9357	.9370	.9382	.9394	.9406	.9418	.9430	.9441
1.6	.9452	.9463	.9474	.9484	.9495	.9505	.9515	.9525	.9535	.9545
1.7	.9554	.9564	.9573	.9582	.9591	.9599	.9608	.9616	.9625	.9633
1.8	.9641	.9648	.9656	.9664	.9671	.9678	.9686	.9693	.9700	.9706
1.9	.9713	.9719	.9726	.9732	.9738	.9744	.9750	.9756	.9762	.9767
2.0	.9772	.9778	.9783	.9788	.9793	.9798	.9803	.9808	.9812	.9817
2.1	.9821	.9826	.9830	.9834	.9838	.9842	.9846	.9850	.9854	.9857
2.2	.9861	.9864	.9868	.9871	.9874	.9878	.9881	.9884	.9887	.9890
2.3	.9893	.9896	.9898	.9901	.9904	.9906	.9909	.9911	.9913	.9916
2.4	.9918	.9920	.9922	.9925	.9927	.9929	.9931	.9932	.9934	.9936
2.5	.9938	.9940	.9941	.9943	.9945	.9946	.9948	.9949	.9951	.9952
2.6	.9953	.9955	.9956	.9957	.9959	.9960	.9961	.9962	.9963	.9964
2.7	.9965	.9966	.9967	.9968	.9969	.9970	.9971	.9972	.9973	.9974
2.8	.9974	.9975	.9976	.9977	.9977	.9978	.9979	.9979	.9980	.9981
2.9	.9981	.9982	.9982	.9983	.9984	.9984	.9985	.9985	.9986	.9986
3.	.9987	.9990	.9993	.9995	.9997	.9998	.9998	.9999	.9999	1.0000

INDEX

Undergraduate Texts in Mathematics

Probability Theory

Independence, Interchangeability, Martingales

by **Y.S. Chow**, Columbia University and **H. Teicher**, Rutgers University

1978. xv, 455p. cloth.

The measure theoretical foundations of probability theory and the main laws of theorems which emerge therefrom are the subject of this text. The main topics treated are independence, interchangeability and martingales. Particular emphasis is placed on stopping times both as tools in providing theorems and as objects of individual interest. Many exercises and examples are given. No prior knowledge of measure theory is assumed; the book is unique in the combined presentation of measure and probability. For students who have already had a course on measure theory, guidelines for use of the text are found in the preface. The text is suitable for a graduate level course in probability theory.

Special Features

- inclusion of the Marcinkiewicz-Zygmund inequality, its extension to martingales and applications thereof
- a comprehensive treatment of the law of the iterated logarithm
- extension of the central limit theorem to the martingale and interchangeability cases, and
- development and explanation of the second moment analogue of Wald's equation

Springer-Verlag New York Heidelberg Berlin